Encyclopedia of Scientific Principles, Laws, and Theories

Encyclopedia of Scientific Principles, Laws, and Theories

Volume 1: A–K

Robert E. Krebs

Illustrations by Rae Déjur

GREENWOOD PRESS
Westport, Connecticut • London

Library of Congress Cataloging-in-Publication Data

Krebs, Robert E., 1922–
Encyclopedia of scientific principles, laws, and theories / Robert E. Krebs ; illustrations by Rae Déjur.
p. cm.
Includes bibliographical references and index.
ISBN: 978-0-313-34005-5 (set : alk. paper)
ISBN: 978-0-313-34006-2 (vol. 1 : alk. paper)
ISBN: 978-0-313-34007-9 (vol. 2 : alk. paper)
1. Science—Encyclopedias. 2. Science—History—Encyclopedias. 3. Physical laws—Encyclopedias. I. Title.
Q121.K74 2008
503—dc22 2008002345

British Library Cataloguing in Publication Data is available.

Library of Congress Catalog Card Number: 2008002345
ISBN: 978-0-313-34005-5 (set)
978-0-313-34006-2 (vol. 1)
978-0-313-34007-9 (vol. 2)

First published in 2008

Greenwood Press, 88 Post Road West, Westport, CT 06881
An imprint of Greenwood Publishing Group, Inc.
www.greenwood.com

Printed in the United States of America

The paper used in this book complies with the Permanent Paper Standard issued by the National Information Standards Organization (Z39.48–1984).

10 9 8 7 6 5 4 3 2 1

To Carolyn, who contributed much to this encyclopedia as my researcher, pre-editor, constructive critic, proofreader, supportive wife, and friend.

Contents

List of Entries

Preface

The development of universal scientific principles, fundamental physical laws, viable theories, and testable hypotheses has a long history. Humans are unique in that they can think about and contemplate the world around them, conceive ideas to explain natural events and processes, and then make use of what was learned. Early explanations of natural phenomena were interesting but not very reliable. Not until a few thousand years before the birth of Christ would recordings of history provide us with evidence of how humans related to the events and phenomena of nature. Some of the early Egyptian, Greek, Islamic, and Asian theories, as well as those from other cultures, demonstrated great insight into the structure and functions of animals, plants, matter, meteorology, agriculture, Earth, and astronomy. Our ancestors had the curiosity but lacked the means for forming truly accurate explanations and conclusions about nature as the understanding of cause and effect, and objective use of controlled experimentation was yet to be developed.

Included in this volume are a few ancient classical ideas and concepts that were theoretical *descriptions* of nature, often very inaccurate. This ancient "classical" philosophical process resulted in many dead ends. The modern sciences began when people learned how to *explain* nature by objectively observing events, asking questions that could be answered by making reliable measurements, using mathematics, and then considering probabilities to make reasonable predictions. This process led to "operational facts" that continued to be upgraded and corrected by the "self-correcting" nature of modern science. However, testing theories and hypotheses in a controlled situation developed late in human history and thus science, as we think of it, was slow to advance in ancient times. The development and implementation of scientific processes increased the "growth of knowledge," as well as the *rate* of growth of science from the seventeenth to the twenty-first centuries. During this period science accelerated at an astounding exponential pace, and this growth will most likely continue throughout the twenty-first century and beyond, particularly in the biological sciences based on quantum theory. None of our current understanding of the universe, nature, and ourselves would have been possible without men and women using the processes and procedures of scientific investigations.

The purpose of this encyclopedia is to present in two volumes a historical aspect for the important principles, laws, theories, hypotheses, and concepts that reflect this amazing progression of scientific descriptions and explanations of nature. Some more recent theories are also included. These scientific principles, laws, theories, etc., did not just appear out of "thin air." They are related to the period and people who developed these explanations and descriptions of the nature of our universe. The entries are listed alphabetically, in most cases, according to the name of the person credited with formulating the law, theory, or concept. Their names may be familiar to you. Others are less well known. Inventions and discoveries are included only if they contributed to the development or understanding of a particular scientific law, etc. Where appropriate, practical applications of particular laws and theories are included. Only laws, principles, theories, and concepts related to the basic sciences of physics, chemistry, biology, astronomy, geology, mathematics, and related fields such as medicine are included. Social, political, behavioral, and related studies are not included.

This encyclopedia is designed for the high school and college-level student as well as for the general reader who has an interest in science. Following the A–Z portion of the encyclopedia is a glossary of technical terms. The terms contained in the glossary are highlighted in **bold** type the first time the word is used in the text. Following the glossary are four appendices. Appendix A groups the encyclopedia entries by scientific discipline. Appendices B through D list Nobel Prize recipients for Chemistry, Physics, and Physiology/Medicine, respectively. Following the appendices is a selected bibliography containing sources for additional information related to the scientific principles, laws, theories, etc., included in the volumes. A general subject index concludes the set.

The following notations are used in this book:

BCE = Before the Common Era (instead of BC)
CE = The Common Era (Instead of AD)
c. = Approximate Date (e.g., approximate birth or death dates)
$\sim$ = Approximate amount, quantity, or figure
ppm = Parts per million
ppb = Parts per billion
% = Percent
α = Alpha particle (radiation)
β = Beta particle (radiation)
λ = Gamma radiation (similar to high-energy X-rays)

IMPORTANT:

Due to the technical nature of many of the original statements of principles, laws, theories, etc., and to the fact that some of the original statements are lost, are in a foreign language, or include "technical jargon," they have been paraphrased and restated to make them more comprehensible. The paraphrased statements of the principles or laws are printed in italics for easy identification.

Introduction

WHAT SCIENCE IS AND IS NOT

In his book *Asimov's Biographical Encyclopedia of Science and Technology* (1964) Isaac Asimov states "science is a complex skein, intricately interknotted across the artificial boundaries we draw only that we may the more easily encompass its parts in our mind. Pick up any thread of that skein and the whole structure will follow."

That "interknotted skein" may be thought of as a complex interrelationship between the *processes* and *products* of science. The processes are the *verbs* that relate to the so-called methods of science, whereas the products are the *nouns* that represent the knowledge about nature that we gain through the rational and pragmatic uses of scientific processes. Scientific research investigations of nature may lead to the technology/engineering that uses the knowledge gained through applying the processes of science. Thus, an understanding of Earth and universe requires the knowledge of science (the *products—nouns*) and the methods of obtaining that knowledge (the *processes—verbs*).

There are a series of analogies that can also be used to describe this process/product duality of science: basic science versus applied science, research and development, induction/deduction, and experimental investigation/technology innovations.

There is no *one* scientific method. Scientists do not use a cut-and-dried procedure during their investigations of nature. There are at least twelve characteristics and processes of science that make up the so-called scientific method, and in real-life, objective, rational scientific research one does not always start with number 1 and follow in order to number 12. First of all, scientists are skeptics when they investigate nature or hear about some extraordinary new discovery. As the saying goes "extraordinary claims require extraordinary evidence." They are skeptical of new evidence claimed by other scientists and will try to repeat their experiments. They also like to explore new problems, identify patterns or breaks in patterns existing in nature, and, in general, gather information and data to come up with answers to answerable questions and viable solutions to problems. In general, these processes are:

Observing—The "thread" used throughout a scientific investigation
Concentrating—Skepticism, critical thinking, induction/deduction
Recognizing—Scientific problems, questioning the ways things seem
Discriminating—Judging viability of identified problem, stereotyping
Relating—Old knowledge related to new information, being informed
Establishing—Cause and effect, eliminating irrational relationships
Formulating hypotheses—Asking answerable questions, forming answers
Testing—Selecting and using appropriate instruments and equipment
Experimenting—Objective testing, research using controls
Gathering, Treating, and Interpreting—Information and data
Using appropriate statistical instruments—Confirming and predicting
Communicating—Results and conclusions, peer review publishing.

A slightly different list of "techniques of inquiry" are given in Surendra Verma's excellent book *The Little Book of Scientific Principles Theories & Things* (2005):

1. *Observations* and search for data
2. *Hypothesis* to explain observations
3. *Experiments* to test hypotheses
4. *Formulation* of theory
5. *Experimental* confirmation of theory
6. *Mathematics* of empirical confirmation of theory
7. *Use* of this confirmation to form scientific law
8. *Use* of scientific law to predict behavior of nature.

It might be mentioned that these statements, and other similar ones, describe human ways of thinking and acting. The human concept of identifying and using the "processes" relates to a way of thinking and acting we call science. These rational procedures for acquiring new knowledge were developed and applied very late in human history. It is assumed that bits and pieces of these processes were known and used by proto-humans even before *Homo sapiens* arrived on Earth several millions of years ago. For example, as a means for survival early humans learned the difference between poisonous plants and "safe" plants to eat, and animals that humans could kill and eat and those more likely to kill and eat the humans. This required trial and error leading to stereotyping (profiling) varieties of plants and animals as well as other aspects of their environment pointing to ways of making practical use of something that "worked." These are rudimentary process of science. Applying the processes of observing, discriminating by stereotyping, relating new knowledge to what is known, and arriving at conclusions ensured humanity's continued existence, as well as the development of cultures and societies. This resulted in laws, civil and scientific, culminating in our modern use of knowledge gained through the use of the processes of science to sustain and improve human life.

Above all, scientists are critical thinkers and skeptics who do not accept many things and events at face value. They are not convinced just because others find something acceptable or convenient. They do not accept certainties just because others accept them, and they do not trust absolutes including absolute ideologies. Finally they have no "sacred cows," and they carefully select their authorities. They are aware that knowledge and science in general are subject to justifiable change and corrections.

Conversely, science is NOT many of the things often attributed to it.

First, science is not democratic—we do not vote on scientific theories or natural phenomena that scientists wish to explore. A possible exception might be when government "supports" a particular scientific effort.

Second, science is not based on everyone's opinion—some persons are more informed, wiser, better educated, and better equipped to explore natural phenomena or solutions to problems than are many other people.

Third, science is less subjective than other disciplines—although scientists are human and thus, to some degree subjective in their judgments, they are, as a group more objective in their outlook than most people. Claiming to be an authority does not make it so.

Fourth, science is not magic—much of science may seem like "magic" to the non-scientist, but there is no occult influence, sleight of hand, or conjuring related to the use of the processes that make up the scientific method. Pseudoscience and "junk" science are identified as such and exposed.

Fifth, science is not religion—scientists must have some degree of "faith" in the processes they use while conducting scientific investigations, but this is much different than the concept of having faith in spiritual nonexistent things. Although many scientists observe one form of irrational spiritual religious belief or another, to be successful in their particular discipline, they must separate theological ideologies from the rational science parts of their lives.

Sixth, science is not parapsychology—this includes nonsciences and the pseudosciences such as telepathy, clairvoyance, extra-sensory perception (ESP), psychokinesis, and unproven activities such as fortune-telling, tarot cards, reading bones and head bumps, witchcraft, channeling, astrology, etc. These types of activities are irrational and have no bases in facts gained from controlled experiments to verify them. It is said that about 75 percent of the people in the United States believe in angels; it is doubtful that many scientists do because no verifiable evidence of such entities exists.

Seventh, science is not politics—although politics sometimes attempts to dominate science for its own purposes, science is best kept out of the political environment except in the use of science in an advisory capacity to the political operatives. Conversely, politics is best kept out of scientific endeavors. This is difficult because governments of the developed world have increasingly provided funds to support scientific investigations, and many believe that control follows the money, or as the saying goes: He who pays the piper selects the tunes. Peer review panels have been established to ensure the scientific viability of proposed research sponsored by the government. Unfortunately, as with any system of controls there are a few flaws. Furthermore, political science is not a really a science but a study of politics. Just as economics is not a science but rather is more of an "art" that uses some of the processes of science, such as statistics, to study the economics of a society.

Eighth, science is not philosophy—historically there were philosopher/scientists and not much distinction was made until the eighteenth century "enlightenment period" of history when modern concepts of science as a systematic self-correcting process were developed. No other discipline or intellectual/social activity has a natural built-in self-correcting component.

SCIENTIFIC PHYSICAL LAWS

The concept of "laws" is as old as civilization and originally implied a "law giver" that was often interpreted as one type of god or another. The laws of science grew out of the early concept that the universe operated in certain ways and that "gods" established these particular laws that controlled nature. These several gods later evolved into a supreme lawgiving god who controlled nature according to his or her whims. This concept changed as humans became more curious and the "lawgiver" became the "worker of nature," which, in turn, became "natural laws." By the time of the Enlightenment, the concept became the "laws of science." This evolution of "scientific laws" naturally followed as humans learned how to objectively and logically examine and explain the workings of nature. It might be mentioned that while some so-called scientific theories were assigned the title "law" many years ago, many more modern theories, such as the theory of relativity, could just as well be referred to as the "law of relativity" today. Accepted and established *theories* are similar to accepted and established *laws* as far as authenticity is concerned. In other words, a scientific law is derived from a theory that has been proven and repeatedly verified by the application of mathematics and is consistent in predicting specific effects.

Scientific physical laws are statements of fact that explain nature and how it works and are not merely applicable to the nonliving things in the field of physics (matter and energy). Scientific laws also apply to all living things in the field of biology (plants and animals), and all areas of nature. Physical laws do not apply to things that are not natural, for example, the spiritual, philosophical, and parapsychological. Physical laws are generalized factual principles that describe how things behave in nature under a variety of circumstances. Where did these physical laws come from? The answer is obvious—from within the universe itself. We live in an ordered universe, made comprehensible by applying the concept of "rules" and using intelligent powers of reasoning, logic, and critical thinking. Scientific physical laws describe *how* things work and to some degree *what* they are in reality, but not always *why* they work as they do (e.g., gravity—we really do not know *what* gravity is or *why* it works as it does, but it is universal, predictable, and can be described mathematically). Scientific laws are consistent, long lasting, universal, and rational. In other words, a phenomenon, event, or action that occurs and behaves in the same manner under the same conditions, and is thus predictable, can be stated as a scientific law. These are sometimes referred to as "fundamental laws," "universal laws," "basic laws of science," or just "physical" or "scientific" laws.

At least five characteristics apply to all scientific laws: 1) They can be expressed mathematically. (Mathematics is basic to understanding nature and expressing scientific laws—even abstract mathematics.) 2) Physical laws are not always exact (as is mathematics). Scientific laws may need future "adjustments" as more knowledge is gained concerning the natural phenomenon as expressed by the law (but revisions are not made often nor extensive). 3) The natural systems may be complex and contain many "pieces," but the law describing the phenomenon always turns out to be simple. The universal "theory of everything" (TOE) or the "grand unification theory" (GUT) is expected to be an extremely complex composite of matter/energy. Even so, scientists predict that when found, the TOE or GUT will be a very simple mathematical expression. 4) Most important, scientific theories are universal. (Any well-established physical law that applies on Earth also applies throughout the universe.) And, 5) by using statistical probabilities, scientific theories can be used to predict future natural events in the real world.

SCIENTIFIC PRINCIPLES

Scientific principles are similar general statements about nature as are scientific laws. A scientific principle of nature must be objective and universal. Principles are not dependent on the views or statement of individuals but must be true from all points of view and true from all points of reference in the universe. There are only a few general, fundamental, and overriding postulates from which scientific laws and theories are derived. One of the unifying principles is *symmetry*. The scientific definition of symmetry relates to the orientation of something regardless of its orientation in space and time. No matter where in space or at what time in the past, present, or future the event occurs, acts, or reacts, it will do so in the same way. Symmetry also enables an object to rotate on a fixed axis in any orientation regardless of where in the universe it is located, and it will move at a uniform velocity in a straight line regardless of orientation. Symmetry in our everyday lives is considered twofold—bilateral and radial. Rotational symmetry of a sphere does not point to any specific direction in space. Anything that can be reoriented or its position changed while keeping its same basic geometry is considered to be symmetrical. An example is a two-dimensional square drawn on a piece of paper. Despite its orientation, it will look the same (unless you view it from the edge of a piece of paper). Some three-dimensional figures and objects when examined can exhibit two kinds of symmetry—bilateral and radial. For example, a drinking glass can be cut lengthwise (bilateral) or crosswise (radial). The human body is considered to have bilateral symmetry.

Another fundamental principle for all physical laws is *conservation*. The principle of conservation of matter and energy is related to symmetry because everything is in balance—somewhat like the reflection in a mirror or a process that obtains equilibrium. (Note: the fundamental concept of the conservation of energy is also known as the "first law of thermodynamics.") When something comes out the same way or results in the same answer, no matter what takes place during the event, conservation is involved. Another way to look at conservation is the principle (concept) of *cause* and *effect*. All effects (events) have a cause, or possibly several causes may be related to the effect, but there is a symmetrical pattern involved in this natural phenomenon. This is true for quantum mechanics that connects the "minimum principle" with the laws of conservation. There are antiparticles for all particles, and positives (+) for negatives (−). In other words, they are symmetrical but opposites and are conserved (no basic loss or gain). The fundamental property of mass is inertia. Inertia is the property of an object (with mass) that offers resistance to any change in its position or speed (orientation in space) when a force is applied to the mass. For our Earth and solar system this explanation is adequate, but inadequate for the relativity of space and time. Einstein's principle or theory of relativity redefined mass when considering the vastness of space, as expressed in his famous equation $E = mc^2$, which equates mass and energy as being conserved. (See Einstein for more on this principle.)

Humans did not invent the principles and laws of physics—they came from the formation of the universe from the big bang—from the void of the spontaneous formation of matter and energy—not a supreme being.

Other examples of conservation principles are Newton's three laws of motion and the theory of kinetic energy of particles that can be

> It might be mentioned that our recent explorations of space by unmanned vehicles that "fly" by other planets and their moons would be impossible to perform accurately if it were not for the use of the theory of relativity in the mathematical calculation to guide the flight path. Without the use of mathematical constants, scientific principles, laws, and theories by NASA's space exploration program, astronauts would have had a more difficult time landing on the moon and would make the planned trip to the planet Mars more hazardous.

expressed in standard mathematical terms. A different example is the uncertainty principle related to quantum mechanics used to explain the nature of subatomic particles and energy. The uncertainty is based on the inability to ascertain, at the quantum level, the exact position of a subatomic particle, while at the same time trying to determine its momentum energy. The converse is also evident as a subatomic particle's momentum cannot be determined at the same time as locating its position. Treating the quanta data statistically solves this dilemma. Indeterminacy of quantum mechanics is a good example of the use of statistical probabilities to solve a problem (see Heisenberg).

Interestingly, it seems that there is a deep correlation or connection between the concepts of symmetry and the conservation laws. Someday there may be a common mathematical formula or constant that relates the two ideas.

SCIENTIFIC CONSTANTS

Generalized scientific principles involve mathematical rules. In other words, the answer reached will always be the same if the mathematical rules are followed. This concept utilizes what are known as fundamental *constants* that are mathematical expressions that establish a mathematical relationship between two or more variables that never change their values. These constants never change regardless of where or how they are used in the universe and assist scientists in the formation of mathematical expressions of universal scientific laws and theories. The preceding paragraph on Einstein's theory of relativity contained a constant that is universally accepted. It is the speed of light (the c which he squared) in the formula $E = mc^2$. The speed of light is 186,282 miles per second (299,792 km/sec) and applies only when light traverses unobstructed space. Other examples of the many constants used in mathematical expressions in science include Planck's constant ($\hbar$), and the elementary charge of an electron (e^-). (See gravity and Coulomb's law related to electrical charges over distance as examples.)

There are over thirty fundamental physical constants that have specific and universal invariant quantities. These constants are expressed as symbols representing mathematical expressions that have a degree of accuracy as high as possible to obtain. Some examples of more familiar and frequently used fundamental physical constants and their mathematical expressions are:

Planck's constant ($\hbar$) is $6.6260755 \times 10^{-32}$ Joules/second.
Elementary charge (e^-) is $1.60217733 \times 10^{-19}$ Coulombs.
Avogadro's number (N_A) is 6.0221367×10^{23} particles per mol.
Electron's mass at rest (m_e) is $9.11093897 \times 10^{-31}$ kilograms.
Proton's mass at rest (m_p) is $1.6726231 \times 10^{-27}$ kilograms.
Neutron's mass as rest (m_n) is $1.6749286 \times 10^{-27}$ kilograms.
Atomic mass unit (*amu*) is 1.66054×10^{-27} kilograms.
Bohr radius (atomic) (*a0*) is 5.29177×10^{-11} meters.
Acceleration due to gravity (*g*) is 9.80665 meters/sec^2.
Gas constant (*R*) is 8.3145′ m^2 × kg/s^2 × K × mol.

SCIENTIFIC THEORIES

The origin of the word "theory" is from the Greek word *theorein* or *theoria* that means "to look at" or to be "observed" as actors in a theater production are observed. In modern usage it does not mean that a person is seeing something not real or staged, or just as a guess or hunch.

Scientific theories are a type of model designed as general explanatory statements about the workings of nature. A theory might be thought of as a hypothesis that has been tested by experiments. At times it is possible for an exception to be found in a theory, but if this happens the exception is usually found and corrected or the theory is discarded. Theories can be used to predict natural phenomena and lead to more specific laws.

Many people think that a theory and hypothesis are somewhat interchangeable and can be related to a not well defined idea. In other words, a statement may be tested even though it may not be true. This confusion may result when the same scientific description of something appears to be in two different stages that are testing the idea. The hypothesis is a statement that is used in the early development of the research investigation of phenomena. The idea is new, and results have a good chance of being wrong, even though correct data to be analyzed is presented. An example is the theory that electromagnetic radiation (light) requires some form of matter, such as the aether, in space to transmit electromagnetic waves. If the data gathered from related investigations can verify the experiment consistently, then the concept may be considered a theory. Conversely, if further experiments arrive at different and more viable explanations for the phenomenon then the theory may be falsified, which is what happened to the theory of the aether.

Some modern scientists have other ways of looking at this conundrum by thinking about scientific "models" instead of theories or physical laws to describe the reality of nature. Some consider the term "theory" as handicapped with much social and historical baggage and believe that the concept of "model" can be much more productive as a statement for a well-established scientific idea or concept. Even so, the terms "hypotheses," "theories," "laws," and "models" are all important concepts of science; and, at some point, they all started as ideas. They all may be thought of steps in the systematic searches for "truth," which in science means the way things are—not as some humans would like them to be. Also, "truth" is not relative as some advocate. Accepted scientific theories are established facts and truths related to scientific laws and are viable and proven fundamental models (statements) about nature. Many established scientific theories might be considered laws because they both meet the same criteria for proven ideas.

Scientific models of nature are mathematical constructs of step-by-step rules that reflect what happens in actual natural events. Scientific models are often undated and are revised as new evidence become available. Scientific theories may therefore be thought of as types of models designed as general explanatory statements about the workings of nature. As established explanations of how things work in nature, they are the end points of scientifically gathered evidence about specific events that

In his book *The Trouble with Physics*, Lee Smolin (2006) explains why the recent "string theory" was proposed to combine all theories of matter and energy into one simple unified theory, but it has not yet become a proven theory for several reasons, mainly because it cannot, as currently stated, be used to make predictions. Smolin describes a theory as:

> "In science, for a theory to be believed, it must make a prediction—different from those made by previous theories—for an experiment not yet done. For the experiment to be meaningful, we must be able to get an answer that disagrees with that prediction. When this is the case, we say that a theory is falsifiable—vulnerable to being shown false. The theory also has to be confirmable, it must be possible to verify a new prediction that only this theory makes. Only when a theory has been tested and the results agree with the theory do we advance the statement to the rank of a true scientific theory."

may incorporate other laws and hypotheses. Humans derive all theories, and as such scientists make linguistic constructs of assumptions. They are similar to, but much more than, educated guesses because they are the results of crucial observations, experimentation, logical inferences, and creative thinking. They are not the same as what we think of as "guesses" as stated in everyday life. They are neither undocumented statements nor uninformed opinions. People often come up with "theories" based on social or behavioral notions or assumptions, which are often accepted by faith. These nonscientific "theories" are without experimental proof and seldom based on mathematics, verifiable facts, measurable data, and evidence. The main test of validity (truth or correctness) of an idea, concept, or theory (model) is found in the results of the experiment, and when possible, a controlled experiment. Theories advanced by scientists can be described as predictions based on the scientist's knowledge of a probable occurrence within a given set of circumstances or conditions. In fact, the validity of a theory, law, or model is determined by its predictability of one or more events. The nature of science is that exactly the same thing may *not always* occur at exactly the same time in exactly the same way in nature, but this does not negate the requirement of predictability of the law, theory, or model. Therefore, scientists, through experimentations, often seek a statistical average upon which to make their predictions. This being said, a proven fundamental scientific law or theory is as close as humans can get to 100 percent truth. Unfortunately, there are some people who consider everything relative (not in the Einstein concept of relativity as related to one's point of reference) but in accepting that everyone's relativity and often ambiguous "truths" are as authentic as proven scientific laws and theories.

An important characteristic of a theory is that it must be stated in such a way that it is nonambiguous. A vague theory cannot be proven wrong, and it is possible to come up with almost any answer desired for an ambiguously stated theory. Confirming a theory requires specific conditions, an experiment (controlled, if possible), and measured results that are analyzed statistically. If the related facts indicate a high probability for its validity and reliability, the theory can then be said to be justified and acceptable. Even so, there must exist the possibility that the theory may not be quite correct and will require additional testing, etc., and new knowledge related to the basic theory becomes available.

A theory is only as good as the limited number of assumptions and generalizations postulated. The fewer astute assumptions incorporated in a theory, the more likely it will stand the test of time. This is known as Ockham's razor (William of Ockham, c.1284–1349) which is a maxim that states that "*Entities ought not be multiplied, except from necessity.*" In modern

> There are two general classifications of theories. One covers a large range of ideas and concepts, often referred to as "breadth" or "broad" theory. Historically, ancient people attempting to understand their world used "broad theory" such as myths by using stories and folktales to explain observed phenomena. Scientists are still attempting to reduce all scientific theories, principles, and laws into a unified field theory (UFT), grand unification theory (GUT), theory of everything (TOE), or come up with the "final answer." However, theories based on the laws of gravity, time, conservation, symmetry, relativity, quantum mechanics, etc., are not yet explained in ways that can be incorporated into a general, universal theory of everything. Nature is extremely complex, and humans are still attempting to understand it more fully.
>
> In summary, a theory must exhibit the following conditions: 1) It must explain the law from which it was derived; 2) It must in some way be related to the law it challenges; 3) It must be able to predict new, verifiable adjustments to the law or postulate a new law; and 4) it must be stated in such a manner that it can be proven false as well true (validity/true and reliable/repeatability).

vernacular it might be thought of as "KISS" (keep it simple stupid). In other words, the number of unnecessary assumptions should be avoided in formulating theories and hypotheses.

SCIENTIFIC HYPOTHESES

The word "hypothesis" comes from the Greek, *hypo thesis*, meaning "placed under" or "foundation." Somewhat similar to a theory a hypothesis is a more tentative observation of facts. Every scientific law and theory began as an idea, question, or hypothesis. Hypotheses and theories lend themselves to deductions that can be experimentally tested. A hypothesis is a logical and rational explanation of a series of critical observations that have not yet been disproved or proved, nor accepted by the scientific community. Hypotheses are reasonable statements, measurable assumptions, and generalizations drawn from a series of observations and selected facts. In other words, scientists confirm a hypothesis by experiments under controlled conditions, and if the data resulting from the experiments do not support the original hypothesis, it must be altered or discarded.

The origins of hypotheses (or concepts) are immaterial. They can be derived from intuitions, dreams, or as ideas arrived at by scientists who have knowledge and understanding of the subject related to the hypothesis. What matters is that the investigator must be familiar with the processes (methods) of scientific research and that the hypothesis is systematically tested to determine consequences. Like theories, hypotheses are products of the informed imaginations of scientists, but they are not wild speculations. Hypotheses are only accepted when tested and confirmed by additional observations by other investigators who conduct their own related experiments.

A viable scientific hypothesis must be stated in such a way that it has some chance of being disproved, and proved, and that it conforms to the observed facts (just as with a theory). Hypotheses can be "proved" or "disproved" by continuing observations of the phenomena and additional experimentation. It is the responsibility of the person advancing the hypothesis (or theory) to "prove" his or her case—not the responsibility of someone else to disprove it, although others can certainly challenge the results and conduct their own related experiments for another person's hypothesis. As an example, it is the responsibility of someone who states as a fact that angels, ghosts, and spirits exist to prove that they are real and actually exist. It is not the responsibility of science to disprove such beliefs (hypotheses).

Scientific hypotheses can only answer questions for which answers can be achieved by observing, testing, measuring, gathering data, and statistically treating and analyzing the data. The results become valid and reliable only when others repeat the experiment and obtain similar results. This is why science can deal only with "answerable questions." Most questions are not "answerable" in a scientific sense nor can they be stated as a scientific hypothesis because answers for such questions cannot be measured, or effectively analyzed. For example, the questions: "What is the secret of success?" or "How can I become more popular?" or "How can I become rich?" (One wag suggested that that person acquire more wealth.) These are, scientifically speaking, nonsense questions that don't lend themselves to rational answers. Even so these are the types of questions many people ask—the answers just cannot be gleaned by controlled experimentation and the analysis of data. For instance, "success" is subjective and not the same for everyone. "The" implies there is only one secret to success. And, if it is a "secret," no one knows the answer. Being able to formulate questions that can be answered by conducting controlled experiments by all who wish to investigate nature requires a statistical analysis of data. A scientific hypothesis (also theory, principle, or law) must be changed if new observed facts or experiments contradict or even slightly alter the original statement or data. This self-correcting process is one of the basic reasons that science may some day triumph over ignorance.

SCIENCE CONCEPTS

A concept is one step above a *specific idea*. Related to ideas are assumptions (notions) that are based on beliefs in or about something, most often accepted without proof. What counts in science is how ideas are developed into viable concepts, hypotheses, theories, and models that accurately describe nature. Most ideas and concepts people come up with are a "dime a dozen" and seldom result in anything beneficial.

CAUSALITY

There is one more concept that can be confusing. It deals with the causes of events on the micro/quanta and macro/universal levels. Causal laws are considered those that can predict and explain empirical and theoretical laws that describe the nature of our known universe. This relationship is usually referred to as "cause and effect." Some science philosophers classify causal principles as 1) empirical generalizations from facts based on other facts, 2) a rational interpretation of a required connection between two or more events, and, 3) a cause may be a useful or pragmatic explanation of science. This is what we usually consider as a specific cause (or causes) leading to a specific effect (or effects or events) but may not necessarily be a direct observable connection between the two.

During Sir Isaac Newton's time science was considered a system to describe a mechanistic, deterministic, and reductionist world. Determinism and predictability are not the same. Determinism deals with how nature behaves, particularly nature as we know it within the solar system—it depends on the "laws of nature." Predictability is based on what scientists are able to observe, measure, and analyze, as related to outcomes for specific events. To better understand complex nature, scientists attempt to "reduce" its complexity to more manageable and understandable laws, principles, and theories. We live our everyday lives in a Newtonian mechanistic solar system based on logic, physics, and mathematics, whereas the Einsteinian universe is based on theories of relative space and time in particular perspectives to each other that are not easy to apply on Earth as are Newtonian laws of physics. Why? Because the setting for events on Earth is very small compared to the relational aspects for the frames of reference in space and time of the universe. Einstein's relativistic physics rendered some of Newton's laws of motion only approximations. Newton's laws do not hold up when considered for the immense universe consisting of great space, energies, masses, and velocities in different frames of reference over very great distances of space and time.

Historically, all effects or events were assumed to have a cause, or possibly several causes, or to be co-events. We now know that many natural events are described and predicted by statistical probabilities, not mathematical certainties. This is true for very large events in the universe and the very small events as related to subatomic particles and energy. These very small events led to quantum theory and indeterminacy (uncertainty principle) resulting in some problems with the cause-and-effect concept for accepted physical laws. This is one reason scientists think in "statistical probabilities" rather than what is "possible." Probabilities involve measurements to arrive at predictable events, whereas possibilities may or may not lead to an event.

No one ever expects an *effect* to precede a *cause*. Scientists would say that such a situation is unlikely (with a very low degree of probability), but we never know for sure. In other words, the probability of the effect preceding the cause is practically zero. There are some effects (events) in nature where the cause or causes have not been detected and thus not well understood. This is one reason that the cause-and-effect relationship is not highly thought of by some scientists.

The nature of the universe is extremely complex. The universe is about 13 to 14 billion years old, the Earth is about 3.5 to 4 billion years old, and thinking, rational humans have been on Earth for only several hundred thousands of years. Early humans must surely have wondered about the world around them and by observing cycles of nature speculated how things worked and affected them. It has been about five hundred years since humans arrived at a systematic method of "asking" questions of nature for which rational answers were possible. There are many areas of nature that we do not yet understand, including our own nature and role in the scheme of things. No doubt, we will continue exploring the unknown throughout the twenty-first century and beyond.

ABBE'S THEORY FOR CORRECTING LENS DISTORTIONS: Physics: *Ernst Abbe* (1840–1905), Germany.

> *The equation for Abbe's theory is:* $u'/\sin U' = u/\sin U$, *where* u *and* u' *are the angles for the entering and exiting of rays from the object to the image, as in a microscope.*

Ernst Abbe was raised in a poor German family but managed to become a physics professor at the University of Jena (now located in Germany). He also was the director of the university's observatory and a theoretical optical consultant for the Carl Zeiss instrument company. Up to this time the field of optics was an "art," but Abbe brought several viable theories of optics to bear on several problems of the day, for example, chromatic aberrations in lenses. He also worked with Otto Schott, a glassmaker, to improve several optical devices. In 1888 he became the owner of the Carl Zeiss optical company that is known worldwide for its high-quality precision instruments.

The *Abbe sine condition* is a mathematical concept used to make lenses that produce sharper images and less distortion. It is a means to eliminate spherical aberration of an optical system to produce an **aplanatic lens** (corrected lens). In 1886 Abbe used his mathematical approach to develop apochromatic lenses (corrected for chromatic and spherical aberrations) to eliminate primary and secondary color distortions. The U and U' are the corresponding angles of any other rays transmitted.

Abbe's contributions to the field of optics include several inventions, most notably the Abbe condensing lens to focus light onto microscopic slides, the achromatic lens, the Abbe refractometer, and the prismatic binocular, a design for binoculars that is still used today. His contributions led to the improvement of all of optical instruments, including sharper images with less color distortion for cameras, microscopes, refracting telescopes, spectroscopes, and so forth.

See also Newton

ABEGG'S RULE AND VALENCE THEORY: Chemistry: *Richard Abegg* (1869–1910), Poland.

Chemical reactions are the result of electron transfer from one atom to another.

Richard Abegg attended several universities in Poland and Germany, graduating from the University of Berlin in 1891. He was the pupil of Wilhelm Hofmann, the famous theoretical chemist. Although trained in organic chemistry, Abegg was interested in the physical nature of chemistry. He was working as an electrochemist when he published his famous paper, "Die Elektronaffinität," that explained his theory on the combining power of chemicals (**valence**).

Abegg not only became famous for what became known as the *Abegg Rule*, but also he arrived at his rule years before the existence of valence for elements was firmly established. His rule and valence theory predated Dmitri Mendeleev's development of the periodicity of elements. The rule stated that each element has a positive valence and a negative valence whose sum is eight. This predated the "octet rule" for the periodic table. He further theorized that the *attraction of electrons for atoms of all elements has two distinct types of similarities or valences, that is, a "normal affinity" (valence) and a "counter affinity" (countervalence).*

His theory that two related valences always add up to eight, published in 1869, is responsible for the octet rule as related to the **Periodic Table of Chemical Elements** (*see* Figure S2 under Sidgwick). Abegg's early theories of valence became valuable for later chemists and were used to explain chemical reactions and the organization of the periodic table. For instance, elements in the groups with just one or two electrons in their outer orbit (metals), when combining with other elements (nonmetals), tend to give up their outer electron(s) and thus attain an outer octet of electrons. Elements with six or seven electrons in their outer orbits tend to acquire electrons to complete their outer octet orbit.

The octet rule is just a device to remember the position of some elements in the periodic table, and it is only valid for elements located in the higher periods and groups of the periodic table. The concept of valence has become extremely useful in all fields of chemistry.

See also Mendeleev; Newlands; Sidgwick

ABEL'S THEORY OF GROUPS (ABELIAN GROUPS): Mathematics: *Niels Henrik Abel* (1802–1829), Norway.

Following is a list of Niels Abel's notable accomplishments in his short lifetime:

1. In 1824 at age nineteen he proved that there was no general algebraic solution (proof) for roots to solve *quintic equations* by radicals, or in other *polynomial equations* of degrees greater than four.
2. In 1826 he published an updated version of this proof in August Leopold Crelle's new journal on mathematics.
3. He invented a branch of mathematics known as *group theory*, for which he has become famous.
4. He did original research in the theory of functions including *elliptic* and *hyperelliptic functions* that later became known as the famous abelian functions.
5. He is also known for the *abelian group*, *abelian category*, *abelian variety*, and the *abel transformation*. Note: these terms are so common in mathematics that they

are expressed with the lower case (a). This is contrary to what one might expect when naming such important mathematical concepts and theories for a person who has received many of the highest honors in his field.

> Niels Henrik Abel's story is that of a twenty-six-year-old genius's life of "rags-to-rags." He was born into a large family with six male siblings whose father was a poor Protestant minister in a church in Christiansand, a diocese in Norway. At age fifteen Niels proved to have a remarkable knack for understanding complicated mathematical principles. He was only age eighteen when the breadwinner of the family died, leaving the family to live in miserable conditions. A small government pension provided Niels an opportunity to attend the local cathedral school in Christiania in 1821. In 1825 he was awarded another state scholarship that enabled him to visit France for ten months where he became friends with many mathematicians. He briefly visited Germany where he met local mathematicians. As he was unable to secure a paying job, he ran out of money and had to return to Norway. Even after mastering a number of remarkable developments in the field of mathematics that few other mathematicians understood, Niels remained in a life of poverty as he sought, through the assistance of a friend, a position at the University of Berlin. The letter appointing him to this position by his friend, the amateur mathematician August Crelle, to whose journal Abel contributed several publications, arrived two days after his early death from tuberculosis on April 6, 1829.

The Abel Prize was established in 2001 by the King of Norway on the bicentennial of the birth of Niels Henrik Abel. This prize is awarded annually to an outstanding mathematician and is intended to stimulate interest in mathematics among young scientists and increase the amount of research in the field of mathematics, as well as improve the image of Norway as a nation of "learning." The Abel Prize in mathematics is considered on a par with the Swedish Nobel Prize in other fields.

Abel's theorem is proof of a fundamental theorem on transcendental functions. Abel's theorem, and his outstanding theory of elliptical functions, kept mathematicians busy for the latter part of the nineteenth century.

The usefulness of the Abelian group concept in mathematics is based on notations and is sometimes called the "commutative group" that can be additive or multiplicative. For example: if the group is (G *) a * b = b * a, the end product is immaterial. Complete discussion of the Abelian group theory is beyond the scope of this book.

ADAMS' CONCEPT OF HYDROGENATION: Organic chemistry: *Roger Adams* (1889–1971), United States.

1. *Additional hydrogenation will take place when hydrogen is added to double bonds of unsaturated molecules of organic substances such as liquid fats and oils.*
2. *Hydrogenolysis hydrogenation will take place when hydrogen breaks the bonds of organic molecules, permitting a reaction of hydrogen with the molecular fragments.*

The first type of hydrogenation led to the formation of solid or semisolid fats from liquid oils, and the second process is used in hydrocracking petroleum or adding hydrogen to coal molecules to increase its heat output. In the early 1900s Roger Adams' ideas resulted in the successful hydrogenation of unsaturated organic compounds by catalyzing them with finely powdered platinum and palladium metals under heat. It is similar to the process of reduction in organic chemistry that led to the hydrogenation of many of our fuels and foods, where liquid oils are converted to semisolid oils or fats.

The hydrogenation of unsaturated organic compounds is used in industrial processes that result in many of today's most popular products. Some examples are the synthesis of liquid fuels; the production of several alcohols, including methanol; the production of aldehydes and benzene derivatives; the synthesis of nylon; the hydrogenation process of peanut butter to make a smooth spread without the separation of the peanut oil; and hydrogenation of liquid vegetable oils to form margarine. Hydrogenation is also used to form solid petroleum fuels and some semisolid medications. The second type of hydrogenation led to an increase in the production of petroleum products from crude oil, such as gasoline.

ADHEMAR'S ICE AGE THEORY: Astronomy: *Joseph Alphonse Adhemar* (1797–1862), France.

> *Because Earth tilts 23.5% from its ecliptic (the orbital plane of Earth around the sun), the Southern Hemisphere receives about two hundred fewer hours of sunlight per year. Therefore, more ice accumulates on Antarctica than at the North Pole.*

In 1842, Joseph Adhemar proposed his theory, the first to provide a reasonable answer to the questions as to the causes of the ice ages. His theory is based on evidence gained from astronomical events expressed in his book *Les Revolutions da la mer* published in 1842. He was the first to propose that natural astronomical occurrences might be responsible for the ice ages and, by implication, global warming/cooling cycles. He realized that Earth is just one focal point in its elliptical orbit around the sun, which means that Earth is farther away from the sun in July. In addition, Earth's axis does not always point in the same direction in space; rather, it slowly rotates in a small circular orbit approximately every twenty-six thousand years (called *precession*). These astronomical factors, plus the inclination of Earth, cause the winters to be slightly longer for the Southern Hemisphere.

Due to the tilt of Earth on its axis the South Pole receives about 170 to 200 fewer hours of sunlight than does the North Pole. Adhemar claimed these differences were adequate explanations for the extensive ice build-up at the South Pole. It might be mentioned that during the current global warming trend, unlike the North Pole, the South Pole is still building up its ice sheet, which Adhemar claimed contributed to the ice age. However, many scientists today do not completely agree with his theory.

See also Agassiz; Kepler

AGASSIZ'S GEOLOGICAL THEORIES: Geology: *Jean Louis Rodolphe Agassiz* (1807–1873), Switzerland.

Agassiz's glacier theory: *The movement of glaciers created scratch marks in rocks, smoothed over vast areas of the terrain, gouged out great valleys, carried large boulders over long distances, and left piles of dirt, soil, and debris called moraines.*

In 1836–1837 Jean Agassiz arrived at the new idea that glaciers in his native Switzerland were not static when he realized that a hut on a glacier had moved over a period of years. He then constructed a line across a glacier and noted that after a year it had moved. After discovering rocks and scars in the landscape that he found under the glacier, he concluded that much of Northern Europe had at one time been covered with ice. This led to his theory of the cause and effect of glaciers. As a keen observer of geological phenomena he believed there was evidence of glaciers where none now exists. After his theory was published in 1840, he was invited to the United States in 1846 as a lecturer where he speculated that North America had also experienced the effects of glaciers. He further speculated that ice sheets and glaciers formed at the same time in most of the continents.

Agassiz's second glacier theory: *Glacier ice sheets had movements that included advancements as well as periods of retreat, which correspond to the ice ages.*

Jean Agassiz's work led to the concept that glaciers were the result of the ice age. As the ice sheets that covered Earth melted in the warmer zones, great deposits of compact ice were deposited in the colder regions of the Northern and Southern Hemispheres. These ice masses, now called *glaciers*, moved slowly toward the warmer areas, and they continue to move even today. Periodic ice ages, some much smaller than the original frozen Earth period, formed, advanced, and retreated over many millions of years. Ice ages on Earth are considered normal cyclic occurrences. Previous to Agassiz's glacier theories, scientists assumed glaciers were caused by icebergs. In fact, the opposite is true, icebergs are caused by the calving of huge chunks off the edges of glaciers that extend into lakes and oceans. Agassiz, sometimes known as the father of glaciology, also made contributions to evolutionary development through his study and classification of the fossils of freshwater fish.

Agassiz's theory of fossils and evolution: *The lowest forms of organisms were found in rock strata located at the lowest levels of rock formations.*

Before Agassiz devised his theory, William Smith (1769–1839), the English surveyor and "amateur" geologist, proposed that fossils found in older layers of sedimentary rocks were much older than the more modern-appearing fossils located in sediments laid down in more recent geological times. This concept answered some of the questions about the evolution of species because the ages of fossil plants and animals now could be determined by their placement in the rock strata. This information led to Agassiz's more formal theory on the subject.

Agassiz at first did not believe in evolution but later did accept the concept of evolution while still rejecting Darwin's theory of natural selection because it required long, gradual periods of change. Oddly, Agassiz's studies of fossils were used to support Darwin's theory of natural selection. Agassiz accepted and expanded on Georges Cuvier's catastrophism theory of evolution, which postulated periods of rapid environmental changes, not natural selection, as the basis of evolution.

William Smith was born in 1769 on a small farm and received little formal education. His interest was in collecting and studying fossils while he learned geometry. By age eighteen he worked as an assistant to the master surveyor, Edward Webb. This enabled him to travel extensively in England and led to his supervision of the digging of the Somerset Canal in southern England. The digging of the canal provided Smith with an opportunity to observe the fossils found in the sedimentary rock. He noticed that older-appearing fossils were always found in the older, deeper layer of rocks. This led to the principle that "The layers of sedimentary rocks in any given location contain fossils in a definite sequence." In other words, the same sequence can be found in different geographic locations, thus providing a correlation between locations. Smith's career expanded, and he became a well-known engineer who developed a complete geological map of England and Wales. He had difficulty in raising funds to publish his map, but it was finally published in 1815. A biography of William Smith by Simon Winchester titled *The Map That Changed The World—The Birth of Modern Geology* was published in 2001 by HarperCollins. The book's cover is a unique expandable geological map of the below-surface geological structure of Great Britain.

More recent uses of Agassiz's theories of ice ages and stratification of fossils should give pause to many claims that modern society is responsible for global warming and an increase in the extinction of plants and animal. Both processes have been occurring over eons of time as cycles in the cooling and heating of Earth and the rate of plant and animal extinction due to natural environmental changes. Although humans certainly influence and alter their environment, just as do all living organisms, which in turn result in some evolutionary changes, there are multiple causes and influences on these two interrelated natural cycles. These cycles most likely will continue long after the human species is extinct.

See also Adhemar; Agricola; Charpentier; Cuvier; Darwin; Eldredge–Gould; Gould; Lyell; Raup

AGRICOLA'S THEORIES OF EARTHQUAKES AND VOLCANOES: Geology (Mineralogy, Metallurgy): *Georgius Agricola* (1494–1555), Germany.

Agricola's original name was Georg Bauer, but he followed the custom of the time and chose the Latin name, Georgius Agricola. *Agricola* means "farmer" while his original name meant "peasant." In the 1540s Agricola's studies based on his observations of minerals and stratified layers of rock led to several geological theories.

Agricola's theory of stratification: *Stratified forms of rocks are the arrangement and relationships of different layers of sedimentary rocks as formed during earthquakes, floods, and volcanoes.*

The planes between different strata (layers) assist in determining not only the source of the sedimentary rocks but also the area's local history. In addition, the fossil content found in different strata provides a record of the biological and geological history of Earth. Today, the study of stratification is called *stratigraphy*, which is the branch of geology that studies the different layers of rock. Agricola's stratification system, though primitive, proved useful as one means of identifying the location and sources of petroleum.

Agricola's theory of earthquakes and volcanoes: *Earthquakes and volcanoes are caused by subterranean (below ground level) gases and vapors originating deep in Earth, where they are heated and then escape to the surface.*

A keen observer and practicing physician specializing in miners' diseases, Agricola's main interest was minerals, known in those days as metals. In 1546 he was one of the first to classify minerals according to their physical properties, such as color, weight, and texture. He believed these minerals/metals originated deep in the underground and were brought to the surface by earthquakes and volcanoes. Agricola was also known as a paleontologist for his

Although later scientists disputed some of his ideas and subsequently revised his theories, Agricola's publications were used in the field of geology for over two hundred years. His most famous book *De re metallica* (1556) provided much new information on mining and metallurgy, including mine management, how to use windmill-driven pumps, and how to derive power from water wheels. He updated the process required for working with metals. Much of the information in his writings, as was other knowledge of the Middle Ages, was derived from antiquity and Arabic scholars. Interestingly, in 1912 President H. Hoover and his wife Lou translated Agricola's book into English. It is still available. In 1546 Agricola published *De ortu et causis subterraneorum* (Origins and Causes of Subterranean Things) in which he updated the concept of "rock juices" as subterranean fluids. Also in 1546 he published, *De natura fossilium*, which was a new classification system of minerals (called fossils in those days) that was based on a mineral's physical properties such as color, weight, texture, solubility, combustibility, etc.

work with objects found in the soil, including fossils, gemstones, and even gallstones. He is sometimes referred to as the father or founder of mineralogy and the science of geology because he was among the first to describe fossils as once-living organisms.

AIRY'S CONCEPTS OF GEOLOGIC EQUILIBRIUM: Geology and Astronomy: *Sir George Biddell Airy* (1801–1892), England.

Airy, along with other scientists, proposed two theories of equilibrium as expressed in geological structures.

Airy's first theory of geologic equilibrium: *Mountains must have root structures of a lower density in proportion to their heights in order to maintain their isostasy (equilibrium).*

Isostasy is Airy's theory that there is a proportional balance between the height of mountains as compared to the distribution of the root structure or mass underneath the mountain. He claimed this equilibrium resulted in a balance of **hydrostatic** pressure for the formation of mountains. It became an important concept in geology and aided in the exploration of minerals and gas and oil deposits. Although his theory has been revised, it was a step in the right direction in understanding the dynamics involved in geological systems.

Shortly after graduation from Cambridge University in 1819 he wrote several successful textbooks in mathematics and optics. He later became a professor of astronomy and the director of the Cambridge Observatory where he developed several innovative instruments. These include a type of altazimuth device for lunar observations, a transit circle, and a new equatorial telescope with a spectroscope. He also devised an optical device with a central hole (called the *Airy disk*) to examine diffraction patterns of a point source of light.

His second theory deals with internal water waves and is based on ideas expressed by Vagn Walfrid Ekman (1874–1954).

Airy's second theory of equilibrium: *In areas where the sea is covered with a thin layer of freshwater, energy is generated by internal waves and is radiated away from ships, which subsequently produces a drag on ships.*

Because this slows the ship's progress, it is known as *dead water*. One area where freshwater and seawater are at different levels, which causes these internal waves to drag on ships, is the entrance to the Mediterranean Sea at Gibraltar.

> A story from World War II pertains to the fact that the water in the Mediterranean Sea has a higher salt content due to evaporation than does the water in the Atlantic Ocean. Because the Mediterranean's water is saltier and thus denser, it flows out past the Strait of Gibraltar near the bottom of the Strait into the Atlantic Ocean. At the same time, the less salty (less dense) Atlantic Ocean water flows into the Mediterranean near the surface. It was suggested that when submarines wished to avoid detection as they passed through the fortified Strait of Gibraltar they drifted quietly in the less dense water into the Mediterranean near the surface. When submarines wished to leave without being detected, they drifted quietly in the more dense water near the bottom as they were "carried" past the protected Gibraltar and into the Atlantic. By using the difference in density of saltwater in the Atlantic Ocean and the Mediterranean Sea submarines could apply the "Airy waves" concept to their advantage. The success of this tactic remains unknown.

AL-BATTANI'S THEORIES: Astronomy: *Abu Abdullah Al-Battani* (c.858–929), Mesopotamia (present-day Iraq).

Abu Abdullah Al-Battani developed various theories after improving measurements completed by Ptolemy of Alexandria.

Al-Battani's theory of solar perigee: *Solar perigee equations demonstrate slow variations over time.*

Al-Battani determined that the sun's perigee (the point at which the sun is closest to Earth in Earth's elliptical path around the sun) is greater than Ptolemy's measurement by a difference of over 16°47′. Although Al-Battani admired Ptolemy, he improved on several of Ptolemy's calculations, including the ecliptic of Earth's orbit to its equatorial plane.

Al-Battani's theory of the Earth's ecliptic: *The inclination of the angle of Earth's equatorial plane to its orbital plane is 23°35′.*

This figure is very close to the current measurement. The ecliptic for Earth can also be thought of as the apparent yearly path of the sun as Earth revolves around it. In other words, it is the angle of tilt of Earth to its solar orbital plane that is the major cause of our four seasons—not how close Earth is to the sun. The Northern Hemisphere is tilted more toward the sun during the summer, thus receiving more direct sunlight for more daylight hours than in the winter, when the Northern Hemisphere is tilted away from the sun. The situation is reversed for the Southern Hemisphere.

Al-Battani's theory of the length of the year: *The solar year is 365 days, 5 hours, and 24 seconds.*

Al-Battani's calculations are very close to the actual figure of 365.24220 days that is accepted today. This led to more exact recalculations for the dates of the spring and fall **equinoxes**. The dates for the equinoxes and the accuracy for the length of the year were important for various religions of the world that base many of their holy days on the seasons and these particular dates.

Al-Battani's concept of the motion of the moon: *It is possible to determine the acceleration of the motion of the Moon by measuring the lunar and solar eclipses.*

Al-Battani was able to time the lunar and solar eclipses and thus able to extrapolate this figure to calculate the speed of the moon in its orbit. He also devised a theory for determining the visibility of the new moon. Albategnius, an eighty-mile plane surface area on Earth's moon surrounded by high mountains, is named for Al-Battani.

Al-Battani's concepts of two trigonometric ratios: 1) *Sines* (which he formulated) *were demonstrated as more practical and superior to the use of Greek chords.* (A chord is a line segment that intersects a curve only at the end of the curve.) 2) *Cotangents are the reciprocal of tangents.*

Considering the period in history during which Al-Battaini lived, his contributions to astronomy and mathematics are extremely innovative and accurate. Al-Battani devised tables for the use of *sines*, *cosines*, *tangents*, and *cotangents*, which are invaluable for modern algebra and trigonometry. Copernicus credited Al-Battani with advancing astronomy based on his work in trigonometry and algebra. For many years, Al-Battani's contributions to science and mathematics were considered preeminent and advanced the cause of knowledge over the next several centuries. He is sometimes referred to by his Latin name, *Albategnius*.

ALVAREZ'S HYPOTHESES OF SUBATOMIC COLLISIONS: Physics: *Luis Walter Alvarez* (1911–1988); United States. Luis Walter Alvarez was awarded the 1968 Nobel Prize in Physics.

Alvarez K-capture hypothesis: *The vacant inner K shell will capture an orbital electron as the electron moves from an outer shell to the innermost K shell of the atom.*

Luis Alvarez's discovery of the number of high-energy nuclear collisions of subatomic particles was the result of his developing a liquid-hydrogen bubble chamber. This "bubble chamber" was able to detect the decay of nuclei of atoms using the procedure known as K-capture of the orbiting electrons from the innermost orbit (K shell or orbit) of atoms.

He also worked on the World War II Manhattan Project where he developed the device for detonating the atomic bomb. In addition, he held over thirty patents, including those for unique radar systems. His large, seventy-two-inch bubble chamber first used in 1959 contributed to the discovery of many elementary subatomic particles.

Along with his son, Walter Alvarez (1940–), and several other geologists, he arrived at what is known as "the Alvarez asteroid impact theory" in 1980. The theory states that an asteroid about ten miles in diameter struck Earth approximately sixty-five million years ago with an impact that sent debris into the atmosphere that blocked the sun and caused worldwide storms, acid rain, and other chemical changes in Earth's air. Other conditions, such as fires and high winds, along with the many volcanoes present at that time, contributed to the disruption in the balance of the environment resulting in a cooling period during which much of the plant life was wiped out. This loss of plant life killed off the herbivores (animals that survive on plants) and subsequently the carnivores (meat eaters) that needed the herbivores as a food source. Thus, with their food supplies eradicated, all species of dinosaurs suffered extinction, that is the herbivores, carnivores, and omnivores.

> There is considerable evidence for this catastrophic event. A 120-mile-wide crater, known as the "Chicxulub" impact crater is located at the tip of the Yucatan Peninsula in the Gulf of Mexico. There is evidence that this impact caused huge tsunamis in the area. (It might be mentioned that another huge asteroid hit Earth at about the same time in the Arabian Sea off the coast of Bombay, India. It is called the *Shiva Crater* after the Hindu god of destruction and renewal.)
>
> Additional evidence for the Chicxulub Crater is known as the K-T layer of sedimentary deposits. This is a strip of clay with a relatively high concentration of iridium brought to Earth by the asteroid. (Note: the "K" stands for *Kreide,* which is German for "cretaceous," and the "T" stands for "tertiary," that are the two geological periods between which the clay strip of sediment is found.) Other evidences of the ancient event are the discovery of several rare earth elements (siderophiles), particularly the rare earth iridium, as well as tektites that are also found in the region. Tektites are quartz grains that were vaporized by extreme heat and pressure that crystallize into glass beads when cooled. Tektites result from high-impact meteorites, asteroids, and **bolides.** Many tektites are found in the K-T layer. Other evidence of the impact are layers of quartz and glass beads found at the crater sites.

AMBARTSUMIAN'S THEORY OF STELLAR ASSOCIATIONS: Astronomy: *Viktor Amazaspovich Ambartsumian (also known as Ambarzsumian Ambarzumyan)* (1908–1996), Armenia-Russia.

Star systems, including galaxies, form cluster type associations as they evolve.

Ambartsumian was a professor of astrophysics at the University of Leningrad and later at the University at Yerevan in Armenia. He established the Byurakan Astronomical Observatory in 1946 when he did most of his work in the evolution of galaxies and star clusters. His main interest was the **cosmogony** of stars and galaxies and nebulas.

His work on stellar dynamics led to his theory of stellar associations that is based on loose groups of young hot stars that are located near the disk-shaped plane of our Milky Way galaxy. This association occurs only with young stars that are just a few million years old. As they age over many millions of years, the galaxy's gravity will separate them from their relatively close association sending them further apart from each other. This seems to be evidence that new star formation in our galaxy is ongoing.

Ambartsumian was the first to propose that the T Tauri type stars are very young and are found in groupings (clusters) that are expanding. In addition, he demonstrated that as galaxies evolve, they lose mass. During his long career he also explored the nature of interstellar matter and the radio signals emitted by galaxies.

A small planet was recently named for Ambartsumian, as was a dwarf galaxy located in the constellation Ursa Major, referred to as *Ambartsumain's Knot*.

AMDAHL'S LAW: Computer Science: *Gene Myron Amdahl* (1922–), United States.

Amdahl's law of parallel computing can be stated in several ways, but it is basically a law related to the acceleration of computers from using just a single computer to multiple parallel computers. The law is used to find the maximum expected improvement to an entire system when only one part of the system is known. A simple statement of the law is: *Parallel computing that performs speedup of a parallel algorithm is limited by the fraction of the problem that must be performed sequentially, which is to say that a design is only as strong as its weakest link.*

In essence, this means it is not the number of computers involved that is the limiting factor, but rather it is the algorithm that cannot speedup beyond limits of the paralleling. The term "speedup" is defined as the time it takes a computer program to execute a program with just one processor, divided by the time it takes to execute the program in parallel using many processors. The formula for "speedup" is:

$$S = \frac{T(1)}{T(J)}$$

S = the "speedup." The *T(1)* is the time required to execute the program with a single computer, while the *T(J)* is the time required to execute the program using a *(J)* multiple number of computers in parallel.

Amdahl worked for IBM in developing the famous IBM 704, 709, and 7030 mainframe super computers. In

Sandia National Laboratory located in Livermore, California, is one of three U.S. Defense Program Laboratories of the U.S. Department of Energy. The other two are Los Alamos in New Mexico where the Manhattan Project created the first two atomic "nuclear fission" bombs, and the Lawrence Livermore Laboratory in California where the fission-type atomic bomb technology was advanced to "nuclear fusion" used by the much more powerful hydrogen bomb (H-bomb). These labs, although owned by the federal government are managed and operated by contract to private corporations. The Lockheed Martin Corporation operates the Sandia facilities. The Sandia laboratory has four major responsibilities, all related to meeting national security needs. These four functions are 1) ensuring the safety, reliability, and security of the nation's nuclear weapons; 2) reducing the nation's vulnerability to other nation's use of weapons of mass destruction; 3) improving and enhancing critical infrastructures, in particular energy; and, 4) addressing any new or possible threats to the nation's security.

Following World War II these Department of Energy laboratories have expanded their responsibilities and assumed the major role in providing integrated systems and engineering support for nuclear weapons and the explosive core used to fire these weapons, and to ensure the nation's security from nuclear attacks.

1970 when IBM did not accept his idea for a new super computer, he left the company and established the Amdahl Corporation located in California. He later founded other related corporations in California. The Sandia National Laboratories in Livermore, California, question the validity of Amdahl's law when dealing with massive ensembles of parallel computers. The equation for Amdahl's law may need some revision for the new hypercube-type processors.

AMPÈRE'S THEORIES OF ELECTRODYNAMICS: Physics: *André-Marie Ampère* (1775–1836), France.

Ampère came from a wealthy French merchant's family who hired a private tutor for his education. He was mainly a "self-taught" young genius who early in life showed talent for mathematics. He taught mathematics in several universities and was honored by Napoleon who in 1808 appointed him as the inspector general of the newly formed French university system. In 1820 his interest was aroused at a demonstration of the emerging field of electricity. He was inspired to learn more and just a week later was performing his own experiments to explore the nature of electricity.

In the early 1820s Ampère based his theories of electrodynamics on how electric currents influence each other and how they interact with magnetism.

Ampère's theory of flowing current and magnetism: *When two parallel wires carry current in the same direction, they will attract each other. And if the current flows in opposite directions in the two parallel wires, they will repel each other.*

Ampère related this attraction/repulsion concept to the two poles of **magnets**, which led to his famous laws developed in 1825.

Ampère's law, part I: *The force of the electric current between two wires (or conductors) will exhibit the inverse square law, which states that the force decreases with the square of the distance between the two* ***conductors****, and that the force will be proportional to the product of the two currents.*

Ampère's law, part II: *When there is an electric charge in motion, there will be a magnetic field associated with that motion.*

Ampère's law is related to induction, which can be expressed as the equation: $d\beta = k\ A\ dl\ \sin\theta/r^2$, where A is the current and *k* is a proportional factor based on either the cm (centimeter) or m (meter) units of the SI system. $d\beta$ is the increase in the strength of the magnetic field due to an infinitesimal increment *dl* in the length of the current element (a wire carrying the current A); θ is the angle between the current elements; and *r* is the distance from the element of wire to the part where the field is measured. (Note: the symbol for electrical *current* may be A or *I*, depending on the convention used.)

Ampère devised another law related to the magnetic effects of flowing electrical current in curved wires.

Ampère's circuital law: *The magnetic intensity of a curved or enclosed loop of wire is the sum of the current and can be determined by considering the sum of the*

AMPERE'S LAW

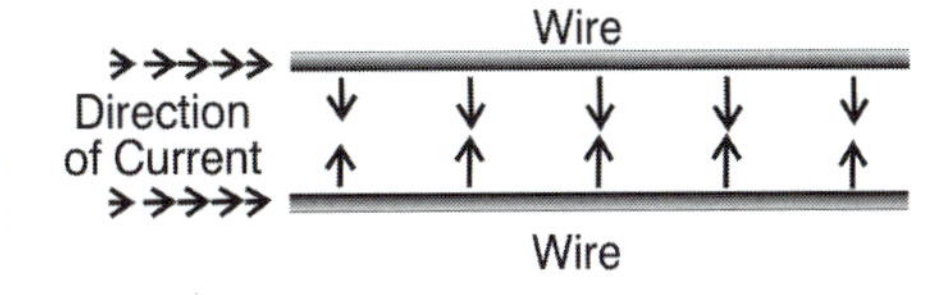

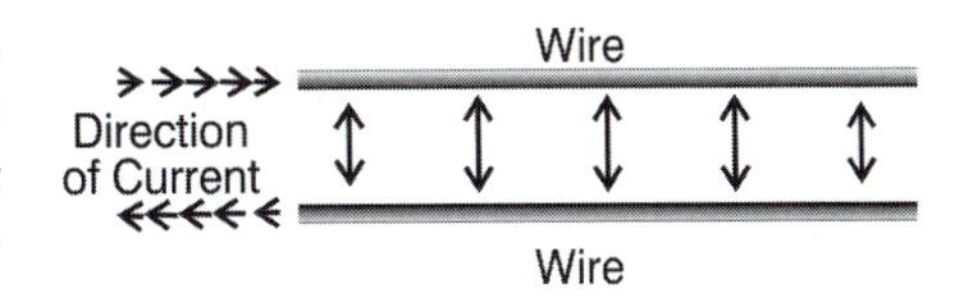

Figure A1. Ampère's Law. For both examples, the strength of the magnetic fields generated between the wires decreases with the square of the distance between the wires.

MAGNETIC FIELD OF CURRENT

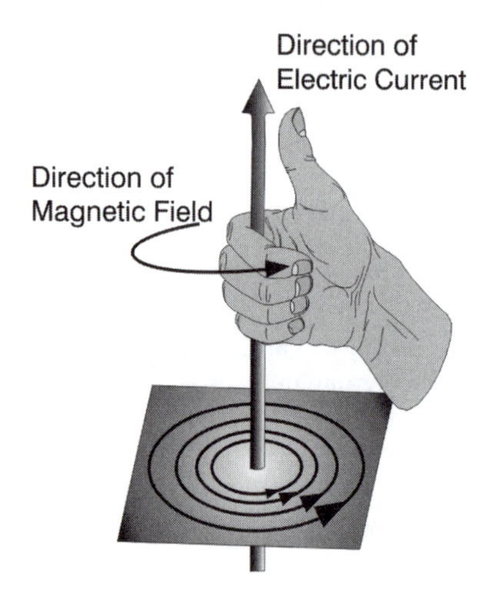

Figure A2. The "right hand rule" indicates the direction of the magnetic field as related to the direction of the flow of current.

magnetic field for each segment of the loop. (This is similar to Gauss's magnetic law for closed surfaces.)

The equation for the circuital law is $B(r) \times 2\pi r = \mu A$ (*B* is the magnetic field at the center of the loop, *r* is the radius of the loop, π and μ are constants, and A [or upper case letter *I*] is the electric current). This law measures the strength of the magnetic field in a **solenoid** and determines the strength of the field in **electromagnets** used in electric generators and motors.

The unit of electrical current known as an ampere or amp was named after Ampère and is given the symbol of A or *I*.

***Ampère* rule:** *The unit of electric current flowing through parallel, straight, long wires in a vacuum produces a force between the wires of* 2×10^{-7} N (newtons) *for every meter of wire through which the current flows.*

In other words, it is the measure of the "amount" of electrical current flowing through a wire. As an analogy, consider amps similar to the "amount" of water flowing through a pipe every second. Another way to think of an amp (A) is to count the number of **electrons** that cross a particular point while flowing through a conductor (wire). The rule also states: *One amp equals* 6×10^{18} *electrons passing this point every second.* Electrical appliances are rated according to the number of amps (current) they use (e.g., a TV set uses 3 to 8 amps, a small motor about 2 to 5 amps, a 100-watt light bulb about 1 amp, an electric stove between 10 and 25 amps, or more) (*see also* Faraday; Galvani; Oersted; Ohm; Maxwell).

There is another Ampère rule that states: *The direction of the magnetic field surrounding a conductor will be clockwise when viewed from the conductor if the direction of the current flow is away from the observer* (also commonly known as the "right-hand grip rule").

> Ampère's contributions advanced the development of many practical industrial devices that make our lives more enjoyable and easier. For example, he suggested his discovery could be used to send signals, which, over time, became a reality (e.g., telegraph, the radio, TV). Others, including Faraday and Oersted, used Ampère's laws to construct the dynamo (electric generator) and the electric motor. A more recent application is the experimental nuclear fusion project to generate heat for the production of electricity. This process requires very strong magnets to produce the pinch effect that will contain and concentrate the hot plasma gases required for the application of nuclear fusion. However, controlled nuclear fusion to produce electricity has yet to be developed to the point where it is practical. Strong electromagnets are also important for the operation of particle accelerators and magnetic resonance imaging (MRI) and positron emission tomography (PET) medical diagnostic instruments.

ANAXIMANDER'S CONCEPTS AND IDEAS: Natural Philosophy: *Anaximander of Miletus* (c.611–547 BCE), Greece and Turkey.

Although Anaximander's writings were lost, his many original concepts and ideas were well known and reported by other Greek philosopher/scientists. Following are some of his original concepts:

1. Anaximander contested his mentor's (Thales of Miletus)

assertion that water was the main substance (element) on Earth. Anaximander stated that basic matter is indefinite. This led to the Greek atomists' concept of the indivisible atom of matter.

2. The Earth does not "rest" on a body of water, as believed by many in his day. Rather, it is not supported by anything, but it is in equilibrium with other bodies in the universe.
3. He was the first to use a sundial (in Greece) to determine the spring and fall equinoxes.
4. He was the first philosopher/scientist to propose a theory for the origin of humans, as well as how Earth was formed.
5. He determined that the surface of Earth was curved but thought that its shape was similar to a cylinder, not a sphere.
6. Earth's axis was oriented east and west within the cylinder. Anaximander was the first to draw a map of the entire world as known in his day.

ANDERSON'S POSITRON THEORY: Physics: *Carl D. Anderson* (1905–1991), United States. Carl Anderson shared the 1936 Nobel Prize with Victor Franz Hess, who discovered cosmic rays.

Cosmic rays passing through a cloud chamber produce tracks of negative particles deflected in one direction (electrons), while producing tracks of particles with equal curvature and equal mass in the opposite direction. Both are deflected by a magnetic field. These particles can be only positive-type "electrons." Thus, they are new elementary particles called a positron.

In 1932 Carl Anderson's concept of a positive electron (positron) was verified by Patrick Blackett (1897–1974) and Giuseppe Occhialini (1907–1993). This conclusion was arrived at by basing it on Paul Dirac's (1902–1984) theory, that the **positron** was the equivalent in mass but opposite in charge to the electron. Thus, it is an antiparticle, and when considered by itself, it is stable. However, when a positron collides with an electron, they annihilate each other to form a photon (quantum unit of light). In the early 1930s this discovery was followed by nuclear physicists realizing that the nuclei of an unstable element (radioactive nuclei) consists primarily of neutrons and protons and will, during radioactive decay, produce four (instead of the previously thought three) basic particles, namely, 1) the alpha particle (similar to a helium nucleus); 2) the beta particle (similar to a negative electron); 3) the positron (similar to an electron, but with a positive charge); and 4) the gamma ray (similar to a high energy X-ray).

The positron is considered **antimatter** instead of normal matter that led to speculations that there might possibly be an "anti-universe" somewhere consisting of antimatter. This is just that—speculation, not a proven fact. Anderson later discovered what is called the *mu-meson* that was predicted by Hideki Yukawa. It is now called the **muon** whose nature and role in nuclear physics are not yet completely understood. Anderson's use of the **cloud chamber** and his discovery of two new particles opened the path to the exploration and understanding of numerous subnuclear particles.

See also Dirac; Yukawa

ANDERSON'S THEORIES AND MODEL: Theoretical Physics: *Philip Warren Anderson* (1923–), United States. Philip Anderson shared the 1977 Nobel Prize in

Philip Warren Anderson was born on December 13, 1923, and raised on a farm in Crawfordville, Indiana (near Indianapolis). He grew up during the Great Depression. His family moved to Urbana, Illinois, where he attended school. His grandparents and parents were educated and had many friends who were scientists. As a young man, Anderson did extensive reading and was influenced by a high school mathematics teacher. After graduation from the university high school in Urbana he received a national scholarship to Harvard. During the war years of 1940 to 1943 he majored in "electronic physics" and in 1943 was employed by the Naval Research Laboratory to work on the then-secret radar project. Anderson attended graduate school in 1945–1949 where he made many friends and studied with some famous physicists. In 1949 to 1984 he worked in Bell Labs (now Lucent Technologies that was acquired by the Alcatel Corporation) and is still a consultant to the lab. In 1967 he became a professor of theoretical physics at Cambridge University, and in 1977, along with two other scientists, was awarded the Nobel Prize in Physics. Philip Anderson is well known for his works in magnetic superconductivity and quantum disorder in systems and related areas. He has received many awards over the years.

Physics with John Van Vleck and Neville Mott for investigations of the electronic structure of disordered magnetic systems as related to semiconductors.

Anderson's "superexchange" theory: *This theory explains how the different spins of magnetic atoms in a crystal interact with nonmagnetic atoms that are between the magnetic atoms within the crystal.*

This theory led to theoretical statements related to superconductors and antimagnetism.

Anderson's model describes what takes place when a metal, particularly a semiconductor, acts when "impure" atoms are present. This is of importance to the modern semiconductor industry because the semiconductor's capacity is dependent on the type and amount of impure atoms present in the semiconducting element such as silicon and germanium.

Anderson's localization is based on the idea that extended states of matter be localized by the presence of disorder in a particular system. Again, this deals with impurities in crystals as related to superfluidity and superconductivity at extremely low temperatures. This is related to the unique characteristic of heavy helium (H-3) at low temperatures that behaves in an unusual fashion by climbing up and over the sides of a beaker when at near absolute temperatures.

The Anderson–Hamiltonian theory describes how electrons behave in a metal undergoing a transition phase.

Anderson's work in quantum relationships as to how the structure of magnetic and electronic structures affect disordered systems provided the information needed for the development of electronic switching and magnetic memory disks used in modern computers.

ÅNGSTRÖM'S PRINCIPLE OF SPECTRUM ANALYSIS AND RELATED THEORIES:

Astronomy and Physics: *Anders Jonas Ångström* (1814–1874), Sweden.

Ångström grew up in a simple home in Sweden where his father was a chaplain. He studied and taught physics at the University of Uppsala. He received his doctorate from the university in 1839 followed by a professorship and in 1858 became chairman of the university's physics department.

In 1853 Ångström published *Optical Investigations* where he compiled a list of his measurements of over one thousand atomic spectra lines that were visible for both the gas and the types of electrodes used in the analysis. This work led to his principle related to spectrum analysis that states: A *hot gas will emit light at exactly the same wavelength at which it absorbs light when cooler*.

This principle was formulated from his work with spectral analysis and led to his analysis of the sun's light spectrum. In 1868 he published *Researches on the Solar Spectrum* based on his observations leading to his conclusion that hydrogen gas is present in the sun because it showed up in his analysis of the sun's spectrum. This work enabled him to demonstrate that the spectra of alloy metals are of a composite nature. In other words, the metals that compose the alloys show up in the analysis as unique individual spectral lines. He was also the first to view and analyze the spectrum of the aurora borealis, also known as the northern lights.

ARAGO'S WAVE THEORY OF LIGHT AND ARAGO'S DISK: Physics: *Dominique Francois Jean Arago* (1786–1853), France.

After discovering chromatic polarization of light in 1811, Arago investigated the idea proposed by the French physicist A. J. Fresnel that light was a wave. This was contrary to the theory of other physicists of the day, including Pierre de Laplace and Jean-Baptiste Biot, that light was of a corpuscular nature and required a medium (the aether or ether) through which to travel. Arago set up an experiment to prove the theory that *light travels through air and media with different densities in waves*. He did this by measuring the speed of light in air and water. Later, after Arago's death, Jean Foucault and Armand Fizeau proved his theory correct. Today, light is considered both a wave and particle (photon), and there is no aether in space.

Arago also discovered how to produce magnetism by wrapping a wire that is carrying electricity around a cylindrical piece of iron. This discovery led to the development of the electric motor, the dynamo, solenoid, and other modern electrical devices.

Arago's disk was a device consisting of a copper disk suspended above, but in close proximity to, a compass. When the disk is spinning, it deflects the compass needle. This is another example of the phenomenon known as Ampère's electric/magnetic induction.

Arago was interested in astronomy and discovered the sun's chromosphere and assisted Urbain Leverrier (1811–1877) in the discovery of Neptune. In his later life instead of continuing his work in physics, he became involved in politics. He was the government official most responsible for the abolishment of slavery in most of the French colonies.

See also Ampère; Fizeau; Foucault; Fresnel

ARBER'S CONCEPT OF THE STRUCTURE OF DNA: Microbiology: *Werner Arber* (1929–), Switzerland and the United States. Werner Arber and two other microbiologists (Daniel Nathans and Hamilton O. Smith) received the 1978 Nobel Prize for Physiology or Medicine.

Arber was born in Switzerland and graduated from two major universities in his home nation before attending the University of Southern California. He returned to Geneva, Switzerland, where he served as professor of microbiology from 1960 to 1970.

He experimented with bacteriophages (a form of virus) that invade bacteria and may cause hereditary mutations in the host bacteria, as well as undergo similar mutations themselves.

Arber's main concept proposed in 1962 relates to the use of a specialized enzyme that can destroy invading phage viruses by cutting up and separating their DNA molecules into smaller pieces. Further, these "restriction enzymes" always attack the DNA at precise and predictable locations on the molecule, which allows the smaller strands of DNA to be reformed in combinations during what is now called "genetic engineering." This is possible because the bits of DNA that are separated are somewhat "sticky" and can be made to combine with other "sticky" bits of DNA at different sites on molecules.

This process is used in the treatment of genetic diseases (although not always successfully), the cloning of plants and animals (also not always successfully), the use of DNA as evidence in legal proceedings, and the mapping of genes in the Human Genome Project. The main benefit of Arber's discovery is its use as a "tool" for further genetic research. The benefits of genetic engineering are mixed, and much research is still needed to make all its promises a reality.

See also Delbrück; Lederberg

ARCHIMEDES' THEORIES: Mathematics: *Archimedes of Syracuse* (c.287–212 BCE), Greece.

Archimedes was an accomplished theoretical and applied mathematician who developed "thought" experiments to test some of his ideas and then expressed their results mathematically. He actually did not conduct controlled experiments as we think of the process today. Only a few of his many theories and accomplishments are explored.

Archimedes' theory of "perfect exhaustion" (calculation of pi): Archimedes was not the first to recognize the consistency of the ratio of the diameter to the circumference for all circles or to attempt the calculation of **pi** as this ratio has became known. Objects of differing shapes and how their dimensions were related intrigued ancient Stone Age people. They realized straight lines do not exist in nature and recognized curved lines in the shape of rocks, plants, animals, and other objects. By about 2000 BCE humans recognized and roughly calculated the relationship of a circle's measurement in the sense that the larger the circle, the greater is its circumference. By the era of the ancient Greeks, mathematicians understood this ratio was consistent for all circles because they measured and compared the diameters and perimeters of various circles. Soon after, this constant irrational number was given the Greek symbol π (pi).

Using his knowledge of the geometry of many-sided plane figures, such as squares and multiple **polygons**, Archimedes proposed his theory of *perfect exhaustion*, which he demonstrated by drawing a circle and inscribing several polygons on the inside and outside of the circle's circumference. At first, he used polygons with just a few sides. Later he used multiple polygons with as many as ninety-six or more sides. This is often referred to as *perfect exhaustion* because Aristotle used polygons with larger numbers of sides. Theoretically, a polygon with an infinite number of sides could be used. Through the use of geometry and fractions, Archimedes measured the inside polygons and compared them with the measured outside polygons. He concluded that the polygons

ARCHIMEDES' METHOD TO APPROXIMATE π

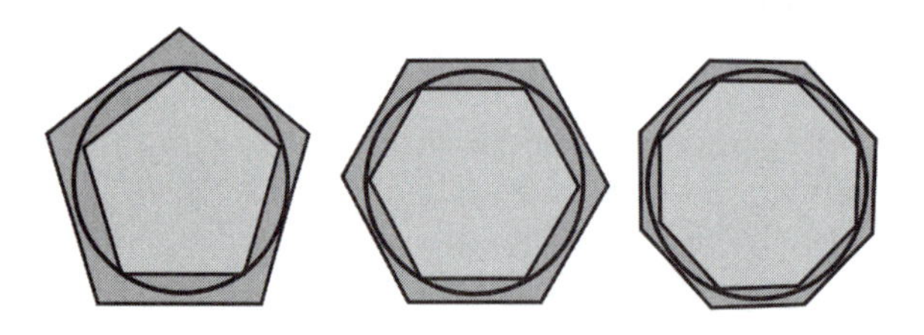

Figure A3. Artist's example of Archimedes using geometry for "Perfect Exhaustion" to estimate pi.

touching the circle on its outside circumference (perimeter) were slightly larger than pi and that the polygons touching the inside of the rim of the circle were slightly smaller than pi. Therefore, pi must be a value somewhere between these two measurements. His value for pi was 3.14163, which he calculated as the figure between the inner and outer polygons ($3_{10/71} < \pi > 3_{1/7}$). His figure for pi has been used for many hundreds of years and was developed by using Euclidean plane geometry. This method of using geometry to calculate pi has physical limitations for arriving at the correct ratio.

Later mathematicians used algebra, which enabled the calculation of a more accurate value for pi. With the invention of fast computers, pi has been run off to several hundred thousand decimal places in a few hours. Yet one could run off pi on a computer forever and still never reach a final number to make pi come out even, because it is an **irrational number**.

Archimedes' theory for the volume of spheres: *The volume of a sphere is two-thirds the volume of a cylinder that circumscribes (surrounds) the sphere.*

It is said that Archimedes wanted this theorem inscribed on his tombstone. Historically, measuring the volume of a sphere was difficult, whereas measuring the volume of a cylinder was easy. Therefore, if one knew the volume of a cylinder that surrounded a sphere, its volume could be determined.

Archimedes' theory of levers: *The mechanical advantage of a lever is due to the ratio of the weight (load) to the action (effort) required to move the load, which is determined by measuring the distance the effort moves from the central point (fulcrum) divided by the distance the load moves from the central point.*

Humans have used the simple lever since prehistoric times. How people learned to take advantage of this simple lever is unknown, but evidence exists that ancient people were aware of the advantage of using sticks for digging and moving heavy objects by prying them with sticks. Archimedes was the first to calculate the ratio of the distance between a force and a weight, separated by a fulcrum. The placement of the fulcrum in relation to the force and weight determined the ratio for the mechanical advantage. Archimedes used his knowledge of geometry and mathematics to calculate the mechanical ratio for several simple machines. For the simple lever, he believed the advantage was the ability to move very heavy loads with little effort. Most of his demonstrations of mechanics dealt with the simple lever. His major demonstration was the raising of a large ship by pushing down on one end of a large lever that he had designed.

Archimedes' concept of the inclined plane: *It is easier to move a load along a long, sloping ascent of a given height than it is to move a load of the same weight along a shorter but steeper ascent to the same height.*

Archimedes knew the mechanical advantage of rolling objects up a long inclined plane of a given height rather than lifting them vertically for the same height. He applied the concept of an inclined plane as a means of raising water in a well up to the surface. He wrapped an inclined plane device around a central shaft to form a "water screw," which was placed with one end in the well and the other on the surface where the water was to be used. When turned by a crank handle, this "helical pump" enabled one man to lift water more efficiently than with any other pump then known. Remarkably, it is still used, twenty-three hundred years later, in Egypt and other parts of the world.

Archimedes also developed catapults, cranes, pulleys, and optical devices that consisted of a series of shiny metallic mirrors that reflected and concentrated rays of the sun. All of the devices are believed to have been used to defend his city of Syracuse

THREE TYPES OF LEVERS

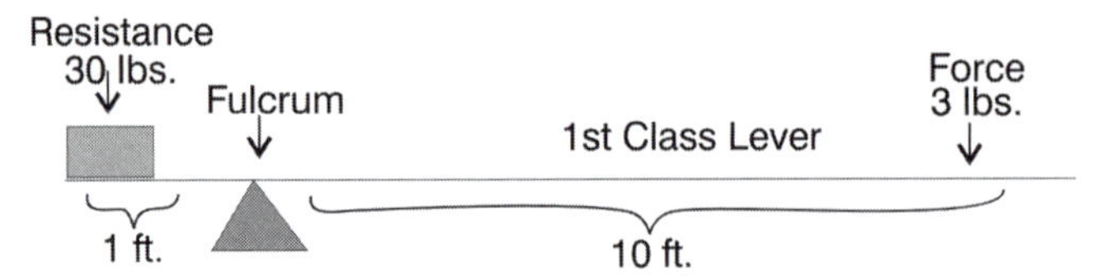

Distance X Resistance = Distance X Force: 1 X 30 = 10 X 3
Mechanical Advantage is the Load Divided by Effort

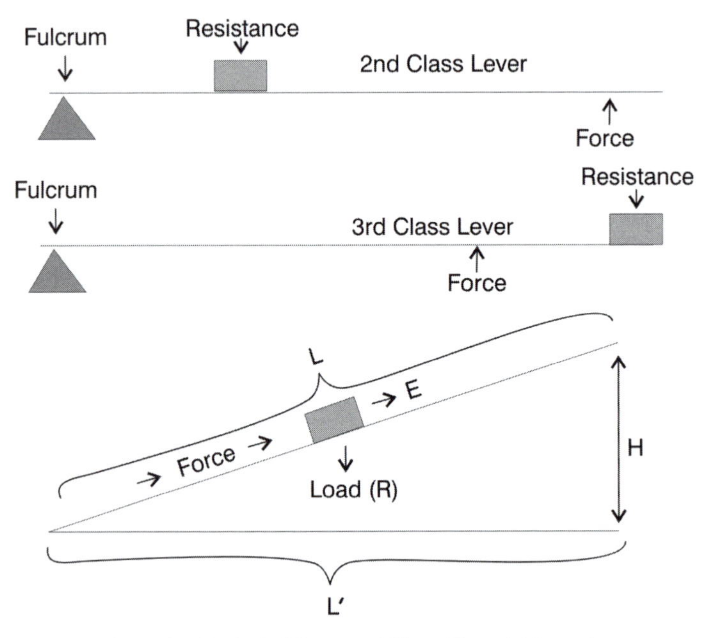

Velocity Ratio is Distance Effort (Force) Moved
Divided by the Height Moved (H) $Vr = \frac{L}{H}$

(The greater L is to L' the greater the mechanical advantage,
but the amount of work accomplished to achieve H
is the same for both L and L'.)

Figure A4. Archimedes' Theory of Levers. Humans used levers for centuries and intuitively knew their advantage, but the theory was not formalized until Archimedes stated it.

from Roman invaders. Although not the first to use his knowledge of physics and mechanics in the name of war, Archimedes was one of the most successful.

Archimedes' concepts of relative density and specific gravity: *The compactness of an object is related to the ratio of its weight divided by its volume.*

Archimedes used his concept of buoyancy to measure the relationship between the weight and volume of an object. No discussion can omit the famous story of how Archimedes gained insight into the concept of density and specific gravity. As the story goes, King Hiero of Greece asked Archimedes to ascertain whether a gold crown he commissioned from a goldsmith was pure gold or whether silver was substituted for some of the gold. Archimedes pondered the question while taking a public bath. He lowered himself into a bath filled to the brim. As he sank deeper into the bath, more and more water spilled over the sides. He immediately grasped the significance of this phenomenon, jumped out of the bath, and ran naked down the street shouting "Eureka! Eureka!" ("I have found it! I have found it!"). He proceeded to fill a bucket to the brim with water into which he lowered the crown, catching and measuring the volume of the water that overflowed. He did the same with equal weights of gold and silver. Because gold has a greater density than silver, the ball of gold was smaller; thus with a smaller volume less water spilled over the edge of the bucket. Once he could measure the volumes of the water representing the volumes of the gold, silver, and the crown, all he needed to do was to divide the figures obtained for the weight of each item by their volume of water and calculate a ratio representing their comparative densities. He could then determine how much of the crown was gold and how much was silver. (Supposedly, the crown was not pure gold, and the goldsmith was executed.)

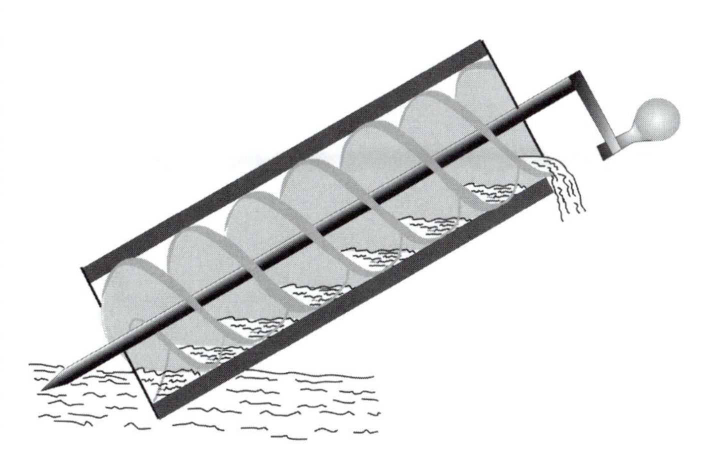

Figure A5. Archimedes' "Screw" used the concept of an inclined plane wrapped around a central shaft to form a "helical pump" that one person could use to raise water from a well.

His principle led to the expression of density as the weight (mass) of an object divided by its volume $(d = m/v)$. Specific gravity is the ratio of an object's density to that of some standard. For liquids, water at 15°C temperature is the standard for specific gravity expressed as 1.0. (Any object denser than water would have a specific gravity greater than 1.0 and would sink, whereas any object with a density less than 1.0 would float in water.) For gases, dry air is used as the standard pressure and temperature. Specific gravity is easier to use than density for making calculations because it is the same value in all systems of measurement.

Archimedes' theories, principles, and concepts have been refined over the ages, but his genius has provided the basis for modern-day machinery and instruments.

ARISTOTLE'S THEORIES: Physics: *Aristotle of Macedonia* (384–322 BCE), Greece.

In the estimation of many historians, Aristotle was one of the most influential humans who ever lived. Although he was a philosopher concerned with classes and hierarchies rather than a scientist concerned with observations and evidence, his philosophy, methods of reasoning, logic, and scientific contributions are still with us and continue to be influential. Much of Aristotle's philosophy is related to his four causes: 1) *the matter cause, which makes up all material, including living organisms*; 2) *the form cause of species, types, and kinds of things*; 3) *the efficient cause of motion and change*; 4) *the*

ARISTOTLE'S LADDER OF NATURE

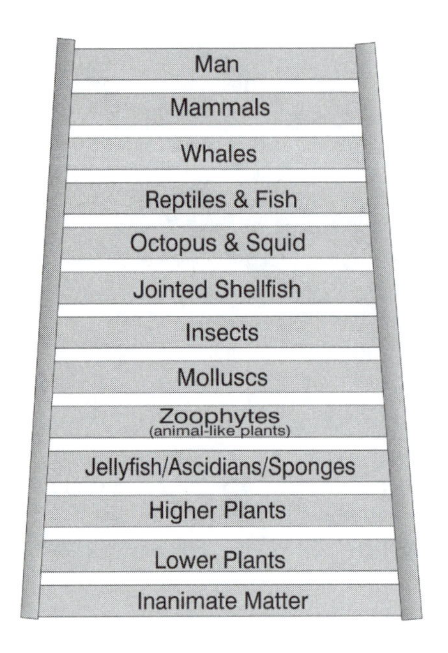

Figure A6. Aristotle's classification (taxonomy) of living things known as the "Ladder of Life"

final cause of development or the final goal of an intended activity (maturity). He related these "causes" to inanimate and animate phenomena. Only a few of Aristotle's theories related to limited areas of science are presented.

Aristotle's topological-species theories: An *ideal form is a living group in which each member resembles each other, but the group is distinct in structure from members of other groups.* His concept can further be stated as *Each living thing has a natural built-in pattern that, through reproduction, growth, and development, leads to an individual type* (***species***) *similar to its parents.* Aristotle also believed that all living species reproduce as to type (e.g., humans beget humans, cattle beget cattle), but he considered a possible exception to this theory when applied to the lowest of species. He organized species from lowly flies and worms at the bottom, then lower animals, up to mammals, and then humans at the top. Aristotle classified everything and endeavored to write all as a unified theory of knowledge, which preceded Einstein's long-sought grand unification theory by many centuries.

Aristotle's theory of spontaneous generation: *Flies and low worms are generated from rotting fruit and manure.* He based this theory on observations, not experimentation. Later scientists, through the use of simple experimentation, demonstrated "spontaneous generation" of life does *not* exist, at least as expressed by Aristotle and some former scientists (*see also* Leeuwenhoek; Pasteur; Redi).

Aristotle's theory of taxonomy (classification) of living things: *Nature proceeds from tiny lifeless forms to larger animal life, so it is impossible to determine the exact line of demarcation. Reproduction identifies those giving live birth (viviparous) as being mammals and humans, whereas those laying eggs (oviparous) are subdivided into birds and reptiles, and fish and insects.* Aristotle developed an elaborate classification system of nature later called *Aristotle's ladder of nature.* It listed inanimate matter at the bottom, progressing upward from lower plants, higher plants, minor water organisms, shellfish, insects, fish, reptiles, whales, mammals, and finally on the top rung of the ladder, humans.

Aristotle expressed his theory in a graphic structure called *scala naturae* or *chain of being*, better known as "Aristotle's ladder of nature."

Aristotle's three classes of living things: 1) *vegetable, which possessed a nutritive soul*; 2) *animals, who were able to move and thus had a sensitive soul*; and 3) *humans who had intelligence and thus a rational soul, and who also possessed souls of all the types of creatures.*

One of Aristotle's classifications was that male humans had more teeth than did females. As a philosopher concerned with the meaning of classes and hierarchies rather than a scientist concerned with observations and evidence, neither he nor anyone else at that time bothered to count the teeth in men and women. Regardless, "man" was at the top class with a rational soul.

Aristotle's concept of reproduction: *An invisible "seed" of the most rudimentary structure was imparted by the male to join a female egg to produce an offspring of the same species.*

Some philosophers and scientists of Aristotle's day (and later) believed the "seed" (sperm) was a tiny, invisible person or animal that grew larger once it joined with the female egg. Aristotle concluded this by observations made while dissecting and studying fertile chicken eggs at different stages of embryonic development. He rejected most theories about reproduction proposed by previous philosophers, including those that posited that the sex of an embryo is determined by how it was placed in the womb, a seed originates as a whole body, and the embryo contains all of the adult and body parts (preformationism).

Aristotle's laws of motion: 1) *Heavy objects fall faster than lighter ones, and the speed of descent is proportional to the weight of the object.* 2) *The speed of the falling object is inversely proportional to the density of the medium through which it is falling.* 3) *An object will fall twice as fast as it proceeds through a medium of half its density. Thus a vacuum cannot exist because the object would proceed at an infinite speed.* This law is one of the few examples indicative of Aristotle's concern with the quantitative nature of things. Unfortunately, he did not verify his insights by experimenting and making measurements. Although his laws of motion were incorrect, they were accepted for many years and provided the background for which Galileo and Newton revised Aristotle's original concepts.

Aristotle's concepts of motion were rather simplistic. When asked why things move, he responded simply that it is because something moves them. He considered three types of motion on Earth: 1) the motion of living things that is voluntary; 2) objects that are moved tend to return to their a natural position of rest; and 3) when something is set in motion its motion will cease once the "mover" is eliminated. These ideas are expressed as Aristotle's laws of motion: *Violent (forced) motion will always be displaced by natural motion that ends in a state of rest. The speed of a moving object is directly proportional to the force applied to it.* In simple language this means if you cease pushing an object, it will stop moving. Philosophers and scientists, before as well as after Aristotle's time, could not accept the concept of action (force) at a distance, such as gravity. There had to be something in contact with the object that would force the object to move, and it could not just be "spirits," as some believed, but rather something physical. To Aristotle, all motion was self-explanatory because all bodies sooner or later came to their natural place of rest in the universe. He explained his theory somewhat in this way: Once "impulse" was given to a stone by throwing it up in the air, this impulse was transferred to the air in tiny increments, which kept pushing the stone up. These "air impulses" pushing the stone upward became weaker as the stone rose, and now the natural motion of the stone returned it to the ground in a straight line, and finally to its natural state of rest. When the "impulse" completely stopped, so did the object's motion.

Aristotle applied his concepts of motion to his observations of heavenly bodies. His theory states: *Heavenly bodies move in perfect circles rather than in straight lines as bodies do on Earth. Thus, heavenly bodies are not composed of the four earth elements but rather a fifth element called aether.* This concept that heavenly bodies and bodies on Earth obey separate laws was followed by scientists until Newton's time. Celestial bodies were "pure," whereas those on Earth were subject to death and decay. Aristotle's theory of the prime mover, impulse, and motion came very close to the modern physical law of conservation of momentum (*see also* Galileo; Newton).

Aristotle's concept of infinity: *Because the universe is spherical and has a center, it cannot be infinite. An infinite thing cannot have a center, and the universe does have a center (Earth). Therefore, infinity does not exist.*

Most philosophers of Aristotle's day believed the universe was composed of crystalline concentric spheres with the earth at the center; therefore, the universe was finite.

Aristotelian logic is still taught in high school and university courses in dealing with reasoning and logic and is often used by debaters. The word "logic" is derived from the Greek word *logos,* meaning a form of reasoning using speech. Several Greek philosophers before Aristotle had developed forms of logic, but it was Aristotle who advanced the study and whose writings still exist. It is from these records that we have learned about the basic logic called "Aristotelian syllogism." A syllogism is a form of verbal deductive reasoning that contains three parts, 1) first major premise; 2) second premise (both are assumptions); and 3) a conclusion drawn from 1) and 2). For example: 1) All warm-blooded land animals with four legs are mammals. 2) Horses are warm-blooded land animals and have four legs. And, 3) therefore, all horses are mammals. Note that both 2) and 3) are consistent with and contained within 1), the major assumption. Syllogistic logic can be either positive [*all* 1) are 2)] as is the above syllogism, or negative [*no* 1) are 2)] as follows: 1) All warm-blooded mammals can run. 2) Birds are warm blooded and can run. And, 3) Therefore, birds are mammals. Thus, a negative syllogism is flawed logic. It is a type of fallacious statement often used by people with the intent to deceive.

Somewhat the same argument was used to negate the existence of a void, or vacuum. Accepting concepts such as infinity and vacuum was beyond the philosophical reasoning of people in Aristotle's time. It was not until the sixteenth century, when Copernicus provided credible evidence that the earth was not the center of the **universe**, that this geocentric concept was overcome.

Aristotle's theory of the matter and the aether: *Because all celestial bodies move in perfect circles, there must be a perfect medium for this to occur.*

This perfect medium that enables circular motion is known as the **aether**, which also has circular motion. Aristotle accepted the classification of elements as devised by Empedocles (c.490–430 BCE) and others that placed all things into four elementary groups: earth, water, fire, and air. He saw the need for a fifth class of matter when addressing the heavens. Until Newton's time, scientists continued to accept the concept of aether (or *ether*, the Greek word for "blazing"). In its more sophisticated form, it was referred to as the "fabric of space." The concept of an ether existed into the days of early radio. It was popular to believe that radio signals (and other electromagnetic waves) were transported by something in space similar to the way sound is carried by air. In Aristotle's time, people did not believe the sun's heat could reach Earth without some form of "matter" transporting it.

See also Maxwell

ARRHENIUS' THEORIES, PRINCIPLES, AND CONCEPTS: Chemistry: *Svante August Arrhenius* (1859–1927), Sweden. Arrhenius was awarded the Nobel Price for Chemistry in 1903.

In 1883 Svante Arrhenius proposed two related theories of dissociation. One deals with what occurs when substances are dissolved in solutions; the other explains what happens when a current of electricity is passed through a solution.

Arrhenius' theory of solutions: *When a substance is dissolved, it is partly converted into an "active" dissolved form that will conduct a current.* Arrhenius' theory is based on the concept that **electrolytes** in solution dissociate into atoms (*see* Figure A7, and *see* Faraday).

Arrhenius' theory of ionic dissociation: 1) *When an electric current is passed through molten salt (sodium chloride, NaCl), it dissociates into charged ions of* Na^+ *and* Cl^-. 2) *Positive charged* Na^+ *ions are attracted to the negative pole* (**cathode**) *and deposited as neutral atoms of sodium metal, and the negative charged ions of* Cl^- *are attracted to the positive pole* (**anode**) *and changed back to neutral atoms of chlorine, as the gas molecule* Cl_2.

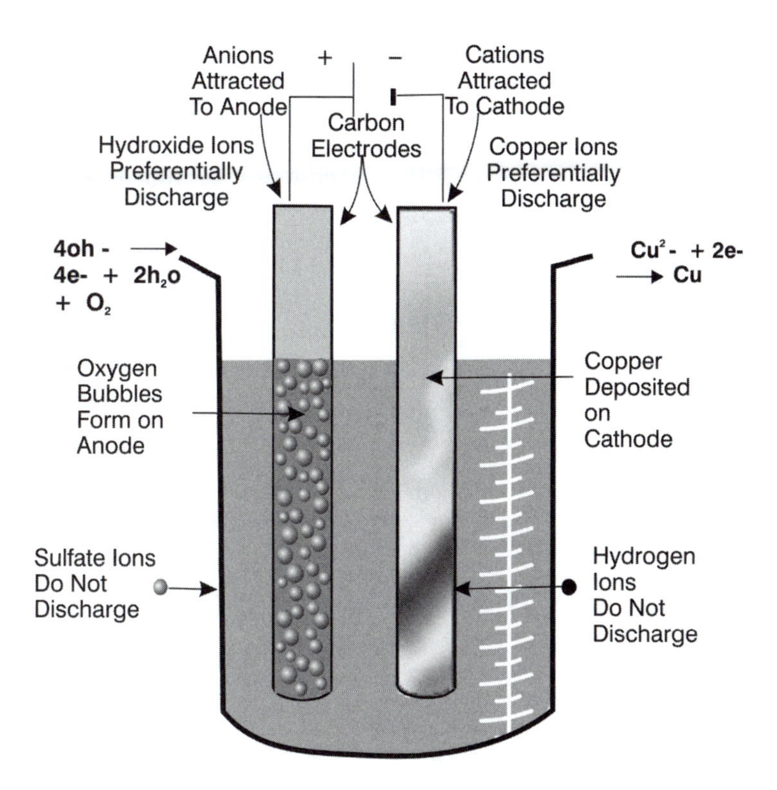

Figure A7. The positive ions are attracted to the negative cathode, and the negative ions are attracted to the positive anode. The ions lose their electrons to form neutral atoms through a discharge of the current between the electrodes and through the electrolyte, which is an ionized compound in solution that can carry electricity.

Once the atoms are "dissociated" into **ions** in a liquid and become an electrolyte solution, an electric current can pass through the solution, producing a completed electric circuit.

These theories are related to electrolysis and are important to many industrial processes today, including electroplating. For instance, ionic dissociation is one way to produce chlorine and sodium from common salt (NaCl). The dissociation of substances, such as $NaNO_3$, separates the compound molecule into Na (sodium metal) and NO—(the negative nitrate ion). A similar process occurs when electroplating gold and chromium, and other metals (*see also* Faraday).

Arrhenius' principle of acid-base pairs: *When an acid splits, it will yield hydrogen ions* (H^+). *When a base breaks apart, it will yield hydroxyl ions* (OH^-). Arrhenius extended his concepts about ionic dissociation to include his theory related to acids and bases. In general terms, this principle has been broadened to make it more useful for chemists, who still speak of *acid-base pairs*. This broadened principle refers to the transfer of a proton from one molecule to another: *The molecule that gave up the proton is the* ***acid****, and the one receiving the proton is a* ***base***. The process is used in industry to produce acids and alkaline (basic) chemicals.

Arrhenius' rate law: *The rate of a chemical reaction increases exponentially with the absolute temperature.*

Arrhenius and others based this phenomenon on their observations that when the temperature rises, the rate (speed) of chemical reactions increases; conversely, when the temperature cools, reactions slow down. They noted this for such things as spoilage and decomposition of fruits and vegetables in hot climates, though their usefulness can be extended by refrigeration. Also, bread rises faster in a warm environment, substances dissolve faster in warm water, and so forth. Arrhenius' rate law can be expressed mathematically: Rate = A exp(−B/T), where A and B are constants that differ from one reaction to another, and *T* is time. The value of this equation is its use in generalizing the concept based on the fact that most chemical reactions occur at room temperatures of about 20°C, and a rise of 10°C will double the rate of the reaction.

Arrhenius' theory of panspermia: *Life came to Earth as a bacterial spore or other simple form from outside the solar system.*

After Redi and Pasteur discredited the theories for spontaneous generation, other theories for the origin of life were postulated, including theories of life arriving on Earth from outer space. Arrhenius proposed a theory, called *panspermia*, for the beginning of life on Earth from extraterrestrial origins. This theory is gaining some new proponents since it was reported that simple organic molecules were found in some meteorites, possibly from Mars, that landed on the ice sheets of Antarctica. A modern version of the theory is that prebacteria-type organisms capable of reproducing are universal and develop within a suitable environment anywhere in the universe (*see also* Hoyle; Pasteur; Redi; Spallanzani; Struve).

Arrhenius' theory for the greenhouse effect: *The percentage of carbon dioxide in the upper atmosphere regulates the temperature, which may be the cause of the ice ages.*

In 1967 when the unmanned lunar lander *Surveyor 3* landed on the moon, it contained a TV camera that was recovered more than two years later by two astronauts. They brought the camera back to Earth for testing. When it was examined in a sterile environment, scientists were surprised to discover that specimens of the bacteria *Streptococcus mitis* were still alive and active. At first, they were puzzled about the origin of these bacteria and then realized they must have been in the camera before it was launched to the moon. It seems amazing that bacteria can exist in the cold/hot vacuum of space, but this does not seem to be a problem for bacteria that are found living in extreme environmental conditions on Earth. When some species of bacteria are faced with lack of water and are exposed to extreme temperatures for extended periods of time, they break open and release proteins, sugars, and other chemicals that act to protect some of the surviving bacteria. If there are enough bacteria in the colony and enough protective substance is released, some of the protected bacteria will survive. These protective substances are called *cryoprotectives.* Some surviving bacteria form an endospore into which the original cell reproduces its chromosomes. These inner endospores formed by the original cell are protected by a surrounding wall, while the outer original cell may perish. The protected endosperm will survive most conditions, including boiling water. There are many cases where bacteria have survived for thousands of year on Earth in very inhospitable environments. One example is the bacteria found in the gut of a bee that was preserved in amber as a fossil for about forty million years. Actual bacteria, spores, or viruses (or any other forms of life) from distant outer space have yet to be found, in spite of advocates of extraterrestrial life found on Earth by people who believe in science fiction.

The theory of panspermia as the basis for the origin of life and its evolution on Earth is now considered a hypothesis that might be tested. The revival of this old theory of life on Earth originating from outer space as proposed by Arrhenius and others is based on some unusual discoveries.

Over eighty years ago, Arrhenius was one of the first to relate carbon dioxide to global climate changes. Although he was unable to establish an exact relationship between carbon dioxide and atmospheric temperatures, he considered the cooling and warming effects of CO_2 as evidence for the cause of the past ice ages. His theory is based on the belief that CO_2 in the atmosphere does not absorb the energy from the sun that arrives on the surface of Earth in the form of light and infrared (heat) radiation. In addition, the energy radiated from Earth is in the form of infrared radiation that is absorbed by CO_2, acting as a blanket thus creating a *greenhouse effect.* This theory is still controversial; however, evidence indicates that there is a slight increase in the levels of atmospheric CO_2 that may, or may not, have a slightly more warming than cooling effect on Earth (about 1.5°C in the twentieth century). The increase in water vapor plus methane from industry, decaying of organic matter, and animal flatulence also contribute to a greenhouse effect. Another theory states the massive ice age six hundred million years ago, when Earth was a complete ice planet, ended as trillions of simple organisms produced enough carbon dioxide to create a greenhouse effect. This may have resulted in the melting of the ice that covered Earth, thus permitting the evolution of higher forms of life. Carbon dioxide also increases plant growth on Earth, which provides food for a more diverse animal kingdom.

See also Rowland

ASTON'S WHOLE NUMBER RULE: Chemistry/Physics: *Francis William Aston* (1877–1945), England. Francis Aston received the 1922 Nobel Prize in Chemistry.

Aston began his career in chemistry and physics just before the turn of the twentieth century. He devised a new and improved pump used to create a vacuum inside glass tubes that were used in gas discharge experiments. One area he explored was *Crookes dark space* found in these discharge tubes. He discovered a phenomenon now known as *Aston dark space.* This work led to his discovery of two isotopes of neon gas, each with a different atomic mass. Further work led to his invention of the mass spectrometer that uses electromagnetic focusing to detect any slight difference in the mass of atoms of the same element (isotopes). This work led to Aston's formation of the whole number rule, that in essence, states: *The mass of the oxygen isotope is defined as a whole number, and all the other isotopes of elements have masses that are very nearly whole numbers.*

Aston invented the mass spectrograph, an instrument that uses electromagnetic focusing to separate **isotopes** of the same element by slight differences in their **atomic weights**. The **nuclei** of atoms of a specific element are composed of both positive **protons** and neutral **neutrons**. The number of protons determines the chemical identity of the element, which does not change, whereas the number of neutrons can be more or fewer than the number of protons. This explains the formation of isotopes of the same element with different atomic weights. This difference in weight is very slight and was not successfully detected until the development of the spectrograph. Aston used his instrument to measure this minute difference in atomic weights and to separate and identify 212 isotopes of nonradioactive elements. From his research he devised the *whole number rule*, which advanced the fields of inorganic and nuclear science.

Aston's invention of the mass spectrometer and identification of isotopes for atoms has been invaluable to science. In his personal life he excelled in such sports as tennis, swimming, rock climbing, and skiing. He was also an accomplished musician playing the violin, cello, and piano.

ATOMISM THEORIES: Physics. Theories related to the nature of atoms and related scientists listed chronologically.

Theories of atomism date back to the fifth century BCE, when philosophers conceived the idea that all matter was composed of tiny, indivisible particles. In ancient times these ideas were classical philosophical theories deduced by reason and logic, not by empirical or experimental evidence. The word *atomos* is derived from two Greek words: *a*, which means "not," and *tomos*, which means "cut." In other words, you cannot cut it, or it is indivisible. Several examples of atomic theories follow:

Leucippus' atomic theory: *Leucippus of Miletus* (c.490–430 BCE), Greece. *All matter is composed of very minute particles called atomos. They are so small that there cannot be anything smaller, and they cannot be further divided.*

Not much is known about Leucippus, but he was the first to be credited with originating the atomic theory, giving the concept the name *atom* and describing the indestructible nature of atoms. One of Leucippus' students was a philosopher named Democritus, who is also credited with the "atomic concept."

Zeno's paradox: *Zeno of Elea* (c.495–430 BCE). One of Zeno's theories stated that *conclusions could be reached by reason even when there was contradictory sensory evidence.* He used paradoxes to present his hypotheses that motion and distance could be divided into smaller units *ad infinitum*. His most famous paradox was used by other philosophers and scientists to explain the concept of the division of matter into smaller and smaller particles while never reaching a final indivisible particle (*see also* Zeno).

Democritus' theory of atoms: *Democritus of Abdera* (c.460–370 BCE), Greece. It is assumed that Democritus and others who followed questioned Zeno's paradox as a rational way of looking at nature in the sense that the division of space and motion could be divided indefinitely, and perhaps there was a final limit to the point of indivisibility. He further developed the atomic theory of his teacher, Leucippus. Democritus and other philosophers considered what would happen if a person took a handful of dirt and divided it by half, then divided that half into half, and continued dividing it by halves. Eventually a point would be reached at which a single tiny speck of dirt that could no longer be divided was all that remained. The result was considered, on a philosophical basis, to be the indivisible atom of dirt. Democritus also theorized *these tiny "atoms" of matter unite to form larger masses, and the large mass could fly apart and the smallest particles would still be the tiny atoms.* This led to his theory that *nothing can be created out of nothing*, which was the precursor to the basic physical law of conservation of matter and energy.

Aristotle's theory of the atom: *Aristotle* (384–322 BCE), Greece. Aristotle recorded much of the philosophy of Democritus. He also credited Democritus with the concept of the indivisible atom and accepted it as a rational, logical, philosophical explanation (*see also* Aristotle).

Epicurus' theory of the atom: *Epicurus of Samos* (c.341–270 BCE), Greece. Epicurus kept the atomism theory current by demonstrating how it could be the basis for perceiving reality and eliminating superstitions—the Epicurean concept of "just being happy" and living a good life without fear. Later, the Romans adopted this philosophy of the "good life." The modern word *epicurean* is derived from Epicurus, whose theory stated that *atoms were forever in constant motion, perceivable, and thus "deterministic."* Although he disputed Democritus' concept of atoms as having "free will," Epicurus was the first to suggest atomic or molecular motion, which later developed into the concepts of kinetic energy, heat, and **thermodynamics**.

Lucretius' theory of atoms: *Titus Lucretius Carus* (c.95–55 BCE), Italy. Lucretius was a follower of Epicurean philosophy that life's goal should be to avoid misery. His theory was the last of the ancient classical period: *There is a natural origin of all things in the universe, including the heavens, physical objects, and living things, and all things, including living organisms, are composed of atoms of different substances.* Although atoms and molecules could not be seen at this time, his ideas preceded the cell theory and the theory for the chemical basis of **metabolism**. He also preceded Charles Darwin by many centuries with his philosophy that all living things struggle for existence, which is one of the principles of evolution, more accurately stated as "natural selection." Up to this time, the indivisible atom was a concept that usually included inorganic matter. Lucretius is credited as being one of the first to write about the atomic structure of living things, including humans, based more on divine knowledge and his philosophy than on **empirical** evidence.

These ancient classical theories, concepts, and philosophies of atomism were mostly ignored and unexplored for over fifteen hundred years. The more modern atomic theories that developed during later periods are presented alphabetically under the names of the scientists.

See also Bohr; Boyle; Gassendi; Heisenberg; Rutherford; Thomson

THE AUGER EFFECT: Physics: *Pierre Victor Auger* (1899–1993), France

Born and educated in France, Pierre Auger was a professor of physics at the University of Paris who, after World War II, became the director general of the European Space and Research Organization.

Auger discovered the "effect" or "process" that was named after him in the early 1920s. In essence it is a two-stage process that can be stated as: *When an electron absorbs energy from an X-ray photon it will lose that energy as an electron is emitted from an inner shell (instead of a photon) as the atom reverts to a lower energy state.* This results in the emission of an electron representing the energy difference and is known as the Auger effect.

It is well known that the various energy levels of electrons in different shells are discrete and unique to the atoms for each individual element. Thus, the Auger process is the identification of the energy levels, which are signatures of the atoms that emit quanta units of energy. Auger developed a spectroscope capable of using this effect to measure this phenomenon. This spectroscope is useful in the laboratory to provide information about the electron structure of ionized atoms with different atomic numbers (protons). The Auger process is used to identify the "signature" of specific atoms that are emitting quanta of energy, including atoms that make up crystals.

AVOGADRO'S LAW, HYPOTHESES, AND NUMBER: Chemistry: *Lorenzo Romano Amedeo Carlo Avogadro* (1776–1856), Italy.

Avogadro's hypothesis: *If the density of one gas is twice that of another, the atomic mass of particles of the first gas must be twice that of the particles of the second gas.* This relationship between the density and the number of particles in a given volume of gas opened the field of quantitative chemistry, which became a more exact science because it involved the analysis of measurements made of observed phenomena. It enabled molecular weights of different substances to be compared by weighing and measuring the combining substances. Using his hypothesis to compare molecular weights of the oxygen and hydrogen molecules, Avogadro established that it required two hydrogen atoms to combine with one oxygen atom to form a molecule of water. Avogadro also hypothesized that *gases such as oxygen and nitrogen must be composed of two atoms when in their gaseous phase*. He named the particle, which is composed of more than one atom, the **molecule**, meaning "small mass" in Latin. This concept led to the structure of the diatomic molecule for gases (e.g., H_2 O_2 Cl_2).

Avogadro's law: *Equal volumes of gases at the same temperature and pressure contain the same number of molecules, regardless of the physical and chemical properties of the gases*. This is true only for a "perfect gas." Avogadro knew that all gases expand by equal amounts as the temperature becomes greater (assuming that the pressure on a gas remains unchanged). Through some insight on his part, he realized that if the volume, pressure, and temperature were the same for any type of gas, the number of particles of each of the gases, existing under the same circumstances, would be the same. His reasoning that all gases under the same physical conditions have the same number of molecules was based on the fact that all gas molecules have the same average kinetic energy at the same temperature. Other physicists of his day called this unique law *Avogadro's hypothesis*.

The scientists of this period of history did not completely understand this concept and its relationship to the atomic weights of elements. This delayed the use of Avogadro's theories and principles for about five decades until they were rediscovered and applied to modern chemistry. In 1858 the Italian chemist Stanislao Cannizzaro used Avogadro's hypothesis to show that molecular weights of gases could be definitely determined by weighing 22.4 liters of each gas; thus the results could explain molecular structure (*see also* Cannizzaro).

Avogadro's Number: *6.023×10^{23} is the number (N) of atoms found in 1 mole of an element.*

In other words, 1 **mole** of any substance, under standard conditions, contains 6.023×10^{23} atoms. Avogadro's number provided scientists with a very easy and practical means to calculate the mass of atoms and molecules of substances. As an example, the number of atoms in 12 grams of the common form of carbon 12 equals the *atomic weight* of carbon 12 (6 protons + 6 neutrons in the nucleus); thus 12 grams of carbon is equal to 1 mole of carbon. This is expressed as the constant *N* and applies to all elements and also to molecules of compounds.

Scientists assigned the simplest atom, hydrogen, an atomic weight of 1, which then results in the weight of 2 for diatomic molecules of hydrogen gas (H_2). It was determined that at 0°C, under normal atmospheric pressure, exactly 5.9 gallons (or 22.4 liters) of hydrogen gas weigh exactly 2 grams. (In other words, 22.4 liters of any gas, under the same conditions, equals its atomic or molecular weight in grams.) This established that the atomic weight of any element, expressed in grams, is 1 mole. Avogadro's number is one of the basic physical constants of chemistry. Thus, 22.4 liters of any gas weighs the same as the molecular weight of that gas and is considered 1 mole. Two other examples: one molecule of H_2O has a weight of 18 (2 + 16), so 18 grams of water is equal to 1 mole of water; sulfuric acid, H_2SO_4 has the molecular weight of 98 (2 + 32 + 64), so 1 mole of sulfuric acid equals 98 grams of H_2SO_4.

Using this constant makes chemical calculation much easier. All that is needed to arrive at a mole of a chemical is to weigh out, in grams, the amount equal to its atomic or molecular weight. These examples can be changed to kilograms, by multiplying grams by 1000, but they are still equivalent as a molar amount.

Using the *kinetic theory of gases* and the *gas laws*, it is now possible to calculate the total number of molecules in 22.4 liters (1 mole) of a gas. The figure turned out to be six hundred billion trillion, or 600 followed by 23 zeros, or more exactly 6.023×10^{23}.

B

BAADE'S THEORIES OF STELLAR PHENOMENA: Astronomy: *Wilhelm Heinrich Walter Baade* (1893–1960), United States.

Baade's theory of stellar populations: *Population I stars are like our sun and are found in the disk portions of galaxies. Population II stars are found in the "halo" region of galaxies.*

As a result of his observations at the Mt. Wilson Observatory located in Pasadena, California, Baade developed the concept of two different types of stars and his theory of galactic evolution, which was based on the following characteristics of the two star populations:

Population I Stars

1. Population I stars are younger halo stars (formed more recently than Population II).
2. Thus, Population I stars have more "heavy metals." "Heavy metals" is a descriptive term for all elements heavier than hydrogen and helium.
3. Population I stars have lower velocities as compared to our sun (disk stars).
4. Orion Nebula is an example of a younger Population I disk with more metals than the sun.

Population II Stars

1. Population II stars are older disk stars (formed early in galactic history). Population II stars have fewer "heavy metals."
2. Population II stars have random orbits and higher velocities than Population I stars.
3. Stars in the "galactic bulge" are old but received heavy elements from **supernovas.**

Baade's theory of star luminosity: *The period/luminosity relationship is valid only for Population II-type Cepheid stars.*

Baade's theory is based on the work of Henrietta Leavitt and Edwin Hubble who determined the relationship of periodicity and luminosity of Cepheid-type stars that vary in brightness. Baade's work, combined with the results of other astronomers, led to methods for determining the distance (in light-years), size, and age (in billions of years) of Andromeda and other galaxies. Baade concluded that the Milky Way **galaxy** was larger than the average galaxy, but, by far, it is not the largest (or oldest) galaxy in the universe.

Baade also theorized that luminosity is related to the mass of stars. In other words, 1) *there are more low-mass stars than bright high-mass stars;* 2) *this determines the "mass function" when referring to the number and density of stars;* and 3) *this is similar to the "luminosity function" that relates to stars with different luminosities.* This theory eliminates the evolutionary processes of stars and just considers the initial mass function for stars.

In addition, Baade, as well as several other astronomers, developed several theories, hypotheses, and opinions concerning *dark matter.* The phenomenon referred to as the massive amounts of "dark matter" in the universe may be described as two major types: 1) massive compact halo objects (MACHOS) identified as brown dwarf stars, extra large planets, and other types of very compact matter and 2) weakly interacting massive particles (WIMPS) that have not yet been discovered.

Following are some possibilities of why it is believed there is so much dark matter in the universe that falls within the two general categories of dark matter: MACHOS or WIMPS:

1. "Failed stars" called brown dwarfs or large planets similar to Jupiter that have a mass-to-light (M/L) ratio less than the sun. (The sun is considered to have a M/L of 1.)
2. "Stars with low luminosity" have a M/L less than 1 and are thus less massive than the sun. (They most likely make up most of the disk of dark matter in the universe.)
3. "Compact objects" consisting of neutron stars, dwarfs, and black holes that have a high M/L ratio and thus not as many are required to form dark matter.
4. "Strange new massive particles" that are odd, not-yet-discovered, strange bodies that in some way weakly interact with normal types of matter to form dark matter.

Baade's theory of gravitational microlensing: *Gravitational microlensing occurs when one star happens to be positioned in front of another star.*

Baade based this theory on work done previously by Albert Einstein. Depending upon the distance between the two stars and on the positioning of the foreground star in relation to the background star, the background star's image is magnified due to the "gravitational lens" effect of the gravity of the foreground star. This is called *microlensing* because the difference (increase in size) in the image of the background star is often too minute to be observed with a telescope. The chances of actually observing gravitational microlensing are relatively small. Millions of stars must be observed to find a situation that correctly aligns one star of the right type in front of another star. In addition, their masses, distance from each other, velocities, and brightness are limiting factors. Therefore, this phenomenon is more likely to occur when looking edgewise at a galaxy that appears as a flattened disk with a bulge at its center. This "bulge of stars" provides more of an opportunity for seeing stars in alignment and thus the microlensing phenomena. Note: When viewing a galaxy from above or below, it appears as a rather

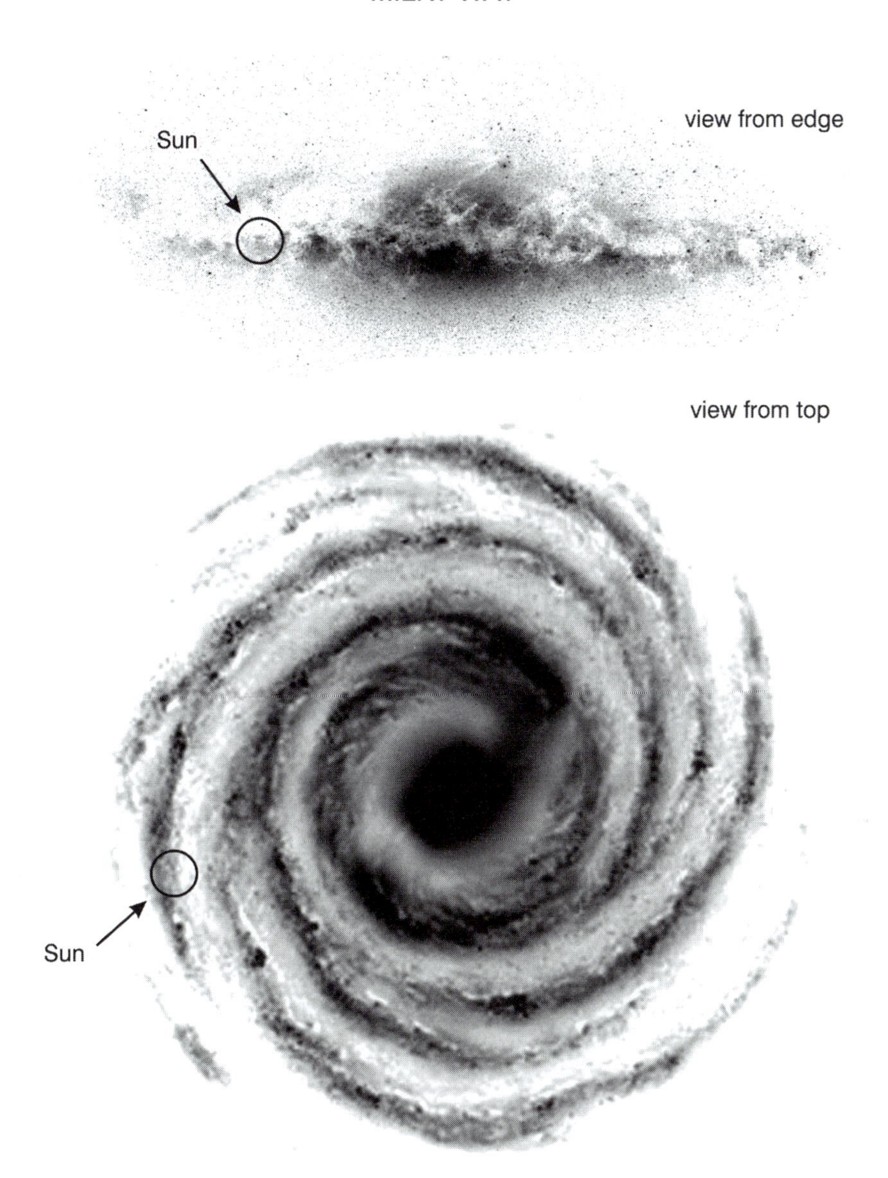

Figure B1. Two views of the Milky Way: Top View and Edge View.

flat, spiral shape structure with a massive cluster of stars at its center and with stars dispersed outward on its spiral arms.

Baade is also credited with discovering two minor planets (more likely **asteroids** than planets) that follow very elliptical orbits. One extends from the asteroid belt (between Mars and Jupiter) to beyond Saturn. He named it Hidalgo, which is Spanish for a person of noble birth. The other he named Icarus after a character in Greek mythology. The "planetoid" Icarus has an elliptical orbit inside Mercury's orbit and sweeps past Earth.

See also Zuckerandl

BABBAGE'S THEORY OF COMPUTING: Mathematics: *Charles Babbage* (c.1791–1871), England.

> *A machine can be built to calculate a series of values of polynomial functions automatically by using finite differences.*

Charles Babbage's birth date is uncertain. It was either the 26th of December 1791 or January 6, 1792. His father was a banker in London who had the means to send him to a private academy and then to continue his education with an Oxford tutor. In 1810 he entered Trinity College in Cambridge. Babbage became familiar with the works of several scientists/mathematicians, including Liebniz's and Newton's calculus. He helped form the Analytical Society whose purpose was to study abstract algebra and improve mathematics in Great Britain. He also helped found the Royal Astronomical Society in Great Britain.

Charles Babbage, known as the father of computing, is also credited with inventing the first ophthalmoscope in 1847, although another inventor in Germany, Hermann von Helmholtz, was unaware of Babbage's instrument and designed his own version in 1850. Babbage gave his ophthalmoscope to a physician to test, but it was neglected for a number of years until later revisions and improvements resulted in the modern instrument that is used to examine the retina, optic nerve, optic disc, and blood vessels of the eye.

Astronomers and mathematicians used other people, usually women, called "computers" to do the complicated calculations involved in their theoretical work, despite the fact that inaccuracies often beset some of their calculations. Thus, Babbage was inspired to develop a machine that could perform complex mathematical computations without errors. In 1822 he outlined his first plan for a "difference engine" that is the forerunner of today's modern computer. It was designed to more accurately calculate star tables to improve navigation. Because of its potential of saving lives at sea, the British government was interested and helped fund Babbage's research. This was most unusual because, in those days, governments did not fund scientific research. However, the promise that his difference engine would improve navigation was a "selling point." Babbage's first attempt was more complicated than he expected. He suffered a nervous breakdown in 1827, the year in which his wife, two sons, and his father died. He ceased work on the project in 1833. Some years later, when he still had no success, the government, as governments are bound to do, stopped funding his research. Although the cessation of funding by the British government effectively ended the practical application of the Babbage "computer," the fact that he could not afford to build the actual machine exclusively with his own funds did not end the theoretical nature of his work. Beginning in 1833, Babbage worked on a different concept that he called an "analytical engine" for which he is more famous. It was based on the system of punch cards used in looms that wove fabrics. Babbage believed instructions could be built into these cards for the loom operators to follow to design the final product. The famous IBM punch card computer of the mid-twentieth century used a similar system.

Babbage started construction on his difference engine in 1823. It is still in existence. In 1991 a working model number 2 was built from Babbage's drawings of his original model. The "first difference engine" stood 8 feet high, had 25,000 parts, and weighed 15 tons. Today, the model can be seen in London's Science Museum. Babbage also designed a printer that could be used with his engine computer that had many features of modern computer printers.

See also Turing

BABINET'S PRINCIPLE: Physics: *Jacques Babinet* (1794–1872), France.

Babinet's principle, sometimes referred to as "Babinet's theorem," states: *The diffraction pattern from an opaque body is identical to that from a hole of the same size and shape as the opaque body except for the intensity of the diffracting light beam.*

Babinet's principle/theorem is true for light and all other types of electromagnetic radiation and is used to detect the relative sameness of size and shape of materials and objects. Although this theorem is most often used in the field of optics, it holds true for all waves of the electromagnetic spectrum. It finds practical uses in determining the equivalence in the size and shapes of objects. For example, by shining a laser light beam through a small blob of blood cells, the diffraction patterns can be used to determine the size of all the blood cells. Another example is the corona or ring-like haze observed around the moon that is caused by sunlight, and which is reflected from the moon's surface, being diffracted by either clouds or water droplets in the earth's atmosphere. This effect is measured by the intensity of the diffraction pattern of the beams of moonlight entering the earth's atmosphere.

Interested in optics and polarization, Babinet developed several instruments and techniques to measure the properties of rocks and minerals, as well as the optical nature of metrological phenomena. He was the first to measure the polarization of light and the nature of rainbows. He was educated at the École Polytechnique and later in 1812 at the Military School in Metz. In 1820 he became a professor at the College Louis le Grand and in 1840 was elected to the French Academy of Sciences. Throughout his life he had varied interests in the optical nature of minerals, polarization of light, meteorology, magnetism, geography, and cartography. He designed and invented several instruments including the *Babinet compensator* and the *polariscope* that are used to polarize light for microscopes. He also invented the *goniometer* used to measure refractive indices.

He is best known for his standardization of the Ångström unit of light as the wavelength of light emitted by heated red cadmium. He suggested that a particular wavelength of light could be used as a standard to measure length. In 1960 his idea was accepted and was used as a definition for the length of the meter. This standard for the meter was the length equal to 1,650,763.73 times the wavelength of orange light emitted by the gas of the pure isotope krypton-86 when the gas is excited by an electrical discharge. The krypton-86 wavelength standard for the meter was changed in 1983 to equal the fraction 1/299,792,458 that light travels in one second.

BABO'S LAW: Chemistry: *Lambert Heinrich Clemens von Babo* (1818–1899), Germany.

> *The vapor pressure over a liquid decreases proportionally when specific amounts of solute are dissolved in the liquid.*

This was the first quantitative measurement that stated that the vapor pressure of a liquid is decreased proportionally to the amount of solvent added to the liquid, when the liquid's temperature is unchanged.

In 1845 Lambert von Babo was appointed to an assistant professorship at the University of Freiburg in Germany and later advanced to a full professor in 1859. Babo was one of the earliest chemists to do quantitative studies of vapor pressures over water. As a chemist he was aware of works of Charles Blagden (1748–1820), the British physician and scientist, who in 1788 observed that adding a solute to a solvent lowered the freezing point of the solution, as well as that of Michael Faraday who in 1822 determined that adding a solute to a liquid raised the solution's boiling point. In 1882 the French physical chemist François Marie Raoult performed more quantifiable experiments to show how solutes that affect the freezing points of a solution might be used to determine molecular weights. All of these chemists were familiar with Robert Boyle's research that established the relationship between temperature, pressure, and volume of gases.

See also Raoult

BACON'S CONCEPT OF INDUCTIVE REASONING: Philosophy of Science: *Francis Bacon*, 1st Baron Verulam, Viscount St. Albans (1561–1626), England.

Francis Bacon was a philosopher/scientist, politician, and writer, whose book *Novum Organum*, published in 1620, has influenced every scientist since his day by introducing the logic of induction and devising his "scientific method," which in essence proceeds from the specific to the general:

1. *Approach* the problem without prejudices; proceed with inquiry.
2. *Observe* situations accurately and critically.
3. *Collect* relevant facts and data from observations; make measurements.
4. *Infer* by use of analogies based on characteristics of observed facts.
5. *Draw* general conclusions from the specific to the general.
6. *Correct* initial conclusions with new insights. Truth comes from error, not confusion.

Francis Bacon started his career at the age of 12 when he entered Trinity College, Cambridge, although he never graduated. He then pursued a career in law in 1576 and was first elected to the House of Commons of Parliament in 1584. He continued his career as a statesman and was knighted 1603. In rapid succession he became the attorney general in 1613 and was proclaimed a Baron in 1618 and the Viscount St Albans in 1621. But all this ended in the same year he became Viscount when he was removed from office after being convicted of accepting bribes. He was removed from Parliament, fined, and sent to prison. In 1621 King James I pardoned him, but he was not allowed to return to Parliament. (King James I was actually the Scottish King James VI who succeeded the unmarried, childless English Queen Elizabeth I upon her death in 1603.) This crime was unfortunate for his political career but was fortunate for science as the rest of his life, although short, was devoted to efforts related to the philosophy of science.

His early philosophy was concerned with purging the mind of what he called "idols" that are the tendency of humans to believe in things that are not true (errors). (This seems to be as true today as in the 1600s.) His intention was to write a six-volume work called *Instauratio Magna* (Great Restoration) that included 1) a way to classify science, 2) his new inductive science, 3) a listing of facts acquired by experimentation, 4) how to use new approaches to learning, 5) general facts learned from natural history, and 6) his final philosophy of the science of nature. However he completed just two parts of this massive project: *The Advancement of Learning* in 1605, which was a review of the state of knowledge in England at that time in history and which was expanded in Latin as *De Augmentis Scientarum* in 1623, and *Novum Organum* (Indications Respecting the Interpretation of Nature) which was published 1n 1620—the year before his removal from Parliament.

Although Bacon is not considered a great experimental scientist, his philosophy of science, best known for his inductive method of investigation nature (scientific method) was appreciated by later scientists, such as Robert Boyle, Sir Isaac Newton, Voltaire, Robert Hooke, and many others who considered him the father of modern science.

His *inductive method* was a great improvement over the Aristotelian *deductive method* (that proceeds from the general to the specific) and the *philosophical thought* processes many ancients used to arrive at conclusions. (Bacon disagreed with Aristotle's philosophy that was based on "truth is derived from authority," and he believed that Arisotelianism produced only disputes.) Bacon's inductive reasoning improved the way scientists observed and experimented and thus arrived at a more authentic or "factual" understanding of nature. It also improved the process of establishing scientific hypotheses that could lead to new theories, principles, and laws of nature while still leaving room for future corrections. Of great importance was his idea of generating tentative conclusions such as hypotheses and theories that could be addressed and corrected by further scientific investigations.

Bacon was the first to observe that the coastlines on both sides of the Atlantic Ocean (Europe/Africa and North/South America) seemed to fit each other. Years later, this concept was developed into the theory of continental drift by Suess and Wegener.

See also Boyle; Ewing; Hess; Hooke; Newton; Suess; Wegener

BAEKELAND'S CONCEPT OF SYNTHETIC POLYMERIZATION: Chemistry: *Leo Hendrik Baekeland* (1863–1944), Belgium and the United States.

Leo Hendrik Baekeland started his career as a professor of physics and chemistry at the University of Bruges in Belgium in 1887 and later returned to University of Ghent where he received his doctorate. On his honeymoon in 1889 in the United States he realized that this was the place to begin a career as an industrial chemist. More of an entrepreneur than an academic professor, Baekeland started a consulting laboratory to explore possibilities in the field of photography. His first success was the invention of paper called Velox that was coated with light sensitive salts that could be used to produce positive photographs from the projection of the image from a negative of the image. It became a well-known product that he sold in 1899 to George Eastman of the Kodak Company for $1 million. These funds provided independence enabling him to investigate a new field of chemistry known as polymerization.

Baekeland began with experiments to find a substitute for shellac that at the time was manufactured by collecting a natural resin secreted by an oriental beetle that deposited this secretion on twigs of trees in India.

As a chemical reaction, polymerization requires that small molecules that make up larger molecules have at least two points involved in the reaction. The reaction usually requires a catalyst and heat, and often light and pressure, to force the smaller molecules to combine into larger chain-like macromolecules called monomers. There are two major types of polymerization. One, known as *condensation polymerization,* takes place when the growing chain eliminates some of the smaller molecules such as H_2O and CH_3OH. The other is called *additional polymerization* in which the polymer is formed without the loss of other chemicals. Polymerization can occur in nature, but today there are many known polymerization chemical processes used to make synthetic versions of what we now know as plastics.

After collecting the resin, and then undergoing a process of cleaning and purifying, it was formed into thin sheets. When broken into flakes, it was called "orange shellac." Natural shellac is soluble in alcohol but not water and is used as an undercoating on wood before varnish is applied. Because it is a "natural" product, shellac's source and availability was limited. This inspired Baekeland to find a synthetic substitute, although his motives were less than altruistic, as he openly admitted his main goal was to make money. Many scientists by patenting their discoveries have a motive to make money, which in a democracy is considered moral if the goal is to provide a useful product.

As an experimental chemist, he combined the synthetic phenol/formaldehyde resin-like substance that was first produced by Johann Friedrich Adolph von Baeyer in 1871 but had no success. By adding some other ingredients, including wood flour filler, as well as applying heat and pressure, he produced the first synthetic plastic, Bakelite in 1909. It was named after him. Soon after, Baekeland founded the General Bakelite Company that merged with two rivals, the Condensite Corporation and the Redmanol Company in 1922. In 1939, the Bakelite Corporation, the new name, was acquired by Union Carbide and Carbon Corporation.

Bakelite can be formed in molds, machined, and produced in many forms. One of its essential properties is its electrical insulation capability. As the first totally synthetic plastic, Bakelite was used in the manufacture of a variety of toys, kitchenware, telephones, and other electrical related equipment. The complexity and high cost of its production, along with its brittleness and other undesirable qualities, proved its undoing when other superior plastics were able to be produced. Bakelite was soon replaced by other polymer plastics.

Leo Hendrik Baekeland served as president of the American Chemical Society in 1924 and continued producing scientific papers until his death in 1944.

BAER'S LAWS OF EMBRYONIC DEVELOPMENT: Biology: *Karl von Baer* (1792–1876), Germany and Russia.

> *Individuals develop by structural elaboration of the unstructured egg rather than by a simple enlargement of a preformed entity.* This theory, also referred to as *epigenesis*, is based on Baer's four rules formulated in 1828:

1. In a large group, the *general* characteristics of animals will appear early in their embryonic development, whereas more *special* differences will appear later in their development.
2. The *more general* structural forms are formed before the *less general* structural forms are developed. Both forms are followed by the development of the *most specific* structural forms.
3. The more an embryo of a given animal becomes specialized, the more different it becomes from other species of animals as it matures.
4. Therefore, the embryos of higher animals only resemble other animal forms in the embryo stage.

This theory is basic to the field of embryology and, in essence, states that mammal development proceeds from simple (general) to complex (specific)—from homogeneous to heterogeneous. This means that, though all mammal embryos may look similar and have similar rudimentary structures, they grow up to be very distinctly different adult species. Baer's theory made the *recapitulation* theory impossible because young embryos are undifferentiated in form and are not previous adult ancestors. This means that mammals of a higher form of animal never resemble any other form of animal, except in the embryo stage. In other words, animal development proceeds from the general to the more specific. As the embryo matures into a fetus and later grows into an adult changes are not only differentiated but also irreversible.

Although of German descent, Karl von Baer was born in Estonia, where he studied and graduated with a degree in medicine in 1814. However, he was dissatisfied with his medical training and moved to Germany and then Austria for more advanced studies from 1814 to 1817. Beginning in 1817 Baer taught at the University of Königsberg (present-day Kaliningrad) in Russia where in 1826, while studying follicles and eggs (ovum) of mammalian ovaries, he identified the ovum as developing into an embryo. He continued the studies of other biologists in this area and is now known as the father of comparative and descriptive embryology. Baer also corrected some of the misconceptions of the mechanistic view of mammalian development from embryo to fetus to adult. The common belief of many biologists of his day and before, even as far back as Aristotle, was the embryos of one species pass through comparable stages to adults of other species. This was known as "recapitulation theory," or as "ontogeny follows phylogeny," and also as Haeckel's "biogenetics law." These theories all state that the embryos of one species pass through stages comparable to adults of other species.

See also Gould; Haeckel; Russell

In addition to embryology Karl von Baer had other interests. He teamed up with Jacques Babinet, a French physicist, to study factors that influence the directional flow of rivers, as well as currents in other bodies of water. This law is known as the "Baer–Babinet law of current flow." The directional flow of rivers, as well as ocean currents such as the Gulf Stream, are affected by the results of tectonics (movement of continental plates) and the Coriolis force created by the rotation of Earth. Tectonic movements of large plates of Earth have altered land structures and influenced the currents of rivers, lakes, and oceans by uplifting some regions as one giant plate overrode another plate. This geological activity creates an uplift of the lithosphere (Earth's crust) in some areas while submerging other regions, causing water to flow from the higher uplifted areas to the lower submerged region. The other force that affects direction of water flow is related to the physical law of "conservation of angular momentum" that is exhibited by the rotation of Earth on its axis. A river flowing northward will be diverted to the east due to the Coriolis force, whereas a stream flowing southward will be directed to the west. This effect is responsible for the direction of flow of the Gulf Stream northeastward toward Great Britain, and the counterclockwise direction of winds in hurricanes heading out of the South Atlantic Ocean to the north along the East Coast of the United States. *See also* Babinet; Coriolis; Wegener

BAEYER'S STRAIN THEORY FOR COMPOUND STABILITY: Chemistry: *Johann Friedrich Adolph von Baeyer* (1835–1917), Germany. He was awarded the 1905 Nobel Prize in Chemistry for his work on organic dyes and hydroaromatic compounds.

CARBON ATOM

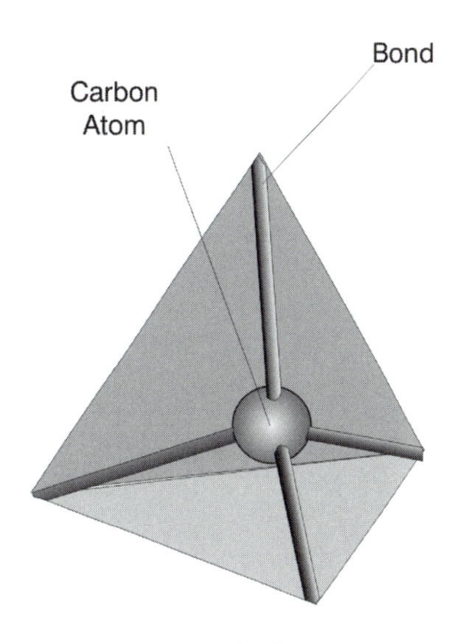

Figure B2. Tetrahedron structure of carbon atom.

After twelve years as a successful teacher of organic chemistry in several schools, Adolph von Baeyer moved to Munich where he spent the rest of his life working on the organic chemistry of dyes. He discovered phthalein dyes, phenolphthalein, and fluorescein as well as a phenol-like formaldehyde resin that was further commercially developed by Leo Baekeland as Bakelite (*see* Baekeland).

Baeyer is best known for his synthetic development of a synthetic indigo dye begun in 1865. Indigo dye known for its distinctive blue color has an interesting history. It is one of the oldest dyes used to color textiles and in paintings and has been used for hundreds of years in Asia and India. It was even known in ancient Greece, Rome, and the Middle East. India was the major supplier of indigo to Europe during the occupation by Roman troops and into the Middle Ages.

Indigo dye originates in several plants including the woad (*Isatis tinctoria*), the dyers' knotweed (*Polygonum tinctorum*), the true indigo plant in Asia (*Indigofera tinctoria*), plus several other varieties from Asia, Central, and South America. Native plants that were used to produce indigo, particularly in India during the 1800s were replaced by the synthetic versions produced in the late 1800s and early 1900s. The development and production of synthetic dyes devastated the indigo business in many countries. In modern times, over 17,000 tons of synthetic indigo dyes are produced in several countries.

Indigo dye is not only used to dye blue jeans. (Cloth dyed with indigo fades when repeatedly washed because indigo is not a "fast" dye. This quality seems to make clothing dyed with indigo more desirable to young people.) It is also used as a food coloring known as Blue No. 2.

The original synthetic indigo dye Baeyer developed was too expensive for commercialization; but later other similar chemical compounds were synthesized, and by 1890 they became inexpensive and thus commercially successful. Baeyer also worked on hydrobenzenes, terpenes, organic explosives, uric acid, and synthesized barbiturate drugs. There are several tales about how he came to the term "barbiturate." One involves it being named after a lady friend of his, Barbara. Another is that he celebrated his discovery on the feast of St. Barbara. Although interesting, both are unproven.

Chemical compounds (molecules) are less stable the more they depart from a regular tetrahedral structure.

A regular tetrahedron is a four-sided (faces) polyhedron. Each face is a triangle. It has four verticals and six line segments that join each pair of verticals. It may also be described as an analog of a three-dimensional triangle. This is the typical structure for some crystals and carbon **compounds**, where the carbon atom provides a **covalent bond** to each of the four corners of the tetrahedral atom to other elements (see Figure B2 tetrahedron, and Figure V3 carbon atom under Van't Hoff.) Think of the tetrahedron structure of the carbon atom as having one electron at each of the four corners. Each of these electrons can be shared with the outer electrons of other carbon atoms and atoms of other elements to form a wide variety of structures, such as long chains of carbon atoms with branches or rings whose "skeleton" is formed by connecting carbon bonds. This unique tetrahedron structure for carbon makes it important for the formation of the many different types of organic molecules that make up plants and animals.

See also Kekule; Van der Waals; Van't Hoff

BAHCALL'S THEORY FOR THE SOLAR NEUTRINO MODEL:

Astronomy: *John Norris Bahcall* (1934–2005), United States.

The sun produces 10^{36} neutrino events every second (solar neutrino units of SNU) at a density (flux) of 8 SNU.

A **neutrino** is an elementary particle classed as a **lepton** (somewhat like an electron of the same class) that has zero mass at rest as well as a zero electrical charge. It has other characteristics that make it useful for studying other minute particles produced in accelerators. The sun's interior is a natural nuclear fusion (atomic) reactor that produces its energy by the proton-to-neutron chain reaction by converting four protons (hydrogen nuclei) into helium, neutrinos, and other forms of energy such as gamma rays as well as the kinetic energy of the neutrinos and other moving particles that travel from the interior of the sun to the earth. The solar neutrino model presents several problems related to the number and types of emissions of neutrinos from the sun. John Bahcall and several other theorists predicted that neutrinos, which are considered weightless, will strike the earth and not be absorbed, as are some of the other heavier particles created by the fusion reaction that takes place in the sun. His theory was tested by others but did not seem to hold up very well. This prompted Bahcall to consider several options to his theory. One was that the sun was going through a passive phase and that only over a long period of time (cycles) would his predictions be accurate. Another consideration was that the neutrinos were decaying before they reached Earth, thus causing a lower count than his predictions. Still another consideration was that perhaps the entire solar neutrino model was wrong, which caused a false count in his predicted neutrino rate and density. The problem is still not settled and is now left up to the development of better instrumentation or revised and improved theories to account for the extent of neutrino production by the sun. More recent speculation involves the vast amounts of dark matter (over 90% of all matter in space) that may be composed of neutrinos left over from the big bang. It is now estimated that one type of neutrino, called the *electron neutrino*, is not exactly massless but has a tiny mass of about 0.5 eV to 5.0 eV, which is less than 1 millionth the mass of a regular electron.

Until improved equipment is developed and additional and improved data is acquired relating to the neutrino problem, it will remain speculation and a problem for future theoretical physicists and cosmologists to solve.

See also Bethe; Birkeland; Fermi; Pauli

BAKKER'S DINOSAUR THEORY: Biology: *Robert Bakker* (1945–), United States.

Dinosaurs were warm-blooded, similar to mammals and birds, and were not related to cold-blooded reptiles.

Robert Bakker based his theory of warm-blooded dinosaurs on the following evidence: 1) Bones of warm-blooded animals, such as mammals and birds (including some dinosaurs), have blood vessels, whereas cold-blooded reptiles' bones exhibit growth rings. Other bone structures also suggest that at least some dinosaurs were warm-blooded. 2) The fossil of a dinosaur heart with four chambers was found that indicates that at least some species were warm-blooded. Other fossilized dinosaur hearts were chambered—similar to reptiles. 3) Cold-blooded animals cannot withstand large variations in climate, such as the cold northern parts of the United States and Canada. Cold-blooded animals are *ectothermic*, which means their bodies cannot self-adjust internal temperatures to react to external temperatures. They do not have an internal thermostatic system to control their internal temperature, thus they need less food and less sleep. Because dinosaur fossils have been found in cold northern climates, it seems that they were *endothermic*. Although warm-blooded animals are endothermic and do have internal thermostatic systems, they need more food and sleep; and 4) warm-blooded animals have a high rate of metabolism. Therefore the *prey ratio* is much higher for warm-blooded mammals than reptiles—that is, the food consumption for warm-blooded animals is many times higher (per unit of body weight) than it is for cold-blooded reptiles. Fossil evidence suggests dinosaurs consumed vast amounts of plant and animal foods.

From these data, Bakker, as well as some others, concluded that, at least, some species of dinosaurs were more closely related to birds than reptiles, both having a common fossil ancestor known as *thecodonts* that means "animals with teeth embedded in the jaws." Bakker's theory created much discussion in the field of paleontology and raised concerns about some of the concepts of evolution. (He is a Pentecostal preacher and a proponent of theistic evolution.) Not all scientists agree that dinosaurs were warm-blooded. His theory is still being debated.

BALMER SERIES: Mathematics and Physics: *Johann Jakob Balmer* (1825–1898), Switzerland.

The Balmer series is the designation of a set of Balmer lines that are lines in the hydrogen spectrum that are produced by changes between $n = 2$ and levels greater than 2 either in emission or in absorption, where n represents the principal quantum number.

Johann Jakob Balmer was born in Switzerland, attending universities in Switzerland and Germany. He received his degree in mathematics from the University of Basel in 1849 and lived there the rest of his life. He began his career teaching in a girls' school and did not make any real contributions to the field of mathematics until the age of 60. In 1885 he devised a rather simple formula that described the wavelength for hydrogen's spectral line. This led to a generalized concept for the Balmer lines and the Balmer series. The formula was limited to the spectral lines of the hydrogen atom but later was expanded to include the spectral lines for all elements.

$$\text{Balmer's formula is: } \lambda = \frac{\hbar m^2}{(m^2 - n^2)}$$

Where λ = the wavelength. $\hbar$ = a constant with the value of 3.6456×10^{-7} meters, or 364.56 nanometers, $n = 2$, and m = an integer when m is greater than n.

Balmer devised the formula by gathering empirical evidence and thus was unable to explain why his formula was correct. (This was due to his and other scientists' lack of knowledge about the structure of the atom at that time in history.) It was later in 1888 when Johannes Rydberg generalized Balmer's formula so that it can be used for all transitions for the hydrogen atom. The four main transitions of hydrogen are based on the principle quantum numbers of the electron in the hydrogen atom. The wavelength and Greek letter associated with the different colors of the electromagnetic spectrum are:

$\lambda = \alpha$ at 656 nm, red color emitted
$\lambda = \beta$ at 486 nm, blue-green color emitted
$\lambda = \gamma$ at 434 nm, violet color emitted
$\lambda = \delta$ at 410 nm, deep violet color emitted.

The Balmer series is important in the field of astronomy because it shows up in many stars due to the abundance of hydrogen in the universe and stars. Starlight can show up as absorption or emission lines in the spectrum depending on the age of the star. Thus, the Balmer series assists in determining the age of stars because younger stars are mostly hydrogen whereas older stars have used up much of their hydrogen due to the fusion process and end up with a higher proportion of heavier elements, thus they are not as bright.

BALTIMORE'S HYPOTHESIS FOR THE REVERSE TRANSFER OF RNA TO DNA: Biology: *David Baltimore* (1938–), United States. David Baltimore, Howard Martin Temin, and Renato Dulbecco jointly received the 1975 Nobel Prize for Physiology or Medicine.

> *A special enzyme, called reverse transcriptase, will reverse the transfer of genetic information from RNA back to DNA, causing the DNA possibly to provide information to protect cells.*

Previous work with **DNA** and **RNA** indicated that genetic information could be passed from DNA to RNA but not the other way around. David Baltimore and Howard Temin independently announced that the enzyme reverse transcriptase enabled RNA to pass some genetic information to DNA, which possibly could aid cells to fight off cancer and other diseases, such as HIV/AIDS.

Baltimore also worked on the replication of the poliovirus and continues research on the HIV **retrovirus**, which was identified by other biologists. Baltimore and other virologists hope their research will lead to a better understanding of the relationship between the HIV retrovirus and AIDS. Biomedical researchers are attempting to find an effective vaccine that will prevent the damage the HIV virus does to the immune system or prevent AIDS by immunization.

See also Dulbecco; Gallo; Montagnier; Temin

> David Baltimore has had a varied career between academic administration and academic research. After studying chemistry at Swarthmore College, and later attending Massachusetts Institute of Technology (MIT) and the Rockefeller University, he changed his field to virology. It was as director of the Whitehead Institute in Cambridge, Massachusetts, that he began his groundbreaking research on DNA and RNA. In 1970 he presented his discovery of the new enzyme "reverse transcriptase" that can transcribe RNA information into DNA in some cancer viruses. This was a unique discovery because, up to this time, it was assumed that the transfer of information could only be by DNA.
>
> Baltimore became president of the Rockefeller University in 1990 where he played an unusual role as university presidents go by combining the careers of administrator, researcher, and fund-raiser. A colleague was accused of falsifying data after submitting a research paper to the magazine *Cell* with Baltimore as a coauthor. As the case developed over a period of years, it became an excellent example of what can happen when politics becomes involved in science. The investigation affected several careers over the years. Baltimore admitted his involvement, removed his name from the paper, and apologized. The researcher was later vindicated of fraud. As a result of the charges of falsifying data, Baltimore's position became difficult resulting in his resignation as president of the Rockefeller University. He returned to MIT in 1994 and in 1997 he became president of the California Institute of Technology where he became appreciative of the great advances being made in all areas of science. He resigned in 2006 after nine years as president of California Institute of Technology. It seems that about ten years is the average tenure of college presidents.

BANACH'S THEORY OF TOPOLOGICAL VECTOR SPACES: Mathematics: *Stefan Banach* (1892–1945), Poland.

> *Banach vector spaces are complete normed vector spaces where the space is a vector space V over a real or complex number and the norm introduces topology onto the vector space.*

Banach's most important work was in function analysis where he integrated related concepts into a comprehensive system of normal linear spaces, which became known as Banach spaces—a type of vector space. In 1924 he and Alfred Tarski (1902–1983) jointly published their theory of "The Banach–Tarski paradox" where they claim that it is possible to dissect a sphere into a finite number of pieces (more than five), which mathematically can be recombined to form two spheres the same size as the original sphere. Banach's major work was published in *Theory of Linear Operations* in 1932. In addition to founding the modern theory of functional analysis, he made contributions to the theories of topological vector spaces, measure theory, integration, and the orthogonal series. In 1979 a two-volume commentary of his works was published.

Stefan Banach was given his mother's surname but never saw her after his birth. His father, Stefan Greczek, gave him his first name, but, because of financial and social circumstances, Banach was raised by another family. Stefan Banach's father contributed to his financial support and maintained a relationship with his illegitimate son. From 1910 to 1914 Banach worked his way through Lvov (or Lwow) Technical University (now the Ivan Franko National University of Lviv in the Ukraine). He taught mathematics in local schools and in 1922 was hired by Lvov University where he did most of his research before dying of lung cancer in 1945.

BANTING'S THEORY FOR ISOLATING PANCREATIC INSULIN: Medicine: *Sir Frederick Grant Banting* (1891–1941), Canada. Frederick Banting was awarded the 1923 Nobel Prize in Physiology or Medicine along with his coresearcher, J. J. Macleod.

> *By ligating the pancreatic duct, it is possible to extract the polypeptide hormone insulin from the islands of Langerhans within the pancreas before the destruction of insulin can take place. The extracted insulin can therefore be administered to a diabetic patient in an effort to regulate carbohydrate metabolism within the body.*

Trysin-secreting cells are produced in the pancreas. When insulin that is produced in the pancreas is destroyed by proteolytic enzymes, the body is unable to metabolize carbohydrates (sugars) correctly. The result is a condition called diabetes.

Frederick Banting was born on a farm in Alliston, Ontario, Canada, where in 1910 he entered Victoria College at the University of Toronto to become a medical missionary. He received his medical degree in 1916 and entered the Canadian Army Medical Corps during World War I. In 1918 he received the Military Cross for heroism after being wounded in the Battle of Cambria in France. After the war he set up a practice related to children's diseases but soon joined the University of Western Ontario at London, Ontario, to specialize in research related to pancreatic cells known as "islets of Langerhans" that were, in some way, related to regulation of sugar metabolism and diabetes.

With the assistance of a professor of physiology, the Scottish physician John Macleod (1876–1935), and a young research assistant, Charles Best (1899–1978) an American physiologist who was educated in Canada, Banting performed a series of crucial experiments in a borrowed laboratory. They tied off the pancreatic ducts of dogs and took samples of the insulin extracted from the islets of Langerhans that was now isolated from other secretions, namely trypsin. They then injected these insulin extracts into diabetic dogs and found that they had some beneficial effects protecting the dogs from diabetes. The trio asked James Collip (1892–1965), a Canadian biochemist, to purify the extract. Soon after, the four of them (Banting, Macleod, Best, and Collip) patented the hormone. It is now known as insulin. They licensed it to Eli Lilly, receiving a royalty that was used to support their research.

It should be noted that at the time of Banting's "discovery," insulin was already identified and named. In 1916 an English physician, Sir Edward A. Sharpey-Schafer (1850–1935) formulated the word "insulin" after theorizing that a single substance produced in the islands of Langerhans in the pancreas is responsible for the condition called diabetes mellitus. Also, a Romanian professor, Nicolae Paulescu (1869–1931), isolated insulin in his lab about a year before the Canadians. He called it pancreatine. However, he only published his findings in French and was never accorded any real recognition for his efforts.

In 1923 the Nobel Prize was awarded to Banting and Macleod, but not to the other two researchers. This infuriated Banting and Macleod who then shared their prize money with Best and Collip. Macleod was Banting's lab supervisor and did little of the actual work involved in this important discovery. Banting was knighted in 1934 and worked with the Canadian and British medical research efforts dealing mainly with the effects of mustard gas and the physiological problems of fighter pilots. In 1941 he died in an airplane crash on his way to England to continue his research.

After eating, carbohydrates and sugars are absorbed by the intestines and then into the bloodstream and finally into the cells. Insulin is secreted by the pancreas as a response to an increase of blood sugar in the system. Cells have insulin receptors with the capacity to bind and absorb the blood sugar (glucose) from the blood into the cell where it is used in the process of metabolism to produce energy. If an individual's body is unable to produce adequate insulin or the cells cannot receive insulin, no matter how much an individual eats, he or she can still "starve." This is why victims of type 1 diabetes become very ill without insulin shots. Whereas, people with type 2 diabetes have developed a resistance to insulin rather than a deficiency of insulin. Type 2 diabetes patients do not respond well to insulin because their cells cannot absorb the sugar from the blood, leading to sugar levels in the blood that are higher than normal. The first insulin used by humans to treat diabetes was purified insulin extracted from cows and later pigs. This nonhuman insulin works well with most people, but some individuals develop allergies and other reactions to animal insulin. By the 1980s researchers developed a method to produce human insulin by using recombinant DNA techniques where the human gene that codes for insulin was copied and then placed inside bacteria. The gene is then "tricked," the end result being the bacteria cells make human insulin constantly. Because all humans have the same insulin genes, sensitive people are not allergic to it nor are humans as likely to reject this "biologically engineered" insulin. Medical research continues to seek a cure for diabetes. Some researchers believe that stem cells may play an important role in this endeavor. In the meantime, there are multiple new drugs on the market that are effective in controlling blood sugar levels in diabetes patients.

BARDEEN'S THEORY OF SUPERCONDUCTIVITY: Physics: *John Bardeen* (1908–1991), United States. John Bardeen is the only person to receive two Nobel Prizes for Physics. In 1956 John Bardeen, Walter Brattain, and William Shockley received the Nobel Prize for the development of the point contact transistor. And in 1972 he shared the Nobel Prize for Physics with Leon Cooper and John Schrieffer for developing the BCS theory of superconductivity.

When electrons interact in pairs in a vibrating crystal lattice, the electrons will cause a slight increase in positive charges in the crystal creating binding energy that holds the electron pair together, except at very low temperatures (near absolute zero), and thus not exhibit electrical resistance.

After graduating from high school at age 15 Bardeen attended the University of Wisconsin where he graduated in 1928, receiving his bachelor's and master's degrees in electrical engineering at that time. He secured a position at Gulf Research Laboratories in Pittsburgh. While there, he helped develop magnetic and geophysical means for oil prospecting, but he decided his interests were really in theoretical physics. He also spent five productive years with Bell Labs in New Jersey, working in the field of solid-state physics. While there, he and his colleagues, Walter Brattain and William Shockley, developed the **transistor**.

Influenced by other outstanding professors and researchers, Bardeen conducted research on the electrical conductivity of metals. After graduating from Princeton in 1936 with a PhD in mathematical physics, he went on to wartime research at the Naval Ordnance Laboratory in Washington, D.C. In 1951 he became professor of electrical engineering and physics at the University of Illinois in Urbana.

Beginning in 1945 his main research interests were in the theoretical effects of quantum mechanics as related to electrical conductivity in semiconductors and metals, which led to the invention of the transistor. In cooperation with two colleagues, Leon Cooper and John Schfieffer, they developed a viable theory of superconductivity, at low and "high" temperatures that is also known as the BCS theory of condensed matter or superconductivity.

John Bardeen's work with transistors and the theory of superconductivity of metals revolutionized electronics. Transistors are a necessity in our modern world. They are used in radio and television transmitting and receiving equipment, telephones, computers, and wherever electrical distribution systems are in place, such as automobiles, airplanes, ships, security systems, and so forth.

BARRINGER'S IMPACT THEORY OF CRATERS: Geology: *Daniel Moreau Barringer* (1860–1929), United States.

Craters were formed on the planets (including Earth) and the moon by the impact of large extraterrestrial objects such as meteors, asteroids, and comets.

Following is the evidence that Barringer developed for his theory:

1. The large amount of silica powder found at crater sites could be formed only by very great pressure.
2. In the past, large deposits of meteoritic iron "globs" were found at the rims of craters, most of which was removed many years ago by humans.
3. Rocks from deep in the craters are mixed with meteoritic material.
4. There is no evidence of volcanoes at crater sites. Therefore, they could be ruled out as a possible cause of impact craters.

Barringer's impact theory for craters is based on his study of the famous meteor crater (also referred to as the Barringer meteorite crater) located near Flagstaff, Arizona. Estimated to be twenty thousand to twenty-five thousand years old, it is almost 1 mile across and 600 feet deep. Compared to other meteorite impact craters, it is considered small.

Barringer was not the first to study this crater or come up with theories of crater formation on Earth. He did, however, establish the impact theory for craters, which is generally accepted within the scientific community. He did this despite having at one time agreed mistakenly with the theory that the meteor crater was the result of the impact of a meteor of the same size as the crater itself. (The current estimation of the size of the meteorite that impacted to create the Barringer crater is about 35 feet in diameter. It was a very dense iron meteorite weighing about 10,000 tons.) After Barringer found small pieces of nickel-iron rocks in the area, he spent a great deal of money establishing a mining company to extract the meteorite iron thought to be at the bottom of the crater. However, he was unsuccessful in finding significant deposits. Today, his theory is still the best explanation for most craters, including the Barringer meteorite crater found in Arizona. It is believed that at one time Earth's surface was pockmarked with craters, as is the current moon's surface, primarily because the moon does not experience extensive erosion. However, the process of weathering and erosion over eons of time has eliminated most of the evidence of the largest craters on Earth.

BEAUMONT'S THEORY FOR THE ORIGIN OF MOUNTAINS: Geology: *Jean Baptiste Armand Leonce Beaumont, Elie de* (1798–1874), France.

> *Mountains were rapidly formed by the distortion of molten matter as it cooled in the earth's crust.*

Jean Beaumont's theory is an explanation for the formation of mountains consisting mainly of basalt rocks, but not sedimentary shales or layered limestone. His theory is still considered viable by some biologists and geologists, particularly by those who believe in the concept of ***catastrophism***—theories that deal with the different types of catastrophic events on Earth that occurred in the past. These catastrophic events on Earth include earthquakes and volcanoes, which possibly are responsible for the formation of mountains, as well as catastrophic meteor impacts/craters and major climate changes. The major evidence in support of Beaumont's theory is that "roots" of mountains are less dense than the rocks found at the mountains' higher elevations.

Modern theory for the origin of mountains is based on the concept of the earth's crust being raised above the surrounding area by the warping and folding of surface rock into layers. Another modern concept is *plate tectonics*: large plates on the ocean floor and under the continents move and crash into each other over eons. This plate movement, at a depth of 25 to 90 miles, has been ongoing for the past 2.5 to 3 billion years and still continues. The crashing together of the edges of these plates cause the development of earthquake fault lines similar to those located in California, Eastern Europe, and Asia. Plate movement is the process responsible for building the global distribution of mountains, as well as resulting in earthquakes and volcanoes. Mountains are formed in either a ring configuration, as in the Olympic Mountains in Washington State, or, more often, in ridges linked together, as in the Sierra Nevada range. A third type is the group of ranges similar to the Rocky Mountains in the western United States, the Andes in South America, the Alps in Europe, and the Himalayas in Asia. Beaumont's theory, though not completely wrong, is too limited as a geological concept for the origin of mountains.

See also Buffon; Cuvier; Eldredge; Gould

BECQUEREL'S HYPOTHESIS OF X-RAY FLUORESCENCE: Physics: *Antoine Henri Becquerel* (1852–1908), France. Antoine Becquerel shared the 1903 Nobel Prize for Physics with Marie and Pierre Curie.

The exposure of fluorescent crystals to ultraviolet light will produce X-rays.

Antoine Becquerel's concept is an excellent example of how his hypothesis, which proved false, later resulted in a discovery of great importance.

Wilhelm Röentgen discovered X-rays in 1895. Becquerel believed he could produce X-rays by exposing his **fluorescent** crystals (salts) to sunlight (**ultraviolet** radiation). He placed his crystals on a photographic plate covered in black paper and then exposed both to sunlight. His original hypothesis assumed that the photographic plate had been darkened by what he incorrectly thought was exposure to X-rays passing through the paper from the crystals. He inadvertently left an unexposed, wrapped, photographic plate in a desk drawer with some of his fluorescent crystals on top of the plate that had not been exposed to sunlight. To his amazement, when the plate was developed, it was darkened as if it had been exposed to something coming from the crystals—obviously not ultraviolet light, because it was stored in a dark drawer. Because neither the plate nor the crystals were exposed to sunlight, he concluded that his original hypothesis was incorrect. He now hypothesized that the crystals gave off some form of penetrating radiation (later identified as radiation of short wavelengths with an electrical charge such as beta and gamma rays.) He continued to experiment and found that the radiation could be deflected by a magnet and thus must consist of minute charged radiation particles. Becquerel is credited with discovering *radioactivity*.

See also Curies; Röentgen; Rutherford

BEER'S LAW: Physics: *August Beer* (1825–1863), Germany. (Note: This law is also known as the **Beer–Lambert–Bouguer law** because all three independently discovered variations of the law at about the same time.

Their law states: *There is a logarithmic dependence between the transmission of light that shines through a material and the density of the material as well as the length of the material that the light is traveling through.*

In 1729 Pierre Bouguer's theory was published that defined the amount of light that was lost by passing it through a given amount of atmosphere. Pierre Bouguer (1698–1758), a French mathematician, also determined that the sun's light was 300 times brighter than the moonlight reflected from its surface that originates from sunlight. Johann Heinrich Lambert (1728–1777), a German mathematician, physicist and astronomer, published a book in 1760 on how light is reflected from different surfaces. He coined the word "**albedo**" (the reflection factor of light or other forms of radiation from a surface). Lambert also presented a hypothesis that the planets near the sun were part of a system that traveled within the Milky Way galaxy, and that our solar system is just one of many found in the galaxy. He also presented a hypothesis for the nebular (interstellar cloud of gas) origin of our solar system.

In 1852 August Beer expressed his law in several elaborate forms of common logarithms and in exponential equations. In essence, the results may vary according to the ability of the material to absorb light and the material's wavelength. In other words, if the material is very dense or opaque the law does not apply because little or no light can transverse the material. This law is expressed in logarithms when applied to **spectrophotometry.** When used with regular optical equipment, it is expressed in exponential form.

Little is known about August Beer's life. He was born in Trier, Germany, in 1815 where he studied mathematics and the natural sciences. He worked for the famous mathematician and physicist, Julius Plücker (1801–1868), in Bonn, Germany. Beer eventually received a PhD in mathematics and published a book in 1854 titled *Einleitung in die höhere Optik*. His work, along with that of Lambert and Bouguer, the German and French eighteenth-century mathematicians, constitutes Beer's Law.

A version of the Beer–Lambert law is also used to describe the absorption of solar radiation as it travels through the atmosphere. Its application and relevance is dependent with respect to the degree the sun's light is perpendicular to the observer on Earth's surface.

See also Ramsay; Tyndall

BEHRING'S THEORY OF IMMUNOLOGY: Biology: *Emil Adolph von Behring* (1854–1917) Germany. Emil von Behring was awarded the first Nobel Prize for Physiology or Medicine in 1901 for his discovery of antitoxins that are produced by humans to counteract the toxins (poisons) produced by bacteria in the body.

> *The blood of animals will produce substances that can neutralize toxins, that is, poisons caused by invading organisms such as bacteria, and that antitoxins similar to antibodies will fight the disease-causing organisms.*

Emil Behring, the son of a small town schoolteacher with a large family, was a brilliant child who, with the assistance of the local preacher, was able to attend the Gymnasium (high school) in Hohenstein in Saxony. He then attended the Academy for Military Doctors at the Royal Medical-Surgical Institute in Berlin. After receiving his medical degree, he entered the Army Medical Corps where he served as a troop doctor and later became a lecturer in the Army Medical College in Berlin. Following his military service he was employed at the Hygiene Institute of Berlin and became an assistant to Robert Koch (1848–1910), the well-known German physician and bacteriologist. At this point in his career, Behring studied and experimented with the development of a therapeutic serum that led to successful treatments for **diphtheria** and **tetanus**. His first successful therapeutic serum treatment took place in 1891. It involved a child who was suffering from diphtheria. These first treatments were not successful because the antitoxins were not strong enough. After more research, an improved protocol, using a mixture of the toxins along with the antitoxins now derived from larger animals such as sheep, and later horses, proved successful. As with many such inoculations, there were adverse reactions to the treatment serum, but in the long run diphtheria and tetanus as devastating diseases have been conquered. Behring received many awards during his long career and went on to develop treatments for other diseases including a vaccine for the immunization of calves against tuberculosis. He spent the end of his career attempting, unsuccessfully, to develop a vaccine for human tuberculosis.

Behring, along with Shibasaburo Kitasato (1856–1931), a Japanese bacteriologist, proposed a serum theory that led to their development of an antitoxin for diphtheria and tetanus (a form of blood poisoning). They demonstrated that giving animals graduated doses of tetanus bacilli caused the animals to produce in their blood substances that could neutralize the toxins that these bacilli produced. These were called *antitoxins*. They also demonstrated that the antitoxins produced by one animal could immunize another animal and that it could also cure an animal showing signs of diphtheria. Their research was confirmed and replicated by others.

Using this information Behring collaborated with Ehrlich to develop similar antitoxin immunity for diphtheria, a major killer of children at that time in history. It might be of some interest that the best antitoxin was made from injecting horses and then using blood serum from the infected horse as the source of the antitoxin. It was also discovered that by mixing a small amount of the original toxin (poison) with the serum antitoxin, the treatment for tetanus and diphtheria was more effective.

See also Ehrlich; Jenner; Koch

BELL'S LAW (ALSO KNOWN AS THE BELL–MAGENDIE LAW): Medicine: *Sir Charles Bell* (1744–1842), Scotland.

> *The anterior spinal nerve roots contain only motor fibers, and posterior roots only sensory fibers.*

Bell's major work in 1811 was the first to refer to the motor functions of the ventral (abdominal) spinal nerve that established the sensory functions of the dorsal roots. This, along with the discovery made by the French physiologist, François Magendie (1783–1855), that damage to the dorsal root and anterior root in spinal nerves destroys both the sensory and motor activity, enabled Bell to arrive at Bell's law that is based on anatomical evidence. This discovery is considered one of the greatest in the history of physiology. He demonstrated that the spinal nerves were able to transmit sensory and motor functions. In addition, he found that the sensory nerves traverse the posterior roots whereas the motor nerves go through the anterior section. Evidence of Bell's law was confirmed by experiments conducted by Magendie, who is considered a founder of experimental physiology, and the German physiologist and anatomist Johannes Peter Müller (1801–1858) who used frogs to demonstrate the theory because it was easier to extract their spinal cords than it was for small mammals. Magendie believed living organisms were merely complex systems that could be subjected to all types of experimentation with impunity. He used living cats, dogs, and rabbits without seeming to care about their pain or discomfort.

Sir Charles Bell received his medical degree in 1799. In 1824 he became the first professor of anatomy and surgery of the College of Surgeons in London. In cooperation with his brother John Bell (1763–1820), also a surgeon, they wrote and illustrated a two-volume medical text titled, *A System of Dissection Explaining the Anatomy of the Human Body*. Bell wrote an earlier book in 1811 called *An Idea of a New Anatomy of the Brain* in which he describes his various experiments with animals and his ability to distinguish between sensory and motor nerves, which was a first. Many physiologists considered the work described in Bell's 1811 text to be the foundation of clinical neurology. In 1826 he was elected a Fellow of Royal Society and was knighted in 1831. Bell established a new hospital and medical school in 1828.

Bell may be best known to the general public for discovering the paralysis of facial muscles caused by a lesion of the facial nerve—known as Bell's palsy, as well as his discovery of a related problem, Bell's spasm, which is the involuntary twitching of the facial muscles.

BERGERON'S THEORY OF CLOUD PROCESSES: Meteorology: *Tor Harold Percival Bergeron* (1891–1977), Sweden.

> *Water vapor is formed as a result of water evaporating from supercooled drops that then are attached to ice crystals that either fall as snow or melt and fall as cool rain, depending upon local temperatures.*

Between 1925 and 1928 Bergeron worked at the Geophysical Institute in Stockholm and, after teaching at Oslo University, was elected as the head of the Meteorological Institute in Uppsala. As a meteorologist, Bergeron collaborated with his German colleague Walter Findeisen (1909–1945), thus, the theory is also known as the "Bergeron–Findeisen theory." The theory is based on their discovery of the mechanism for the formation of precipitation (rain, snow, and ice) in clouds. In 1935 Bergeron wrote a paper titled "On the Physics of Clouds and Precipitation" that documents the change in state from a vapor to a liquid water, which is called condensation. He arrived at the Bergeron process while walking in the mountains where the humidity was high and it began to rain.

We now know that there is more to the process than Bergeron's idea that the saturation vapor pressure with respect to ice is less than the saturation vapor pressure with respect to water. There are several other conditions that result in condensation besides the Bergeron process as follows: 1) when the relative humidity on the surface of Earth reaches 100% and 2) when vapor pressure is the same as the saturation vapor pressure. The primary difference is that, in clouds, the water will not condense until the saturation point reaches a level of supersaturation of about 120%. This level is required for the cloud droplets to overcome the natural surface tension of drops of water. Oddly, the pure water in the atmosphere requires some air pollution (called "aerosol") for the vapor to have something on which to condense and form droplets. Contrary to common beliefs, water vapor in clouds does not condense on itself when the temperature drops in rain clouds, but rather it condenses on tiny particles of matter (less than 2 microns) called nuclei that are forms of air pollution. These nuclei must be below freezing for this process to occur. Once the droplets are formed on the nuclei, air currents cause them to collide and coalesce with each other to form snowflakes—or raindrops carry them. These processes are referred to as ccn (cloud condensation nuclei).

Note: Pure water does not freeze at 0°C, and at temperatures below this point, liquid water is called supercooled water. This is important for another method of forming rain or snow. It is not until the temperature drops lower than freezing that the aerosol pollutants (called freezing nuclei) are cold enough to allow the supercooled cloud droplets to freeze onto these nuclei. This process does not occur until the temperature of the freezing nuclei reaches about −10°C.

BERNOULLI'S LAW OF LARGE NUMBERS: Mathematics: *Jakob (Jacques) Bernoulli* (1654–1705), Switzerland.

> *The average of a random sample from a large population is likely to be close to the mean of the entire population.*

This law of large numbers is a fundamental principle of statistical sequences that can be expressed in another way: The probability of a possible event (no matter how likely or unlikely) occurring at least once in a series of events increases with the number of events in the particular series. Jakob also is known for his work in permutations and combinations. The law of large numbers is sometimes referred to as the principle of probability, meaning that the probability of an event occurring at least once increases in likelihood if the number of events is large enough or approaches infinity. This can be interpreted in several ways. For example, when an increasing number of lottery tickets are sold, the odds increase that there will be at least one winner; whereas if only a few tickets are sold, the odds of a winner decrease drastically. Another way the law of large numbers may be interpreted is related to the differences in the concepts of "possible" and "probable." Anything might be considered possible. Even so, the term "possible" is not quantifiable, whereas the concept of "probable" is quantifiable in a statistical sense ranging from 0.1 to 1.0 or a scale of one (or it may be thought of a scale of 10 or 100%). The probability of an event occurring or not occurring in a nonstatistical sense may be thought of as "likely" or "unlikely" or "reasonably" or "unreasonably." There is no statistical scale for a "possible" event to occur or not occur even for an infinite number of events. This means that a "possible" event is not likely to occur or that it is most unlikely to have occurred in the past. This is the reason why the term "possible" should not be used in a courtroom because the term has no quantifiable or definitive meaning in determining guilt or innocence. After all, according to the law of large numbers, it is also possible for anyone to have committed the crime unless there is evidence establishing a high "probability" of a specific suspect.

Jakob Bernoulli and Johann Bernoulli (1667–1748) were brothers. (Daniel Bernoulli was the son of Johann Bernoulli.) Jakob's and Johann's father, Nicolaus Bernoulli (1623–1708), discouraged their ambitions as mathematicians and encouraged them to have careers in medicine instead. Jakob had degrees in mathematics and theology, Johann in iatromathematics and medicine. The brothers were rivals with each other throughout most of their lives. Part of this rivalry started as a disagreement over how to solve the problem of finding the shortest path between two points of something moving all by itself as influenced by the force of gravity. This problem led several mathematicians, including Leonhard Euler, in the direction of the field that became known as calculus. Their rivalry was not limited to their related work in the field of mathematics but continued in all aspects of their relationship and that of Johann's son, Daniel.

Jakob was involved in the development of and popularization of the new field of integral and differential calculus. Calculus, a field of mathematics dealing with differentiation and integration, was constructed by Sir Isaac Newton and Gottfried Leibniz. Jakob's contribution was in demonstrating how calculus could be used in practical ways in various fields of mathematics, whereas Johann's interest in this field was to support the version of calculus proposed by Gottfried Leibniz. Johann claimed that Leibniz's version had priority over Newton's calculus. Johann's interest extended beyond mathematics and included the fields of astronomy, chemistry, and physics. Even today the controversy of who invented calculus is not settled.

See also Bernoulli (Daniel); Euler; Leibniz; Newton

BERNOULLI'S PRINCIPLE: Physics and Mathematics: *Daniel Bernoulli* (1700–1782), Switzerland.

The sum of the mechanical energy of a flowing fluid (the combined energy of fluid pressure, gravitational potential energy, and kinetic energy of the moving fluid) remains constant.

Daniel Bernoulli's principle is related to the concept of energy conservation of ideal fluids (gases and liquids) that are in a steady flow. This principle, used by mathematicians and engineers to explain and design many machines, further states that if fluid is moving horizontally with no change in gravitational potential energy and if there is then a decrease in the fluid's pressure, then there will be a corresponding increase in the fluid's velocity (or vice versa). One example of this aspect of the principle is the design of airplane wings. The air that flows over the upper curved surface of the wing moves faster than the air that passes past the underside of the wing, thus creating a pressure differential. In other words, the air must travel faster over the curved top of the wing and thus the pressure is less (i.e., the air molecules are spread further apart). While on the underside of the wing, the air flows slower (molecules closer together) and thus exerts greater pressure. This causes the wing to be "pushed" up, keeping the moving aircraft in flight. This upward pressure on the wing is called *lift*, but it might be more appropriate to call it *push*.

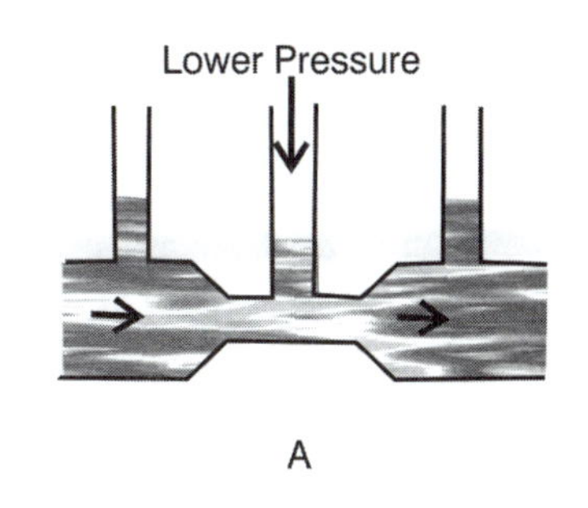

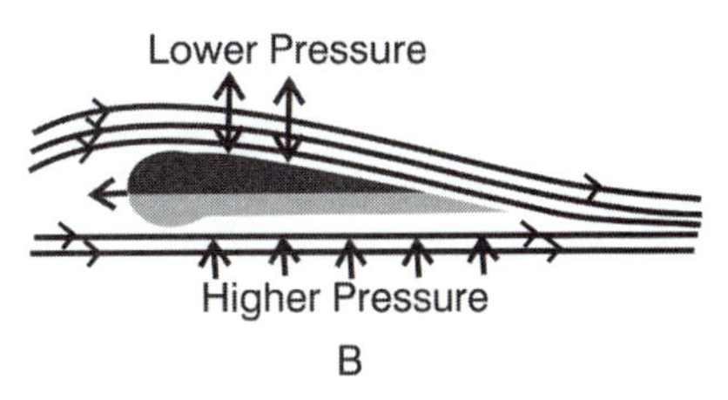

Figure B3. As the speed of the flow of a fluid (air or liquid) increases, the pressure decreases (and vice versa).

A similar part of the Bernoulli principle states, for example, that if there is a partial constriction in a pipe or air duct, the velocity of the fluid (gas or liquid) will increase as the pressure increases. This is known as the *Venturi effect* and can be demonstrated by the narrow nozzle of a garden hose that constricts, and thus speeds up, the flow of water as the water pressure forces the same amount of water through a smaller opening. A spray bottle or atomizer works on the same principle. It is named after the Italian physicist Giovanni Battista Venturi (1746–1822) who first described the effect by constrictions on water flowing in channels.

Bernoulli laid the groundwork for Robert Boyle's work in gases when he proposed a model for gases that consisted of many small atoms that were in constant motion and that exhibited elasticity as they bounced off each other as well as the sides of its container.

Daniel Bernoulli was interested in philosophy, logic, medicine, and mathematics. Bernoulli developed an equation that oddly received the name "Bessel functions." Named after the German mathematician, astronomer, and systematizer, Friedrich Bessel, Bessel functions are solutions involving cylindrical or spherical coordinates and are important for solving problems of wave propagation and signal processing. In 1724 Daniel Bernoulli published work in differential equations, which was well received throughout Europe. Following this publication he was appointed professor of mathematics at the University of St. Petersburg and later as professor of physics at the University at Basel in Switzerland where his father, Johann Bernoulli, formerly held the chair in mathematics. He also made contributions to probability theory, the kinetic theory of gases, electrostatics, and was a pioneer in the field of hydrodynamics.

See also Bernoulli (Jakob); Bessel; Boyle

BERZELIUS' CHEMICAL THEORIES: Chemistry: *Baron Jöns Jakob Berzelius* (1779–1848), Sweden.

Berzelius' electrochemical theory: *Molecules that make up compounds carry either a negative or positive electric charge.*

Baron Berzelius' work with the electrochemical nature of chemical reactions led to his concept of **catalysts** being related to the speeds at which chemical reactions take place. He not only gave this concept the name "catalyst" but also coined the names "protein" and "isometric." However, his positive-negative or, as it became known, his dualistic view of compounds did not hold up very well for future theories related to organic chemistry.

Berzelius' theory for chemical proportions: *The proportion of chemicals in a reaction is related to the masses (atomic weights) of the molecules involved in the reaction.*

His theory of chemical proportions allowed Berzelius to develop an accurate table of atomic weights for elements and molecules. Somewhat oddly, despite understanding the proportional relationship of atomic weights in chemical reactions, he did not accept Avogadro's hypothesis or number (*see* Avogadro). Even so, his accurate measurements of atomic weights of chemicals were one of his most important contributions to chemistry. His table of atomic weights was a precursor to the establishment of John Dalton's atomic theory.

Berzelius' theory of radicals: *Groups of atoms can act as a single unit during a chemical reaction and have at least one unpaired electron.*

Baron Berzelius named these groups of atoms "radicals" because of their nature to act as a singular electrically charged unit (ions), for example, OH, NH4+, $SO4^-$, and NO_2^-, all of which have a charge. They are short-lived, highly reactive charged particles that can initiate a chemical reaction by splitting molecular bonds. It was later discovered that ionizing radioactivity causes illness, including radiation poisoning, and death. Other causes for the formation of free radicals in the tissues of living organisms are not completely understood, but their formation is related to normal metabolism within organisms. The role of free radicals as they affect cells and accelerate the aging process in humans continues to be studied.

Berzelius had a difficult time in securing an education. He was forced to leave the University of Uppsala in Sweden to secure employment to finance his further education. He worked as a chemist testing the local water supply. Although he finally received his medical degree, he chose to work with Wilhelm Hisinger (1766–1852), the Swedish mineralogist, in the field of mining chemistry. Together they discovered the elements cerium, selenium, and thorium and assisted in the discovery of lithium and vanadium. (The German chemist Martin Klaproth (1743–1817) is also given credit for the discovery of cerium.) Berzelius was a meticulous researcher who kept accurate records and published over 250 papers. His textbook on chemistry published in 1808 was translated into several languages. Later, in 1818 he published an important paper titled "Essay on the Theory of Chemical Proportions and on the Chemical Effects of Electricity." Berzelius also developed the system for abbreviating the elements by the first one or two letters of their names (e.g., *H* for hydrogen, *O* for oxygen, *S* for sulfur, *Na* for Natrium [sodium], *Co* for cobalt, and *Ra* for radium).

Berzelius was not as successful in the field of organic chemistry as he was in inorganic chemistry. His insistence on the importance of his "dualistic theory" became an obstacle for future development in that field. Even so, his contributions to the field of organic chemistry were significant. His introduction of the concept of organic radicals (molecular fragments with an electrical charge) was a large step in the understanding of organic chemistry, particularly when it led to the discovery of the benzol radical. (Note: The term "benzol" is an archaic term for "benzene.")

As mentioned, his textbook on chemistry published in the early 1800s was well accepted in universities. It went through many printings and was translated into many languages except English. Berzelius was creative and invented a number of the chemical supplies, equipment, and terms used in today's laboratories. These included rubber tubing, protein, isomerism, catalyst, radical, and filter paper.

See also Avogadro; Dalton; Dumas

BESSEL'S ASTRONOMICAL THEORIES: Astronomy and Mathematics: *Friedrich Bessel* (1784–1846), Germany.

A star's distance from Earth can be determined by measuring its parallax.

Bessel cataloged the positions of over fifty thousand stars up to the ninth magnitude and was able to determine the parallax of 30 seconds for the star 61Cygni. From this calculation he was able to measure the star's distance from Earth as over 10+ light-years, which is now correctly determined to be 11.2 light-years distant. In 1838 Bessel proclaimed that he had determined the parallax for the star 61Cygni. It was difficult to measure the parallax for stars because they are so far from Earth and the instrumentation at the time was inadequate. Also, he was not the first to do so. That honor goes to Thomas Henderson who measured the parallax for the triple-star system Alpha Centauri in 1832. (A simple experiment to recognize parallax involves holding your index finger about 12 inches in front of your face, then closing one eye and then the other. Notice that the finger seems to jump to the left and then to the right. This phenomenon becomes less obvious and less measurable as distance increases.)

Bessel's theory of double star systems: *When the orbit of a bright star is displaced, the displacement is caused by a "dark companion" star that forms a two-star system.*

In 1844 after careful observations of wave-like motions of Sirius, the brightest star in the sky, he correctly determined that it was a double star system and its mate was a dark body that, though not visible, was the cause for the slight displacement of Sirius. Using the same rationale he employed for his theory of the movement of Sirius, he determined that the gravitational motion displacement for the planet Uranus indicated the presence of an unknown planet beyond Uranus. After his death and as he predicted, the planet Neptune was discovered in 1846 by the German astronomer Johann Gottfried Galle (1812–1910). However, Galle used calculations provided by Urbain LeVerrier (1811–1877), the French astronomer.

As a young man Bessel was trained as an accountant and was employed by an import-export firm. His interest in navigation eventually led to his pursuits in astronomy and mathematics. He refined the calculation for Halley's Comet, which started his success in the field of astronomy. After a friend secured him a position as an assistant at the Lilienthal Observatory in Saxony, he made accurate observations for the positions of over three thousand stars. At one time, the Lilienthal Observatory was the best-equipped observatory in the world. This success led to a commission to build an observatory at Königsberg (now Kaliningrad, Germany) where Frederick William III of Prussia appointed him as the observatory's director. He held this position until his death in 1846. Bessel had no university education but was highly respected by his

> Bessel was one of many scientists interested in calculating the shape of Earth. Back in the sixth century BCE Pythagoras calculated that Earth was a sphere, and three hundred years later Eratosthenes calculated the size of Earth. In the late 1600s Jean Richter (1630–1696), a Frenchman, determined that his pendulum time machine was 2 minutes slower when he measured time at the Equator than it was in Paris. Isaac Newton claimed that Earth was flattened at the poles and thus was shaped as a "rotating spheroid." Others came to similar conclusions. Bessel used the imaginary meridian arcs in the sky that correspond to Earth's longitudinal lines in his 1832 calculations to determine that the shape of Earth was an elliptical spheroid.
>
> In mathematical terms a spheroid is a quadric surface in three dimensions that can be formed by rotating an ellipse about one of its major axes; if the ellipse is rotated around its minor axis (rather than its major axis), the surface of the ellipse will take the shape of an *oblate spheroid.* This shape is similar to the "pancake" shape of Earth where the diameter from the North Pole to the South Pole is less then the distance of the diameter of Earth at the Equator. The rotation of Earth on its north/south axis creates the bulge of about 25 miles greater diameter at the Equator than is the diameter from pole to pole.

peers. The largest crater in the moon's Mare Serenitatis (Sea of Serenity) is named after him.

Not withstanding Bessel's contributions to astronomy, he may be better known for "Bessel functions," which is a mathematical theory to determine celestial motion that is influenced by gravity that causes perturbations. Oddly, Bessel functions, which were actually discovered by the mathematician Daniel Bernoulli, refer to a method for solving Bessel's differential equations. (Note: A detailed analysis of differential equations and the various form of Bessel functions of the first and second kind, as well as other applications, is beyond the discussion in this book.) The Bessel function is also useful in finding solutions to Laplace's equations and the Helmholtz equation in cylindrical and spherical coordinates related to wave propagation in the field of communications.

Bessel used his theory of functions to determine the motions of two or more bodies under the influence of mutual gravitation. His work in mathematics and astronomy enabled future astronomers and physicists to arrive at new astronomical observations. Bessel functions also assist in the study of the flow of heat and electricity through cylinders and spheres, as well as solving problems related to the wave functions in electricity and hydrodynamics.

See also Bernoulli (Daniel); Halley; Laplace

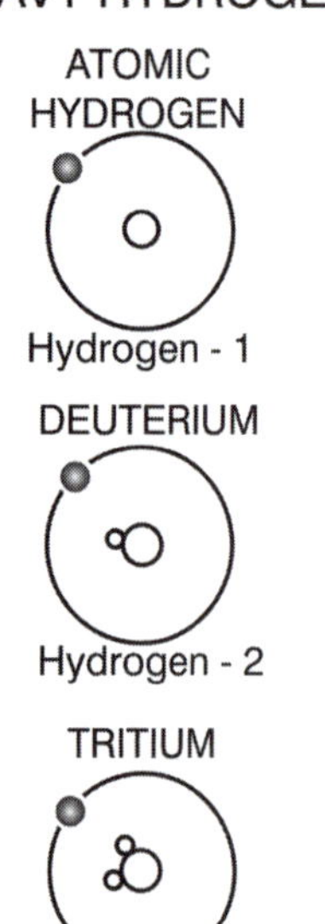

Figure B4. Structure of the three isotopes of hydrogen atoms.

BETHE'S THEORY OF THERMONUCLEAR ENERGY: Physics: *Hans Albrecht Bethe* (1906–2005), United States. Hans Bethe was awarded the 1967 Nobel Prize for Physics.

> *Carbon-12 atoms found in all stars undergo a series of catalytic reactions that convert hydrogen nuclei into helium nuclei through the process of a thermonuclear reaction, releasing 17.5 million electron volts of energy (17.5 MeV).*

Bethe's theory became known as the *carbon cycle*. Others had previously determined the sun is composed of about 85% hydrogen and 10% helium. Bethe postulated his thermonuclear theory as the explanation for the tremendous, long-lived source for the energy produced by the stars, including our sun. Although the thermonuclear reaction involving just one carbon-12 atom and a few hydrogen nuclei will not produce much energy, the stars have an enormous quantity of hydrogen. This reaction has continued over billions of years and produces prodigious amounts of energy. One unsolved problem was why the reaction did not take place faster and blow up the stars similar to a hydrogen (thermal fusion) bomb. Herman Ludwig Ferdinand von Helmholtz suggested that gravitational forces slowed the contraction of hydrogen to keep the system running. This theory did not hold up. Sir Arthur Stanley Eddington suggested that the hydrogen-to-helium reaction could be sustained in the stars if their centers contained very high-temperature gases that would force the nuclei together. Experiments on Earth with high-pressure and temperature-heavy hydrogen plasmas (highly ionized gas) to replicate the fusion process of the sun indicated that Bethe's and Eddington's theory is correct.

The essence of this thermonuclear reaction is that four protons (4 H^+) are converted into helium nuclei ($2He^{++}$), with the carbon-12 atom acting as a catalyst that is not consumed. The hydrogen protons involved are isotopes—deuterium (D) and tritium (T)—which are forms of heavy hydrogen. The reaction can be written as: $D + T + e \rightarrow {}^4He + n + 17.5$ MeV of energy, where *e* is an input of energy required to start the reaction and *n* is radiation. This process led others to develop the "fusion" H-bomb, which is many times more destructive than the nuclear "fission" atomic bomb but produces less harmful radiation. For the past half-century, research had attempted to achieve a similar controlled thermonuclear reaction to produce prodigious amounts of controlled energy for the production of electricity.

See also Gamow; Hoyle; Teller

BIOT–SAVART LAW: Physics: *Jean-Baptiste Biot* (1774–1862) and *Felix Savart* (1791–1841), France.

> *The intensity of the magnetic field set up by a current flowing through a wire varies inversely with the distance of the field from the wire.*

The Biot–Savart law means that the intensity of a magnetic field that is set up by an electric current flowing through a conductor (wire) will vary inversely with the distance of the magnetic field from the wire.

Jean Biot and Felix Savart were French physicists at the Collège de France in Paris where they studied the relationship between the flow of electrical current and magnetism and arrived at the law that is named after them. This law follows other laws of electricity such as Coulomb's law, Ampère's law, and Gauss' law. As a practical statement, it is an analogy between magneto-statics and fluid dynamics that is used to calculate magnetic responses and current density (amperes) in fluids. More recently the Biot–Savart law was used in the calculation of the velocity of air induced by vortex lines in aerodynamic systems because a vortex in fluids (air) is analogous to the velocity of a current's (amps) strength through a magnetic field.

In addition to electromagnetism, Biot was interested in other areas of physics. He was the first person to determine the optical properties related to the polarization of light as the light passes through a solution. He also was the first to determine the optical properties of mica, which is found in the mineral *biot* that was named after him. For his work on polarization he was awarded the Rumford Medal of the Royal Society. He later wrote a book in which he proposed that the shape of Earth might be based on its rotation on its axis. In 1804, along with Joseph Gay-Lussac, he flew a hot-air balloon to an altitude of 5 kilometers as they explored the changes in Earth's atmosphere. A small crater on the moon is named for Biot.

Felix Savart (1791–1841) actually discovered the law related to how the flow of current through a magnetic field is related inversely to the distance of the field from the magnet in 1820 during his collaboration with Jean Biot. Jean Biot was Savart's mentor and was senior in age and academic position at the Collège de France. Savart was also interested in acoustics. He invented the Savart disk that, using a cogwheel as a measuring tool, produced a sound wave of a known frequency. In addition, the *savart* which is a unit of measurement for musical intervals is named after him although it was actually invented by the French mathematician and physicist Joseph Sauveur (1653–1716).

See also Ampère; Coulomb; Gay-Lussac

BIRKELAND'S THEORY OF THE AURORA BOREALIS: Physics: *Kristian Olaf Bernhard Birkeland* (1867–1917), Norway.

The aurora borealis is caused by rays (charged particles) from the sun that are trapped in Earth's magnetic field and concentrated at the polar regions.

The aurora is a curtain-like, luminous, greenish-white light produced by upper atmospheric atoms and molecules that become ionized after being struck by electrons, thus emitting radiation. It is a large-scale electrical discharge affected by the solar wind and Earth acting as a **magnetosphere** "generator" that concentrates the aurora at the polar regions.

Kristian Birkeland studied this phenomenon for some time and arrived at his theory from his knowledge of cathode rays recently produced and named by the German physicist Eugen Goldstein (1850–1930). He recognized the relationship of the glowing charged particles in cathode rays whose directions could be altered by magnetism.

Birkeland made another important contribution involving the great worldwide demand for nitrogen fertilizer. The major supply of fertilizer was limited to guano (bat dung found in caves) and some natural nitrogen compounds. However, the atmosphere is about 78% nitrogen and could be an almost unlimited source of nitrogen fertilizer. In 1903 Birkeland and the Norweigian engineer and industrialist Samuel Eyde (1866–1940) developed a process by which air was passed through an electric carbon arc and produced nitrogen oxides, which were then dissolved in water to form nitric acid. The nitric acid reacted with lime to form calcium to produce calcium nitrate, an excellent fertilizer. It required great amounts of electricity to operate the electric arc. As a consequence, the commercialization of the process led to the development and the growth of hydroelectric power in Norway. This process became known as the Birkeland–Eyde process that produced fertilizer for export worldwide before World War I. About the same time another process that involved a catalytic reaction with hydrogen was developed that used the "free" nitrogen from the atmosphere and converted it into "fixed" nitrogen. The resulting nitrogen oxides, the basis for fertilizer, were less expensive to manufacture than those produced by the Birkeland–Eyde process. This alternate process became known as the Haber process using an electric arc as a means of fixing atmospheric nitrogen. The process was commercialized by the German chemist and engineer Carl Bosch (1874–1940) and is still used today to produce the worldwide demand for fertilizer.

See also Haber

BJERKNES' THEORY OF AIR MASSES: Physics and Meterology: *Vilhelm Friman Koren Bjerknes* (1862–1951), Norway.

The thermodynamic properties of an air mass will determine the weather factors for the area that is covered by the air mass.

Essentially, this means that the physical nature of an air mass is partly determined by the nature of the surface region over which it develops.

Bjerknes was the head of the geophysics department at the University of Leipzig in Germany before establishing the famous Bergen Geophysical Institute in Bergen, Norway, in 1917, which laid the foundation for the Bergen School of Meteorology. In collaboration with his son Jacob Bjerknes (1897–1975), a renowned meteorologist in his own right, they established a series of weather stations in Norway. As a result of the data and research gathered from these stations, Vilhelm Bjerknes developed his theory of air masses and cold (polar) fronts. He recognized that there were at least four types of air masses: equatorial, tropical, polar, and arctic (and antarctic). An air mass is a very large dome of air, which internally has similar factors of temperature, humidity, and pressure. Some fifty air masses exist over the surface of Earth at any one time, and their nature reflects the region from which they were spawned. Bjerknes used these and other factors to develop a system that distinguished properties that determine the weather, such as humidity, temperature, and visibility (based on the amount of dust in the atmosphere). He recognized that the mass movement of air could better be predicted when the hydrodynamics, such as polar fronts, squall lines, and low-pressure areas (cyclonic regions) of large massive weather system were understood. The movement of air masses, related hydrodynamics, and the refinement of his classification system for air masses in the atmosphere are the bases for today's weather predications and reports.

As a young boy Vilhelm assisted his father, Carl Bjerknes, in setting up and conducting experiments in hydrodynamics. He continued this work until he entered the University of Kristiania. (The city of Kristiania was renamed Oslo in 1925.) He wrote his first scientific paper "New Hydrodynamic Investigations" in 1882 when he was just twenty years old.

Prehistoric humans, no doubt, were aware of weather and changes in atmospheric conditions that affected their lives. They could tell if a weather front was upon them by the mere fact that they became colder or hotter, or that it was humid or wet, as well as windy. During the ice ages they were well aware of the seasons and prepared for long winters. People always could, and still do, recognize repeated patterns of the weather and sense the difference between hot/cold, wet/dry, cloudy/bright, and calm/windy. There are records that indicate the ancient Greeks understood that weather changed when masses of air passed through their region. In about 400 BCE, Hippocrates wrote a piece "On Air, Water, and Places" that describes changes in weather and how directions of the wind entering the city could affect the health of citizens.

Until about the seventeen century observing the weather was accomplished by folk tales, myths, and legends. Some of the tales still exist. For instance, the amount of hair on a certain type of caterpillar will predict how cold the winter will be, and the predictions of the famous Punxsutawney Phil groundhog that comes out of his hole on the second day of February each year are but two examples. If the sun is shining and "Phil" sees his shadow, there will be six more weeks of winter; and if he does not see his shadow, spring will come early.

Despite all the knowledge, equipment, computers, satellites, and theories at hand today, weather forecasting is still somewhat unreliable due to the enormous amount of constantly shifting variables involved that are beyond the capabilities of our best efforts to incorporate them into a forecast that is accurate for more than a few days.

He then moved to Germany to continue his studies, returning to Norway in 1907 where he remained for the next five years before going to the University of Leipzig as the head of the geophysics department. He presented his theory in a 1921 paper titled "On the Dynamics of the Circular Vortex with Application to Atmosphere and to Atmospheric Vortex and Wave Motion." His theory is based on the direct analogy of the flow of fluids (air and water), turbulence, and whirlpools in water to the behavior of air masses. In other words, he applied his father's work on hydrodynamics and electrodynamics to air masses.

BLACK'S THEORIES OF HEAT: Chemistry: *Joseph Black* (1728–1799), Scotland.

There are three aspects to Black's theory. One deals with solids changing to liquids (fusion), one involves the change of liquids into gases (vaporization), and the third relates the capacity of heat required to a specific temperature change of a given mass.

Black's theory of latent heat of fusion: *The heat of fusion is the heat capacity required to change 1 kilogram of a substance from a solid to a liquid without a temperature change.*

Black's theory of latent heat of vaporization: *The heat of vaporization is the heat capacity required to change 1 kilogram of a substance from a liquid to a gas without any temperature change.*

Black's theory of specific heat: *Specific heat is the amount of heat required to raise 1 kilogram of a substance by 1 degree Kelvin.*

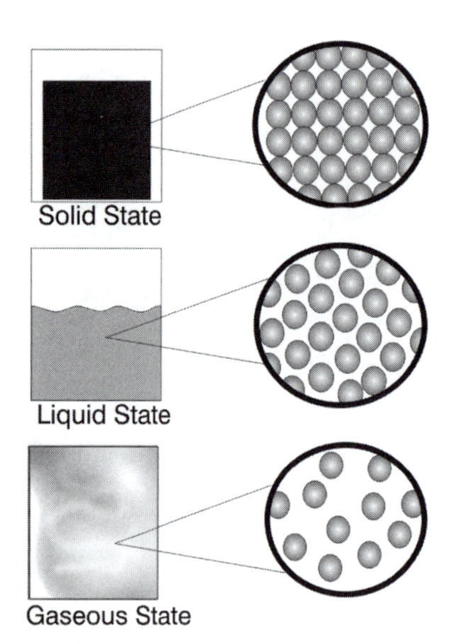

Figure B5. States of matter: A solid state exists when the substance has a definite shape and volume and tends to maintain its shape and volume. A substance in the liquid state has a definite volume but no definite shape; it flows and takes on the shape of it container. The liquid state is between the solid and gaseous states. A substance in the gaseous state has a lower density than solids and liquids, and it will expand to fill the extent of its container. The states of matter are related to the densities of matter and the kinetic energy of their constituent particles.

Joseph Black proposed these theories after experimenting and making many measurements involving changes in the states of matter (e.g., water to ice and boiling water to steam; see Figure B5). He was the first to distinguish between temperature and heat, a distinction that many still confuse today. *Temperature* is based on the law of thermodynamics and is the degree of hotness or coldness transferred from one body to another as measured in degrees Celsius, Kelvin, or Fahrenheit. Temperature is a measure of the average (mean) energy of the motion of molecules and atoms in a substance in internal equilibrium. *Heat*, on the other hand, is a form in which energy is transferred from one body to another. Heat always flows from a substance that contains more **energy** to one with less. Thus, the temperature of the first substance is reduced, and the second substance increases until equilibrium between the two substances is established. In other words, at equilibrium they are at the same temperature.

During the late eighteenth and early nineteenth centuries, a number of physicists developed the science of heat, later named *thermodynamics*, the second law of which states that heat flows naturally from hot to cold, but never the other direction. The second law of thermodynamics describes *entropy*, which is an increase in disorganization (of molecules) within a closed system until equilibrium is established.

In several ways, Black's work was the beginning of modern chemistry. More important, he was one of the first to measure chemical reactions quantitatively. For example, the reaction of limestone with acid that exhibited an effervescence that he called "fixed air," and that later proved to be carbon dioxide, was one of his more famous experiments.

See also Carnot; Clausius; Joule; Kelvin; Maxwell; Mayer; Thomson; von Helmholtz

BODE'S LAW FOR PLANETARY ORBITS: Astronomy: *Johann Elert Bode* (1747–1826), Germany.

Bode's law is a numerical system for determining the average radii (distance) of a planet from the sun, calculated in astronomical units (AU). An **AU** is the average distance of Earth from the center of the sun—approximately 93 million miles.

Start with a series of numbers where each number is twice the preceding number, namely, 3, 6, 12, 24, 48, 96–.–.–.–, then add 4 to each number, namely, 7, 10, 16, 28, 52, 100–.–.–.–, then divide the sum of each by 10. The answer is the mean radii of the planetary orbits in astronomical units (AU), which is the planet's mean distance from the sun—for example:

$$3 + 4 = 7 \div 10 = 0.7 \text{ AU}$$
$$6 + 4 = 10 \div 10 = 1.0 \text{ AU}$$
$$2 + 4 = 16 \div 10 = 1.6 \text{ AU}$$
$$24 + 4 = 28 \div 10 = 2.8 \text{ AU}$$
$$48 + 4 = 52 \div 10 = 5.2 \text{ AU}$$

Bode's law is really the mathematical expression of a concept proposed by the German astronomer Johann Titius (1729–1796) in 1766 or 1772. It is based on Titius's idea that a simple numerical rule governs the distance of planets from the sun. A few years later Bode proposed a useful combination of simple numbers that he claimed could predict the location of unknown planets. It is unknown if this is some true relationship of the nature of the solar system or just coincidence. Most astronomers of his day were unimpressed with his number sequence because the rule did not apply for the planets Neptune and Pluto. Bode's law predicted a planet between Mars and Jupiter, but none was found until the Italian monk, mathematician and astronomer Giuseppe Piazzi (1746–1826) discovered a very small (about 650 miles in diameter) asteroid-like planet, Ceres, in 1801. Ceres was located at 2.55 AUs from the sun in an area with many, many asteroids (also known as planetoids). Bode's law was finally accepted when Bode accurately predicted the location of a yet-to-be-discovered planet. Using a telescope, William Herschel located Uranus in 1781, exactly where Bode's numbers indicated it should be, at 19.2 AUs. Bode was given the privilege of naming this new planet, calling it Uranus after the Greek god of the sky.

BOHM'S INTERPRETATION OF THE UNCERTAINTY THEORY FOR ELECTRONS: Physics: *David Joseph Bohm* (1917–1992), United States.

> *The electron has a definite momentum and position and is thus a real particle, with wave and particle characteristics, but this duality is the result of new "pilot waves" that connect the electron with its environment.*

David Bohm did not completely agree with the Heisenberg uncertainty principle (indeterminacy principle) or with then current interpretations of **quantum** theory. He considered Heisenberg's theory as presenting only a *description* of the behavior of an electron and not a *view* of the electron because it stated neither the position nor momentum (mass multiplied by velocity) of the electron could be determined at the same instant. Bohm claimed that this uncertainty does not represent the **deterministic** nature of reality—that is, an event cannot precede its cause.

Bohm's pilot wave is not the classical or traditional explanation of quantum or indeterminacy theories. It can be measured only by complex mathematics, not by experimentation. Although Bohm's pilot wave interpretation maintains the concept of "real nature" as being deterministic, it is not as well accepted, as is Bohr's interpretation. Today, most physicists accept the latter's theory.

See also Bohr; Dehmelt; Einstein; Heisenberg; Planck

BOHR'S QUANTUM THEORY OF ATOMIC STRUCTURE: Physics: *Niels Hendrik David Bohr* (1885–1962), Denmark. Bohr was awarded the 1922 Nobel Prize for Physics.

Bohr's quantum mechanics theory for atoms: 1) *Electrons reside in discrete energy levels (similar to the shells or orbits of Rutherford's model) in which they move. As long as they remain in their orbit, they do not emit radiation. Therefore, these energy levels (orbits) are stable and are always whole number multiples of Planck's constant as 1h, 2h, 3h, and so on, and the orbiting electrons are limited to a discrete series of orbits.* 2) *Electrons move in stable orbits because they can only emit or absorb discrete radiation "packets" of energy that are equal to the difference between the original and the final energy levels of the electrons. The quanta "packets" of energy are absorbed or radiated when electrons change from one orbit to another.*

Niels Bohr based his theory for the structure of the atom on Ernest Rutherford's famous experiment demonstrating that atoms comprised very small, heavy, dense, positively charged central nuclei surrounded at some distance by very light, negatively charged particles, referred to as *electrons*. This concept of the negative electrons orbiting the positive nucleus was somewhat similar to planets orbiting around the sun. This classical mechanical-electrodynamic concept presented a problem in the sense that electrons carry a negative electrical charge, and according to the laws of physics, they should radiate energy as they orbit the nucleus, which would result in instability and cause them to spiral into the positively charged nucleus. Thus the conservation of momentum would be violated. Bohr solved this problem of the atom's potential instability by postulating that the circumference of the orbit must be equal to an integral number of wavelengths. The extension of this idea led to the development of quantum mechanics.

To account for the conservation of momentum (mass times speed), Bohr assigned specific values to orbits and later to suborbits. This led to his concept that when an electron emitted a quantum of energy (photon), it would move to a lower orbit (lower energy level). Conversely, when an electron absorbed a quantum of energy, it would move to an outer or higher orbit (energy level). This became known as a *quantum leap*. By using Planck's constant ($\hbar$) he measured the difference in radiation for these energy level changes by $\hbar v$, where v is the frequency of the radiation. These developments led Bohr to another principle.

Bohr's correspondence principle: *The quantum theory description of the Bohr atom relates to events on a very small scale but corresponds to the older classical physics, which describes events on a much larger scale.*

This principle is based on electrons' obeying the principle of quantum mechanics but with limits corresponding to and similar to Newtonian classical mechanics. Thus, his model of the atom could exist only if electrons exhibited both wave and particle properties. This explains how electrons, as standing waves, could move in orbits without emitting radiation but still have particle characteristics. Bohr's next principle is related to the quantum nature of photons and electrons.

Bohr's complementary principle: *The electron can behave in two mutually exclusive ways. It can be either a particle or a wave.*

The wave–particle duality was demonstrated by others and is accepted today as the duality nature of quantum particles. Bohr also was the first to theorize that an electron could enter a nucleus and cause it to be excited and unstable. This led to his next contribution.

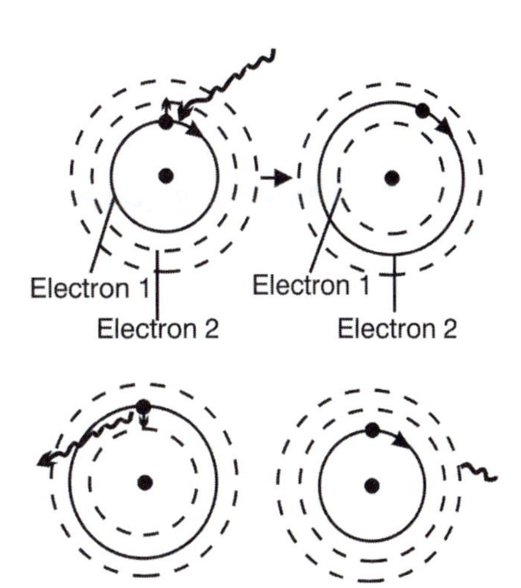

Figure B6. When electrons move from a stable energy level of atoms, they emit discrete packets of energy.

Bohr's theory of a compounded nucleus: *The nuclei of atoms are compounded or composed of distinct parts. The heavier the nuclei the more "parts" they contain and more likely to be unstable and break up.*

This led to his next theory.

Bohr's "droplet model" theory: *The impact of a neutron (corresponding to a "droplet") on a very heavy nucleus can cause the heavy nucleus to be compounded and become unstable and fission or split into two parts, whose total mass almost equals the mass and charge of the original heavy nucleus.*

Later, Otto Hahn who discovered *protactinium* (Pa), and German chemist Fritz Strassman (1902–1980) chemically identified fragmentary decay particles of uranium predicted in Bohr's model but did not identify it as **fission**. This decay reaction, called *nuclear* fission by Lise Meitner and Otto Frisch, occurs when "compounded" heavy nuclei break into two or more lighter nuclei. These experiments were the first evidence for fission of the rare uranium radioactive isotope U-235, which ended as a small radioactive isotope of barium-56. This led to the use of another fissionable element, plutonium, used in atomic fission bombs, which Bohr assisted in developing. Among Bohr's other contributions was his early (1920) theoretical description of the Periodic Table of Chemical Elements that he based on his theories of atomic structure.

See also Bohm; Dehmelt; Frisch; Hahn; Heisenberg; Meitner; Planck; Rutherford

BOK'S GLOBULES THEORY OF STAR FORMATION: Astronomy: *Bart Jan Bok* (1906–1983), United States.

> *The small, circular "clouds" of matter that are visible against a background of luminous gas or by the light from stars are actually massive "globular-like clouds" of dust and gas that are in the process of condensing to form new stars.*

Bart Bok identified these interstellar dark globules near the **nebula** Centaurus, referred to as IC2944. Bok's theory is used to explain one of the concepts for the origin of our universe. Recently his theory has been reexamined as a possible model for the creation, regeneration, or rebirthing of stars to explain the idea for the ever-continuing **universe**.

After receiving his PhD in the Netherlands, Bart Jan Bok arrived in the United States in 1929. He became a citizen in 1938 and was appointed professor of astronomy at Harvard University in 1947. From 1957 to 1966 he was director of the Observatory in Canberra, Australia. After returning to the United States in 1966 he became director of the Steward Observatory in Arizona and professor of astronomy.

In cooperation with his wife, Priscilla, they published a paper "The Milky Way" in 1941 that explained his theory for the spiral structure of the Milky Way. Walter Baade who identified hot young O and B type stars in the Andromeda galaxy further explored Bok's theory. These bright O and B stars act as identifiers in the arms of the spirals of the galaxy. In the early 1950s William Morgan and Hendrik van de Hulst produced data from radio-astronomy experiments that contradicted the circular spiral nature of galaxies. Their data identified them as being more elliptical in shape. Bok attempted to join the two different types of structures (circular vs. elliptical) for galaxies but was not successful. He is now best known for what are called *Bok globules* that are the small dark circular clouds that are visible against the background of stars.

See also Baade

BOLTZMANN'S LAWS, HYPOTHESES, AND CONSTANT: Physics: *Ludwig Edward Boltzmann* (1844–1906), Austria.

Boltzmann's law of equipartition: *The total amount of energy of molecules (or atoms) is equally distributed over their kinetic motions.*

In other words, on the average, the energy of molecular motion is distributed with discrete degrees of freedom within an ideal gas. This led to Boltzmann's description of how *the total energy of a gas is distributed equally among the molecules in the gas, namely, heat*. This became known as the Maxwell–Boltzmann distribution equation, which is based on the Boltzmann constant: $k = R/N = 1.38 \times 10^{-23}$ J/k, where *k* is the Boltzmann constant, *R* is the universal gas constant, *N* is the number of molecules in 1 mole of gas as per Avogadro's number, and *J/K* is joules per degree of Kelvin.

Boltzmann distribution equation: *The probability exists that a molecule of a gas will be in energy equilibrium with the position and movement of the molecule and will be within an unlimited range of values.*

This is another way of stating the energy distribution of gas molecules. It states that atoms and molecules should obey the laws of thermodynamics.

Boltzmann's entropy hypothesis: *The entropy (the measure of disorder in a closed system) in a given state is directly proportional to the logarithm of the number of distinct states available to the system.*

Entropy was the term given to the concept of the second law of thermodynamics. It is based on the fact that unless energy is added to a closed system, the system will always proceed to a state of disorganization and finally to a state of energy or heat equilibrium. Boltzmann supplemented the mathematics related to thermodynamics using a statistical treatment to interpret the second law of thermodynamics, which in essence states that heat can only move toward cold, or to a region of less heat, never the other way around. This hypothesis can be stated as the Boltzmann constant equation: $S/k = \log p + b$, where S is entropy and k is the Boltzmann constant, which has the value of 1.380×10^{-23} joules per degrees Kelvin. Boltzmann thought so highly of this equation that he had it inscribed on his tombstone. Most of Boltzmann's theoretical work contributed to the science of statistical mechanics. He is also known as the father of statistical mechanics.

See also Carnot; Clausius; Kelvin; Maxwell; Rumford

BONNET'S THEORIES OF PARTHENOGENESIS AND CATASTROPHISM: Biology: *Charles Bonnet* (1720–1793), Switzerland.

All organisms are preformed in miniature (homunculi) as little beings inside the female eggs of the species, and the "germ" of a species is constant over time, thus no male of the species is required for propagation of the species.

Charles Bonnet developed this theory of parthenogenesis after discovering the female of a species of a tree aphid reproduced without the aid of male sperm (thus, parthenogenesis, or "virgin birth"). To overcome the objections to his theory, which implied all living organisms were unchanged from the beginning of time, he proposed the concept of catastrophism. Although he was the first to explain evolution in a biological context, and the first to use the term "evolution" in a biological sense, he did not accept the extinction and changes of species as a gradual process.

Bonnet was one of the first to propose catastrophism as the cause of changes (evolution) in biological species. According to his theory, catastrophic events on Earth result in great extinction of most species and that that new species are created from the few individuals that survived catastrophic events. He believed that catastrophism was responsible for apes becoming humans and that the next step is for humans to evolve into angels. Catastrophism as a concept, with modifications, is still accepted by a few biologists today.

The concept of "preformation" of humans in either the female egg or male sperm as a homunculus (small human form) persisted since the days of ancient Greek philosophers. Catastrophism influenced many biologists until Darwin's concept of a more gradual type of biological evolution, but even Darwin was aware that evolution was not "smooth" over periods of time and that there were "interruptions" in the process.

See also Aristotle; Buffon; Cuvier; Eldredge–Gould; Gould; Swammerdam

BOOLE'S THEORY OF SYMBOLIC LOGIC: Mathematics (Logic): *George Boole* (1815–1864), Ireland.

A mathematically (algebraic) logical construction is based on one of the following operators: AND, OR, or NOT, or is based on a construction that may be expressed by all three of these operators.

Symbolic logic, and in particular Boolean logic, is indispensable for use in developing computer programs and computer-based research engines. This system is based on what mathematicians refer to as "elements" of the system and "sets." A set is a collection of things (such as a group of numbers) that have the characteristic that can be identified as being included in a collection of other things (other groups of numbers).

When using the English terms of OR, AND, and NOT, Boolean logic is the basis for computer programming languages as well as computer search engines, such as Google and Ask, although not all computer search engines use the same syntax. As an example of a common engine syntax:

Search for Students with last name = Jones AND first name = James;
Search for Students with last name = Jones OR first name = James;
Search for Students where NOT is last name = Jones

The Google search engine uses the AND syntax as a default in their program as a way to link two related or different items, e.g., "search Jones" "search James." In other words the AND is understood in their program and automatically gives you students with both names Jones and James. If you want to refine the search OR is used as in "search Jones OR James", and they use the – sign for the logical use of NOT, as in – "search Jones." Actually the Google program is more elegant than this simple Boolean logic suggests, but these techniques seem to speed up a more comprehensive search.

George Boole is sometimes thought of as mathematics genius because of his work in abstract algebraic operations that are expressed in his work on Boolean algebra and Boolean logic that are the bases for the programming language of computers. After his shopkeeper father taught him what math he knew, George continued to learn on his own all he could in mathematics as well as other fields, including Latin and Greek. When he was sixteen years old, he added to the family income by teaching school, and four years later in 1835 he opened his own school. He studied the works of other seventeenth-century mathematicians and just four years later published his first paper in *The Cambridge Mathematical Journal* in 1839 that discussed differential equations and invariance. On the basis of this paper and other works he became a professor of mathematics at Queen's College in Cork, Ireland. In 1844 he received The Royal Society Award for his work on differential equations. In 1847 he published *Mathematical Analysis of Logic*, and in 1854 he worked on the idea of applying mathematical approaches to symbolic logic, which he published in *An Investigation into the Laws of Thought*. This last publication before his death from pneumonia in 1864 established the field of symbolic logic later refined by the German mathematician, logician, and philosopher Gottlob Frege (1848–1925), the famous Welsh philosopher Bertrand Russell (1872–1970), and Alfred North Whitehead.

BORN–HABER THEORY OF CYCLE REACTIONS: Physics: *Max Born* (1882–1970), Germany. Born shared the 1954 Nobel Prize for Physics with Walter Bothe.

The sequence of energy involved in the chemical and physical reactions that form lattice ionic crystals is related to the crystal's initial state (zero pressure at zero kelvin), and to the crystal's final state, which is also at zero pressure and zero K (e.g., for a gas of infinite dilution).

The Born–Haber cycle is better known by the early work of Max Born, which resulted in the mathematical theory referred to as the cycle explaining how chemical bonds are the result of sharing or transferring electrons between atoms. As a result, several scientists applied quantum mechanics to the concept of chemical **bonding**. Born and others used the hydrogen atom as a model, and it was soon obvious that quantum mechanics could explain almost all aspects of chemistry, including the different types of reactions between atoms of different elements and the probability of where to find an electron within its orbit surrounding the nucleus.

Max Born is better known for his work in the field of quantum mechanics. In cooperation with German physicist Ernst Pascual Jordan (1902–1980), Born refined Werner Heisenberg's concept of *matrix mechanics* by developing the mathematics that explained the theory in 1925. Born was among the first to provide the mathematics to explain the possibility that particles can also behave like waves. This was about the time that the duality concept of light (photons) and other types of radiation was being discussed and debated.

> Fritz Haber (1868–1934) was born in Breslau, Germany, now Wrolcaw, Poland, and studied at the University of Heidelberg under Robert Bunsen. He and Carl Bosch developed the Haber process for the catalytic formation of ammonia from hydrogen using the nitrogen found in the free atmosphere. The Haber–Bosch process (not to be confused with the Born–Haber cycle) was a great development in the field of industrial chemistry because it greatly increased the supply of nitrogen for use in producing cheap nitrogenous fertilizer and WWI explosives for which he received the 1918 Nobel Prize in Chemistry. He also developed deadly war gases and was personally involved in their release. He also developed an effective gas mask during the war years. Haber converted from Judaism to be more acceptable to the German government. Even so, the Nazis forced him to emigrate because he was still Jewish by the Nazis' definition.

See also Bohm; Bohr; Dehmelt; Frisch; Haber; Hahn; Heisenberg; Meitner; Schrödinger

BOYLE'S LAW: Chemistry: *Robert Boyle* (1627–1691), Ireland and England.

> *There is an inverse relationship between the pressure and volume of a gas when the temperature remains constant. The equation for this law is written* as $P \times V = c$ (pressure times volume equals a constant inverse relationship).

Robert Boyle, an Irish chemist who later worked in England, used air pumps developed by Robert Hooke to experiment with the physical conditions of gases under differing pressures while maintaining constant temperatures. In 1661 he published the results in his book, *The Sceptical Chymist*. In 1662 Boyle discovered air could be compressed, and as the pressure increased, the volume decreased. He demonstrated that if the pressure on a gas doubled, its volume would be just one-half its original volume; if the pressure was increased by one-third, the volume would decrease one-third. Boyle also noted the opposite inverse relationship existed when he used a vacuum pump to decrease the pressure that increased the volume of air. This proved to be a classical inverse relationship, which seems to be a universal constant. It was an important conclusion because it helped explain the atomic (particle) nature of gases—that is, atoms (or molecules) of gases would spread farther apart when the pressure was decreased. Conversely, the atoms would be forced closer together if the pressure increased.

Boyle was an atomist who supported the original concept of matter first proposed by the ancient Greek Democritus. It took more than a century after Boyle's work before the modern atomic theory of matter was fully developed. It might seem ironic that a scientist who is considered one of the founders of modern chemistry was also an **alchemist** who spent much of his time attempting to transmute base metals into gold.

Robert Boyle was born into a wealthy aristocratic family who lived in a castle in Ireland. Like the sons of many aristocratic families, as a young boy he was sent off to schools in Switzerland and later to Italy. It is said that he developed an interest in science while at school when he found copies of Galileo's writings describing his work and ideas on astronomy. After his schooling, Boyle's family moved to their estate in England where in 1644 Robert retired for ten years, during which time he became interested in studying the nature of gases, then called pneumatics. He then moved to Oxford where he met Robert Hooke who had constructed an air pump that he used to demonstrate how air is important for respiration, combustion, and the transmission of sound. An important phase of Boyle's life was when he joined with Francis Bacon who supported the idea that science should be based on empirical observations and controlled experiments. This group originally known as the "Invisible College" became the famous Royal Society for the Improvement of Natural Knowledge in London in 1662.

In his famous book *The Sceptical Chymist* Boyle attacked the age-old concepts of Aristotle's four elements (earth, air, fire, and water) by proposing another ancient concept that matter is composed of basic particles and that different types of matter are identified by their number and position as well as the motion of these particles. He also stressed the importance of accurate empirical observations, planning and conducting well-designed experiments, and keeping records of the results. In addition, an important concept was Boyle's theory that heat was the result of the motion of the particles of matter. Later this idea was expanded into the kinetic theory of matter and the second law of thermodynamics. Boyle's law is a rather simple inverse relationship between the pressure and volume of a gas, assuming no change in the temperature of the gas. Boyle considered this law as the compressibility of air. In 1660 it was published in *New Experiments Physico-Mechanicall, Touching the Spring of the Air and its Effects*. The law is known as Boyle's law in the United States and Great Britain and as Mariotte's law in most of Europe. Edme Mariotte (1620–1684) was a French physicist and priest who recognized the validity of Boyle's law ostensibly before he published his findings in 1676. It is expressed as $p \times V = C$ (when C is the universal inverse constant, p is the pressure, and V is the volume of the gas).

See also Avogadro; Charles; Gay-Lussac; Ideal Gas Law

BRADLEY'S THEORY OF A MOVING EARTH: Astronomy: *James Bradley* (1693–1762), England.

Parallax exhibited by the stars indicates a movement of the earth.

James Bradley was the successor of Edmond Halley as the Astronomer Royal, during which time he erected a telescope in a stationary vertical position to observe the same spot in the sky each night. Over time he observed a slight displacement (**parallax**) of the image of the star Gamma Draconis. At first he thought this parallax of a star viewed over a period of time from the same place on Earth meant that only the star moved in relation to Earth. Bradley later realized it was not really the star's motion causing the parallax because the pattern of the star's displacement was repeated every six months. Thus, it meant the observer and his fixed telescope on a moving Earth caused the change in the star's apparent position. This concept is based on Earth's orbiting around the sun every twelve months, which means Earth is in a very different viewing position in relation to the star every six months. This creates an apparent displacement called *parallax*. The diameter of Earth's orbit around the sun is approximately 186 million miles (twice the radius of 93 million miles). Every six months, Bradley's telescope was 186 million miles from where he viewed the star the previous six months. Parallax seemed to place the star in slightly different positions. Bradley calculated that the aberration or movement he observed was related to the ratio of the velocity of light to the velocity of Earth as it circled the sun. He figured that this ratio was approximately 30 km s^{-1}, which is about 10,000 to 1. From this data he calculated that the speed of light was 3.083×10^8 m s^{-1}. This was the most accurate measurement for the speed of light at that time in history. Bradley's theory that explained the motion of Earth was the first direct evidence for such motion. Up until this time Earth's motion was always inferred from indirect factors.

Today, astronomers using the parallax concept can mathematically determine the distance of the closer stars to our solar system. Parallax does not work very well for distant objects located in deep space. Bradley also was one of the first to conceptualize that light has a finite, not infinite, speed. The idea of light having a finite speed (~186,000 miles per sec.), and the movement of Earth, were important concepts for compiling accurate observations and calculations of stars.

Bradley determined that Earth had a "wobble" (precession) as it spins on its axis. He also calculated the extent of this wobble on the changing gravitational attraction of the moon, which has a slightly inclined orbit around Earth.

See also Brahe; Copernicus; Galileo; Kepler

BRAHE'S THEORY OF THE CHANGING HEAVENS: Astronomy: *Tycho Brahe* (1546–1601), Denmark. (*Tycho Brahe* is the Latinized form of his birth name, Tyge Ottesen Brahe. Universally, he is referred to as "Tycho.")

> *Because a new star does not exhibit any parallax and comets come and go, there must be changes in the heavens, proving Earth with its orbiting moon is the center of the universe.*

Tycho Brahe was an ardent proponent of Ptolemy's concept that Earth was the center of the universe, but he did not accept the idea that the universe was static. Based on his observation, Tycho devised this theory of a changing universe but incorrectly believed Earth was at its center. Tycho's concept of a changing universe was unique; until that time Artistotle's concept of a

Figure B7. Tycho Brahe's 11-foot quadrant is an example of the large-scale astronomic instruments (sextants and quadrants) that he constructed for direct viewing since the telescope had not yet been invented. Some were so large that he was able to climb into them while viewing the heavens.

In 1577 Denmark's' King Frederick II offered Tycho Brahe unlimited funds and the use of the island of Hven for Tycho's lifetime if he would build his proposed observatory that would be large enough to hold Tycho's extra-large instruments for viewing the heavens. These included his large quadrants and sextants that enabled him to observe, with the unaided eye, and with some accuracy events outside of the solar system. This led to the development of his theory. After the death of Frederick II in 1588, his successor, Christian IV, became the king of Denmark. Christian IV and Tycho had a falling out, and Tycho was forced to find a new patron. He succeeded when he met someone a bit more unbalanced than himself, the Holy Roman Emperor Rudolph II, who gave Tycho another castle near the city of Prague to house his enormous instruments. He also acquired a young assistant, Johannes Kepler, who later made a name for himself. They did not get along, but their work was outstanding, and after Tycho's death two years later Kepler made good use of the mass of data Tycho had recorded.

An interesting story about Tycho Brahe's eccentric and argumentative nature related to the duel he fought when he was just nineteen years old. The dispute with a Danish nobleman was about who was the better mathematician. During the fight with swords Tycho lost the end of his nose. It is said that he then formed a false nose out of wax, gold, and silver that he wore for the rest of his life.

permanent, unchanging universe was accepted as fact. This belief changed when Tycho discovered a new supernova (exploding star) known as the "Tycho Star" in 1572. Tycho also discovered a large comet in 1577 that further supported his theory. A large crater on the moon is also named after him. Tycho's theory of a changing universe was not well accepted by other astronomers.

Because telescopes had not yet been invented, Tycho constructed several large sextants and quadrants for his direct sight viewing of the universe. He kept a journal of all his activities as well as astronomical tables, which later proved useful to his assistant, Johannes Kepler. Tycho's most important contribution was the result of the twenty years he spent recording the positions of over eight hundred stars that proved invaluable for the future work of Kepler. These records assisted Kepler in developing his three laws of planetary motion.

See also Copernicus; Galileo; Kepler; Ptolemy

BUFFON'S THEORIES OF NATURE: Biology: *Comte George Louis Leclerc de Buffon* (1707–1788), France.

Buffon's theory of ecology: *The animals of an area (ecology) are the product of the environmental conditions of the land where they developed.*

Comte de Buffon based this theory on his concept that Earth makes and grows the plants on which the animals depend; thus the region's plants determine the geography and geology of the region's animals.

Buffon's theory of natural classes: *Animals were classed not according to genera and species, but rather in a hierarchy of man, domesticated animals, savage animals, and lower animals.*

Taxonomy and species classifications based on structure and functions were not yet fully developed. Therefore, Buffon classed animals according to major categories as he interpreted their status in life. His classification of animals is somewhat similar to Aristotle's "ladder of life" (*see* Aristotle).

Buffon's theory of species: *Animals within a hierarchical group (species), and only those within that group, can reproduce themselves.*

Buffon based his theory on empirical evidence. He observed that animals of one group from his hierarchical classification of animals would breed only with others of their kind. He was unaware of hybrids or mutations.

Buffon's theory of the age of Earth: *Based on a series of stages as evidenced by geological history, Earth is seventy-eight thousand years old.*

Buffon rejected biblical records that contended Earth's age as six thousand years. His estimate of the age of Earth was based on fossils and geology. Buffon believed Earth was originally a hot body that cooled off sufficiently enough for people to exist and would continue to cool, at which time all life on Earth would end. His extension of the age of Earth led other scientists to examine fossils and geological evidence more closely, which provided a more accurate estimate of Earth's age. Today, the universe is considered to be ~13.4 to ~15 billion years old, with Earth being formed about ~3.5 to ~4.5 billion years ago.

Buffon's theory of the origin of the planets: *The formation of planets was the result of a collision between a large comet and the sun.*

Buffon produced numerous writings that influenced other scientists for close to one hundred years. Some of his more famous publications were a translation of Hale's *Vegetable Statistics* (1735) and a translation of Newton's *Method of Fluxions* (1740). From 1740 to the end of his life Buffon worked on a massive forty-four volume manuscript titled *Histoire naturelle*, that included the following:

Volumes I–15 titled *Quadrupeds* (1749–1767)
Volumes 16–24 titled *Birds* (1770–1783)
Volumes 25–31 titled *Epochs of Nature* and *Supplementary Volumes* (1778)
Volumes 32–36 titled *Minerals* (1783–1788)
Volumes 37–44 titled *Reptiles* (including *fish* and *cetaceans*) (1788–1789).

His most influential work was included in Volume I with the subtitle of *Preliminary Discourse of Nature* where his ideas were ahead of his time, nontheological, and very rational but were not always correct. He proposed the division of animals into natural classes, while insisting that only individual species existed in nature and that only two animals of the same species can propagate themselves.

Buffon is credited with providing an important naturalistic history of Earth. He based his concept of the origin of the **solar system** and planets on the more current explanation of natural forces where cosmic "dust" circulated to form the solar system and coalesce into the planets around the sun. He used the mechanics of motion as described by Sir Isaac Newton as well as his own empirical observations.

See also Cuvier; Darwin; Wallace

BUNSEN'S THEORY OF THE SPECTROCHEMISTRY OF ELEMENTS: Chemistry: *Robert Wilhelm Bunsen* (1811–1899), Germany.

Each element, when heated, emits a unique electromagnetic spectrum that can be identified by careful spectrum analysis of the emitted light.

While assisting in an experiment of spectrum analysis, Robert Bunsen, along with his assistant, Peter Desaga, refined Michael Faraday's gas burner by adding a collar that could be adjusted to control the flow of air into the burning gas. This device greatly improved the burner by providing a hotter and steady flame. Since that time, it has been known as the *Bunsen burner*. By using this extremely sensitive instrument (spectroscope) other scientists were able to identify and discover new elements. Bunsen is credited with the discovery of two new elements: rubidium and cesium.

By using the spectra produced by different elements Bunsen and Gustav Kirchhoff are credited with providing the first evidence for the internal structure of atoms that make up all matter. Up to this time it was believed that the "atomic" nature of the structure of the different elements was invisible. The use of his concept that each element has it own unique frequency (wavelength) that can be detected by using his spectroscope was applied to astronomy. Astronomers, using this device, were able for the first time to analyze the light from the stars from which they were able to detect the various chemical elements that composed the stars, including our sun. In addition to the field of **spectroscopy**, Bunsen contributed to the fields of electrochemistry, electrodeposition of metals, and photochemistry.

His career began by experimenting in organic chemistry. The results of his work were published between the years of 1837 to 1842 in *Studies in the Cacodyl Series*. While experimenting with the toxic compounds of cacodyls (tetramethylarsine, a derivative of arsenic), he lost an eye in an accidental explosion. He also nearly killed himself by arsenic poisoning as he worked with poisonous compounds of arsenic, chlorine, fluorine, and cyanide. He soon changed the direction of his work and began experimenting with inorganic compounds, spectroscopy, and electrochemistry. In cooperation with Gustav Kirchhoff, he published the *Chemical Analysis through Observations of the Spectrum* in 1860. His contributions in this field were made possible by the improvement of Michael Faraday's gas burner that he completed in 1855.

See also Berzelius; Faraday; Kirchhoff

THE B_2FH (BURBIDGE–BURBIDGE–FOWLER–HOYLE) THEORY: Astronomy: *Eleanor Margaret Burbidge, née Peachey* (1922–) along with her husband *Geoffrey Burbidge* (1925–) were the lead researchers of this group that included *William Fowler* and *Sir Fred Hoyle*, England and United States.

The theory is known as the B_2FH theory of the formation of chemical elements in the universe. It states: *Chemical elements are produced in the nuclei of stars of during supernova explosions.*

The Burbidges along with Fred Hoyle of Great Britain and William A. Fowler of the United States published the B_2FH theory in 1957. It explained how all the lighter elements from hydrogen to iron began with hydrogen. However, the original theory had some "gaps" that needed to be explained. One problem in their theory was that if all the elements were produced by either supernova explosions or inside stars, the great abundance of helium and deuterium (heavy hydrogen) in the universe could not be explained. George Gamow and other scientists proposed another theory that states that in the very early period after the big bang, all matter was ionized and dissociated forming a type of plasma. In just a few minutes the temperature dropped, resulting in **nucleosynthesis**, at which time a few light elements were created. Protons and neutrons were formed within the first three minutes after the big bang. They collided with each other to produce deuterium (heavy hydrogen), which consists of one proton and one neutron. Then deuterium nuclei collided with additional protons and neutrons to create helium as well as the next element in the periodic table, that is, lithium nuclei consisting of a mass of a total of seven protons and neutrons (which is a combination of one tritium [$_3H$] and two deuterium [2_2H] nuclei). The fact that helium makes up about 25% of all the mass in the universe is evidence that a very hot phase existed shortly after the big bang. Later it was theorized that the heavier elements were created by the fusion reaction within stars. Margaret Burbidge is one of the outstanding women in science and astronomy who can be claimed by England and the United States. She received many academic and scientific honors later in her life.

Margaret Peachey-Burbidge began her career at the University of London where she graduated in 1948. She then joined the University's Observatory and after receiving her PhD served as its acting director from 1950 to 1951. From this position she moved to the Yerkes Observatory at the University of Chicago in 1951 to 1953, followed by research work at the California Institute of Technology from 1955 to 1957. She also did research work at the Cavendish Laboratory in Cambridge, England, before returning to the Yerkes Observatory at Chicago where she served as associate professor of astronomy from 1959 to 1962, at which time she moved to the University of California at San Diego as professor of astronomy from 1964 to 1990. Her career was briefly interrupted when, on a leave, she returned to England to become the director of the Royal Greenwich Observatory where she made improvements in optical astronomy in England. The Royal Observatory located in Greenwich, London, was founded by King Charles II in 1675 and provided astronomical services until 1946. In 1946 the Royal Observatory was moved to the Herstmonceux Castle in Sussex, England. Because this region was located close to the ocean and near sea level, it provided poor viewing conditions for optical telescopes. Also, the castle was very old and in need of many repairs. The castle became a hotel that did not survive and is now a science center. Administration operations, except the telescopes and other equipment, were moved to Cambridge, England, in 1990. After her marriage to the theoretical physicist Geoffrey Burbidge the couple collaborated with other astronomers and physicists to study the physical nature of how chemical elements are synthesized in stars.

See also Fowler; Gamow; Hoyle

C

CAGNIARD DE LA TOUR'S CONCEPT OF "CRITICAL STATE": Physics: *Charles Cagniard de la Tour* (1777–1859), France.

> *The critical state exists at the point where the temperature and pressure create equal densities between a liquid and its vapor. The vapor and liquid can be in equilibrium at any temperature that is below the critical point.*

The term "critical state" is also known as *critical point* (at the critical temperature there is no clear-cut distinction between the vapor [gas] and liquid states) (*see* Figure B5 under Black). Their densities are equal, and their two phases are also equal and considered to be one phase. Cagniard de la Tour Cagniard discovered the concept of the *critical state* in 1822 by heating liquids in a sealed tube until the liquid and vapor were not distinguishable from each other. He referred to this point as the **critical temperature**. His concept was based on the work of the French physicist Louis Paul Cailletet, the English physicist James P. Joule, and the Irish chemist and physicist Thomas Andrews (1813–1885).

Some examples of practical applications are boilers used in home heating and industry, steam engines, and frequently in the generating of electricity by steam turbines. Another example is the pressure cooker. By increasing the pressure, the liquid and vapor can be compressed with an increase in temperature, changing the critical state until pressure is released. Food then cooks faster under the increased pressure, using less applied heat than nonpressurized cooking.

Cagniard is also known as the inventor of the siren used on police cars, fire engines, and so forth to produce a rising, piercing sound. He formed a disk containing one or more holes, and when the disk was spun rapidly and air was passed through the holes in the disk, the typical siren sound was produced. His original, crude apparatus has undergone many improvements in the last century and a half.

See also Cailletet; Joule; Kelvin

CAILLETET'S CONCEPT FOR LIQUEFYING GASES: Physics: *Louis Paul Cailletet* (1832–1913), France.

As pressure on most gases increases, so does the boiling point. Therefore, reducing the temperature while increasing the pressure will liquefy the gas at a lower pressure.

After studying in Paris, Cailletet returned home to assist his father in his ironworks business. He became interested in metallurgy, particularly in the use of forced air applied to the blast furnace. This led him to try to liquefy gases, such as hydrogen, oxygen, and nitrogen that would then be more easily transported to blast furnaces. He failed in this attempt, which led him to identify H_2, O_2, and N_2 as *permanent gases*. It was at this point that he learned about the works of other physicists regarding the concept of critical temperature (*see* Cagniard). This concept gave him the idea that it would require more cooling of his gases as well as greatly increasing the pressure (then releasing the pressure) for the gas to become cool enough to liquefy. This idea is based on the Joule–Thompson effect, which in effect means that when a gas expands, its temperature decreases. Cailletet was the first to liquefy his "permanent gases" by compressing oxygen, hydrogen, nitrogen, and air by rapidly increasing the pressure, and then releasing the pressure of cold gases, which further reduced the temperature by the rapid expansion of the gas. Thus, the gas reached the critical temperature point at which gaseous air turned into liquid air. This process is used today to produce various liquefied gases including air from which other gases are fractionated (e.g., oxygen and nitrogen) that are used in experimental work in nuclear physics, chemistry, and cryogenics. Compressed and liquefied oxygen is essential in the health care of oxygen-dependent patients and in many industrial processes, such as steel production, smelting, rocket fuels, and welding.

Cailletet was also ahead of his time in considering high-altitude flight. He invented the **altimeter,** the **manometer**, and a mask-like breathing device that could be used at high altitudes.

See also Cagniard; Joule; Kelvin

CALVIN'S CARBON CYCLE: Chemistry: *Melvin Calvin* (1911–1997), United States. Calvin was awarded the 1961 Nobel Prize in Chemistry.

The path (route) of carbon dioxide in the chemical and physical reactions of photosynthesis occurs only in the presence of chloroplasts of living plants.

Melvin Calvin began his teaching at the University of California, Berkeley, in 1937. During World War II he worked on the Manhattan atomic bomb project where he conducted research that involved new analytical techniques. He spent the remainder of his career at the University of California where he applied the techniques of ion-exchange chromatography and the use of radioisotopes to the study of **photosynthesis**.

Photosynthesis takes place in green plants when they are exposed to sunlight by absorbing carbon dioxide (CO_2) from the atmosphere and by an intricate process that converts the CO_2 molecules into starch and oxygen (O_2). This process sends oxygen into the air at about 10^{12} kilograms per year on Earth. It is believed that photosynthesis is responsible for supplying the oxygen in the atmosphere of the ancient Earth that enabled primordial life to begin many millions of years ago.

CHROMATOGRAPHIC DATABASE

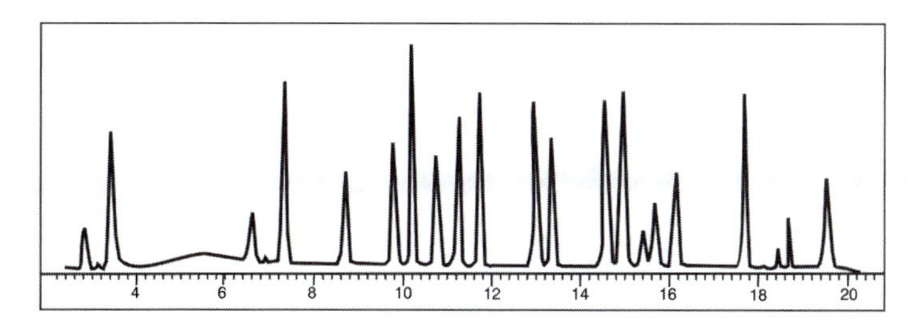

Figure C1. Results from a typical chromatography analysis.

Calvin exposed plants for a few seconds to radioactive CO_2 containing the isotope carbon-14. Carbon-14 has the same atomic number (six protons) and is chemically similar to carbon-12, the more abundant form of carbon. But the carbon-14 nucleus has two more neutrons than the nucleus of C-12. There is no chemical distinction between C-14 and C-12 because they have the same atomic number (number of protons). The radioactive tracer of C-14 was assimilated in plant chloroplasts during the process of photosynthesis and could be traced with radiation detection devices. Because the process is very rapid, he worked quickly to mash the cells to separate the carbon-14 in boiling alcohol. He then separated and identified the components and products of photosynthesis by using paper chromatography. After many experiments, Calvin identified the cycle of absorption and the use of carbon dioxide by plants, leading to a better understanding of the roles of chlorophyll and carbon dioxide in the science of photosynthesis.

Using this technique Calvin identified the cycle for the reductive pentose phosphate reaction that is an important aspect of photosynthesis. This is now known as the "Calvin cycle" for which he won the Nobel Prize.

We now know that most plants increase their rate of growth in an atmosphere rich in CO_2. Carbon dioxide is sometimes added to the inside air in greenhouses to accelerate plant growth. An increase in the growth rate of crops is one of the few benefits of the excessive production of carbon dioxide in modern society. Calvin is also known for his development of the analytical techniques that use radioisotopes for labeling stages in chemical and physical reactions as well as in chromatography methods. They are all important analytical processes used in today's laboratories.

CANDOLLE'S CONCEPT OF PLANT CLASSIFICATION: Biology: *Augustin Pyrame de Candolle* (1778–1841), Switzerland.

> *There is a homologous or fundamental relationship of similarities for the parts of different types of organisms.*

Several classification systems for plants and animals existed at the time of Candolle's work. However, Candolle was the first to use the terms "taxonomy" and "classification" synonymously. His taxonomy (naming system) was based on the recognized similarity of various body parts of near relatives of different species, thus assuming they derived from common ancestors. Candolle also originated the idea of **homologous** parts, which

After studying medicine in Geneva, Switzerland, Candolle went to Paris, France, to study natural sciences as well as medicine. He met and was influenced by several naturalists, including Georges Cuvier and Jean Lamarck, and soon became better known as a botanist than a medical doctor. His many scholarly papers earned him a reputation that resulted in an assignment to conduct a botany and agricultural plant survey of France. In 1813 he published his first original volume titled *Elementary Theory of Botany.* Based on the natural classification systems of Cuvier and other botanists, Candolle introduced, for the first time, the concept of taxonomy as related to classification. Candolle's taxonomy form of classification took the place of the older Linnaeus system and was used for about fifty years before other botanists improved it. Candolle also was the first to realize that geographic location, to some degree, determined what plants were native to a particular region. His concept was based on the premise that various types of soil found in different geographic regions determined the types of vegetation growing in a specific region.

Between 1824 and 1839 Candolle wrote a massive seventeen-volume encyclopedia titled *A Guide to Natural Classification for the Plant Kingdom.* He was only able to publish the first seven volumes before his death; his son completed the publication of the remaining ten volumes after his father's death in 1841.

along with the concept of relating taxonomy to species influenced the British naturalists Alfred Wallace and Charles Darwin in the development of their theories or organic evolution. Only six volumes of Candolle's twenty-one-volume taxonomy project were published before his death. His work, which was superior to that of Carolus Linnaeus, is still used today.

See also Cuvier; Lamarck; Linnaeus

CANNIZZARO'S THEORY OF ATOMIC AND MOLECULAR WEIGHTS: Chemistry: *Stanislao Cannizzaro* (1826–1910), Italy. In 1891 the Royal Society of London awarded Cannizzaro the Copley Medal for his contributions to science.

> *The atomic weights of elements in molecules of a compound can be determined by applying Avogadro's law for gases, which states that gram-molecular weights of gases occupy equal volumes at standard temperature and pressure (STP).*

In essence, Avogadro's law states that all gases that are at the same pressure and temperature will contain the same number of molecules. Cannizzaro's genius was recognizing the utility of this law fifty years after it was proposed by Avogadro and ignored by other scientists. Cannizzaro applied the law as a means for measuring the atomic weights of atoms of elements and the weights of gas molecules. He also determined that the theory could be applied to solids if their vapor density is unknown by measuring their specific heat. Although Cannizzaro credited Avogadro for the basis of his theory, it was Cannizzaro in his 1858 publication *The Epitome of a Course of Chemical Philosophy* that finally convinced the scientific community that molecular weights of gases could be determined by measuring their vapor densities. In essence, Cannizzaro restated Avogadro's theory that definitely established the theory of atoms and molecules in a way that was accepted by the chemists of his day. The field of quantitative chemistry rapidly advanced once atomic and molecular weights could be accurately determined. Cannizzaro's theory provided a means for defining the molecular weights of many organic compounds, as well as clarifying the structure of complex organic molecules. Avogadro's number of molecules in a gram-molecule weight of gases at standard conditions has been determined experimentally as 6.02×10^{23} and is now considered Avogadro's constant.

See also Avogadro

CANTOR'S MATHEMATICAL THEORIES: Mathematics: *Georg Ferdinand Ludwig Philipp Cantor* (1845–1918), Germany.

> Cantor's set theory has a long history although it was essentially created by one person—George Cantor in the late 1800s. The theory's history goes back as far as the Greek philosopher, Zeno of Elea, in ~450 BCE. Since then, ideas related to infinity and cardinal numbers, imaginary numbers, algebra, groupings, classifications, and so on have all led up to Cantor's theory. In mathematics a "set" can be a collection of distinct objects that are considered as a whole (the parts comprise the whole). Although this seems a simple idea, some mathematicians did not accept it. Today it is a fundamental concept in modern mathematics and is now taught at the grade school level. Cantor's concept was that the objects (things) in the set, which are called elements of the members of the set, can be anything such as numbers, pencils, shoes, letters of the alphabet, and so on. Sets are designated by using capital letters (A, B, C, etc,). Two sets of things, A and B are said to be equal if they have the same members (an element or things in each set). A set in mathematics is not the same as what is thought of as a set in real life. In real life it is possible to have a set of something that has multiple copies of the same element (things or objects), whereas in mathematics if two different sets have the same members or items, they are equal even if the members are not in the same order. For instance; Cantor proved that if A and B are sets with A being equivalent to the subset of B, and B is equivalent to a subset of A, then A and B are equivalent. Over the years the mathematical theory of sets and subsets has been refined, and set theory is a fundamental mathematical concept.

Cantor's theory of infinity: *Without qualification, one can say that the transfinite numbers stand or fall with the infinite irrational; their inmost essence is the same, for these are definitely laid-out instances or modifications of the actual infinite.*

This theory is related to Cantor's axiom, which states that if you start with a single point on a two-dimensional surface and continue to add points on each side of the original point, they will continue to extend out in both directions. By adding point-to-point, there will be no one-to-one relationships between the two directions of points, and the lines they form can be extended forever (infinity). Another way to look at this is as follows. If you start from zero, you can progress indefinitely to larger and larger functions, or you can extend indefinitely in the opposite direction to smaller and smaller values.

Karl F. Gauss, the German physicist and mathematician, stated that infinity was not permitted in mathematics and was only a "figure of speech." The concept of endlessness, whether in time, distance (space), or mathematics, is difficult to grasp. Galileo considered the study of infinity as an infinite set of numbers, later defined as Galileo's paradox. However, it was not until Sir Isaac Newton and Gottfried Leibniz developed calculus that it became necessary for a mathematical explanation of infinity. The English mathematician John Wallis (1616–1703), who developed the law of conservation of momentum, also proposed the symbol for infinity (∞), which was known as the "lazy eight" or "love knot." It was Georg Cantor, however, who postulated that consecutive numbers could be counted high enough to reach or pass infinity. His development of transfinite numbers, a group of real numbers (rational and irrational) that represent a higher infinity, led to *set theory*, which permits the use of numbers within an infinite range.

Cantor is better known for his proposed set theory: *The study of the size (cardinality) of sets of numbers and the makeup or structure (countability) of groups of rational or irrational natural numbers.* Following are two examples of sets of numbers:

1 2 3 4 5,–.–.–.–, n,–.–.–.–, ∞
↕ ↕ ↕ ↕ ↕
2 4 6 8 10,–.–.–.–, 2n,–.–.–. ∞

1 2 3 4 5,–.–.–.–, n, –.–.–.–∞
↕ ↕ ↕ ↕ ↕
1 4 9 16 25,–.–.–.–, n^2,–.–.–.–, ∞

Where *n* is a continuation of the sequence or set. By finding a one-to-one match in a set (in the example, a number below the number above it), even if the set is infinite, it is possible to determine the size of the number if the entire structure is unknown. If the set is infinite, some of the numbers can be eliminated without reducing the size or eliminating the structure of the set.

See also Galileo; Gauss; Leibniz; Newton

CARDANO'S CUBIC EQUATION: Mathematics: *Gerolamo Cardano* (1501–1576), Italy.

Definition of cubic equation: *A polynomial equation with no exponent larger than three—Specifically:* $x^3 + 2x^2 - x - 2 = 0$ $(x - 1)(x + 1)(x + 2)$, *which can be further treated by algebraic functions.*

Earlier mathematicians solved equations for x and x^2 but were unable to solve x^3 (cubic) equations. (First-degree equations are linear [straight line], or one-dimensional, involving x; second-degree equations are quadratic [plane surface], or two-dimensional, involving x^2; and third-degree equations are cubic [solid figures], or three-dimensional, involving x^3.) The graphic depiction for the solution of a cube is usually easier to understand than is the algebraic representation. When a graph is used to solve cubic equations, the x-axis is eliminated and reoriented. Cardano was not the first to come up with a solution to cubic equations, having been given the explanation for cubic and biquadratic equations by the Italian mathematician and engineer Niccolo Tartaglia who made Cardano promise not to reveal the secret. After Cardano found that someone previous to Tartaglia had achieved a partial solution, he published the results as his own version. As the first to publish, Cardano has been credited with the discovery of a solution to solving cubic equations. As a result of this controversy, a new "policy" stated that the first person to publish the results of an experiment or a discovery, and not necessarily the first person to actually conduct the experiment or make the discovery, is the one given credit. This is based on the belief that science should be open and available to all rather than kept secret.

See also Tartaglia

CARNOT'S THEORIES OF THERMODYNAMICS: Physics: *Nicholas Leonard Sadi Carnot* (1796–1832), France.

Carnot cycle: *The maximum efficiency of a steam engine is dependent on the difference in temperature between the steam at its hottest and the water at its coldest. It is the temperature differential that represents the energy available to produce work.* $T_2 - T_1 = E$, *where* T_2 *is the higher temperature,* T_1 *is the lower temperature, and E is the energy available to do work. The efficiency E is 1 only if* $T_1 = 0$ *Kelvin.*

The concept of a temperature differential is analogous to the potential energy of water flowing over a water wheel to produce work. This concept enabled inventors to develop more efficient steam engines and locomotives and became known as the Carnot cycle, based on the difference in the temperature of the steam at its highest temperature and the water at its coldest temperature—not on the total amount of internal heat. Carnot pointed out that energy (heat) is always available to do work, but it cannot completely be turned into useful work. Even so, it is always conserved; none is ever "lost." For example, within the steam engine, which is not 100% efficient, some energy is not available to accomplish work due to "lost" heat. The first law of thermodynamics (the word *thermo-dynamics* means "heat-flow" in Greek) states that energy is always conserved, while the second law states there is a limit to how much of the energy can be converted into work. A system cannot produce more energy (work) than the amount of energy that is expended. Perpetual motion is impossible, as some energy is always lost to heat due to friction.

See also Clausius; Fourier; Helmholtz; Kelvin; Maxwell

CASIMIR FORCE (EFFECT): Physics: *Hendrick Brugt Gerhard Casimir* (1909–2000), Netherlands.

> *The Casimir force is the effect of a very small attraction that takes place between two closely spaced plates that, although they have the ability to carry an electrical charge, are uncharged in an environment of a quantum vacuum that contains virtual particles that are in continuous motion.*

These two plates may be composed of a conductor (any substances such as metals that can carry an electrical current), or a dielectric (a substance other than a metal on which an electrical field might be sustained with the loss of minimum power, i.e., insulator).

Casimir's two electron theory: *During superconduction there are pairs of electrons consisting mostly of normal electrons, but some are superconducting electrons.*

Along with the assistance of two colleagues, the German physicist Walther Meiβner and the Dutch physicist Cornelis Gorter (1907–1980), in 1934 Hendrik Casimir explained their "two fluid" model of superconductivity at low temperatures. At near absolute zero some properties of superconductivity are altered. Two types of electrons are formed, most were normal-type electrons and a few were types of superconducting electrons. This explained the physics between the thermal and magnetic properties of superconductors. These electrons form types represented by paired and unpaired electrons referred to as "Cooper pairs."

The Casimir force provided evidence for the concept of a quantum vacuum, which is similar in quantum mechanics to what is described as empty space in classical physics. It is due to quantum vacuum changes in the electromagnetic field between the plates. The Casimir force also proved the viability of the *van der Waals force* that exists between two uncharged atoms (not ions). In addition, the Casimir force affects the ***chiral bag*** (nonmirror image) asymmetric model

An interesting analogy of the Casimir force was evident as far back as the 1700s when sailors observed that when two large wooden ships came closer than 50 feet to each other side-by-side, a calm sea would form between the ships. This "calm" would occur between the ships regardless of how high the waves were on the open ocean side of the ships. As the distance between the ships decreased, they were drawn closer together by the Casimir-like force. If the ships were not forcefully separated, they would soon crash into each other. The solution was to send small boats full of sailors to forcibly keep the ships apart.

A more recent application of the Casimir force is in the field of applied physics (engineering) in the development of the science of **nanotechnologies** where exotic materials are artificially constructed at the atomic and molecular levels.

of the nucleon, which indicated that the mass of the nucleon is independent of the bag radius. Also, the Casimir effect needs to be considered in the field of extremely small electro-mechanical systems including the design of computers and cell phones.

See also Meissner; Van der Waals

CASPERSSON'S THEORY OF PROTEIN SYNTHESIS: Chemistry: *Torbjorn Oskar Caspersson* (1910–1997), Sweden.

Proteins are synthesized in cells by large RNA (ribonucleic acid) molecules.

Torbjorn Caspersson invented a new type of spectrophotometer, enabling him to trace the movement of RNA. He concluded that RNA is involved in the synthesis of amino acids in the production of proteins. RNA is a type of single, long, unbranched, organic macromolecule responsible for transmitting genetic information to deoxyribonucleic acid (DNA) molecules (*see* Figure C5 under Crick). All organisms, except viruses, depend on RNA messengers to carry inherited characteristics to the DNA molecules, which can then be duplicated to pass genetic information to offspring. In RNA-type viruses, the RNA itself acts as the DNA because it contains all the genetic information required for the virus to replicate. Caspersson was the first to determine that DNA had a molecular weight of 500,000 **daltons**. In addition, he discovered a way to dye specimens of DNA so that the nucleotides would appear in dark bands. Caspersson's work provided information used by Crick and Watson in discovering the specific double helix shape of the DNA molecule.

Tests comparing one person's DNA with another person's or their offspring can determine who is genetically related to whom, often years after death. Many police departments and laboratories use this procedure to test and compare human DNA when investigating the crimes of rape and murder.

See also Chargaff; Crick

CASSINI'S HYPOTHESIS FOR THE SIZE OF THE SOLAR SYSTEM: Astronomy: *Giovanni Domenico Cassini* (1625–1712), France.

The mean distance between Earth and Sun is 87 million miles.

Giovanni Cassini developed this figure by working out the parallax for the distance of Mars from Earth, which enabled him to calculate the astronomical unit (AU) for the distance between Earth and the sun. He accomplished this by using calculations attained by other astronomers, as well as his own observations. The figure greatly increased the estimations at this time in history for the size of the solar system. In the late 1500s Tycho Brahe calculated the distance between Sun and Earth at just 5 million miles. A few years later Johannes Kepler's estimation of 15 million miles was better, but still greatly underestimated, whereas in 1824 the German astronomer Johann F. Encke (1791–1865) used Venus's transit with the sun to overestimate the distance as 95.3 million miles. (The correct mean distance from Earth to the sun's center is approximately 92.95 million miles = 1 AU.)

Cassini was the first to distinguish the major gap separating the two major rings of Saturn, now called the Cassini division. He also discovered four new moons of Saturn, determined the period of rotation of Jupiter as 9 hours, 56 minutes, and observed that Mars rotates on its axis once every 24 hours, 40 minutes. Cassini measured Earth's shape and size but incorrectly identified it as a perfect sphere. His work is convincing, inasmuch as Earth is not unique in the solar system and is similar to the other inner planets, with the obvious exception of its human habitation.

See also Brahe; Kepler; Spencer-Jones

CAVENDISH'S THEORIES AND HYPOTHESIS: Chemistry: *Henry Cavendish* (1731–1810), England.

Cavendish's theory of flammable air: *When acids act on some metals, a highly flammable gas is produced, called "fire air."*

Henry Cavendish was one of the last scientists to believe in the **phlogiston** theory of matter, which states that when matter containing phlogiston burns, the phlogiston is released. Because his "fire air" would burn, he called it *phlogisticated air*. It was Antoine Lavoisier who named the new gas *hydrogen*, meaning "water former" in Greek. Cavendish was the first to determine the accurate weights and volumes of several gases (e.g., hydrogen is one-fourteenth the density of air). Until this time, no one considered that matter of any type could be lighter than air (*see also* Lavoisier; Stahl).

Cavendish's theory of the composition of water: *When "fire air" and oxygen are mixed two to one by weight and are burned in a closed, cold, glass container, the water formed is equal to the weight of the two gases. Thus, water is a compound of hydrogen and oxygen.*

Although similar experiments and claims that water is a compound, and not an element (as believed for hundreds of years), were made by other scientists, Cavendish was credited with the discovery of hydrogen, as well as the concept of water as a compound, although he delayed publishing the results of his experiments.

Cavendish developed many concepts but had difficulty making generalizations from his experimental results. Two examples of his "new" discoveries are 1) the distinction between electrical *current*, *voltage*, and *capacitance*, which later led to Ohm's law, and 2) the anticipation of the gas laws dealing with pressure, temperature, and water vapor. He also foresaw the chemical concepts of multiple proportions and equivalent weights for which John Dalton and others were given credit. Cavendish also determined that oxygen gas has the same molecular weight regardless of where it is found and that the percentage of oxygen in ordinary air is approximately the same wherever found on Earth.

Cavendish's hypothesis for the mass of Earth: *Based on the determination of the gravitational constant and the estimated volume of Earth, its mass should be* 6.6×10^{36} *tons, with a density about twice that of surface rocks.*

Cavendish's procedure for determining the mass of Earth is known as the Cavendish experiment. It involved two large lead balls on the ends of a torsion bar and two small lead balls, one on each end of a single rod suspended at its midpoint by a long, fine wire. As the two large lead balls were brought close to the smaller ones from opposite directions, the force of gravity between the balls produced a minute twist in the wire that could be measured. From this twist, Cavendish calculated the gravitational force between the balls, which he used as a gravitational constant for calculating the mass of Earth. His figure was approximately 6,600,000,000,000,000,000,000 tons, very close to today's estimation of 5.97×10^{24} kilograms.

See also Dalton; Ohm

CELSIUS TEMPERATURE SCALE: Physics: *Anders Celsius* (1701–1744), Sweden.

A standard centigrade scale for measuring temperatures with 100 graduation points is needed for scientific endeavors using the metric system.

Invented in 1742, the Celsius temperature scale was originally referred to as the "centigrade" scale from about 1750 until the middle of the twentieth century because the apparatus that Anders Celsius used to boil water was zero and the freezing point was 100. Celsius' scale was the reverse of the modern centigrade (Celsius) scale now in use. The scale was based on 100 units and thus the prefix "centi" for the name of the scale. Its name was not officially changed to "Celsius" until 1948.

Celsius, a young mathematician became a professor of astronomy at Uppsala University in Stockholm, Sweden, where he built and then became the director of an observatory in 1740. He devised his new temperature scale in 1742 that is now the most-used scale throughout the world, not just in science. However, he was not the first to arrive at some type of scale for recording temperatures. Others before, as well as after Celsius, included Daniel Fahrenheit, Lord Kelvin, Christian of Lyons (fl. 1743), William Rankine (1820–1872), Joseph-Nicolas Delisle (1688–1768), Carolus Linnaeus, Per Elvius the Elder (1660–1718), Isaac Newton, Rene-Antoine Reaumur (1683–1757), and Ole Römer.

The Celsius temperature scale was originally based on the freezing point of water as being 100 on the scale and the boiling point of water at 0 degrees, assuming the conditions of standard atmospheric pressure. It was not until after Celsius' death in 1744 at the age of forty-two that the scale was reversed to represent the temperature of boiling water at 100°C and the freezing point of water at 0°C, where it remains to this day.

The temperature at which all molecular motion ceases (except vibrations of individual particles/molecules) is now referred to as absolute zero, which is 0 degrees on the Kelvin scale. The boiling point on the Fahrenheit scale is 212 and the freezing point is 32. The Fahrenheit scale is only used in two countries of the world, the United States and Myanmar (formerly Burma) (*see also* Farhrenheit; Kelvin).

The conversion formulas used to change one scale to another follow:

1. Celsius to Fahrenheit: Conversion Formula: $F = C \times 1.8 + 32$.
2. Fahrenheit to Celsius: Conversion Formula: $C = (F - 32) / 1.8$.
3. Celsius to Kelvin: Conversion Formula: $K = C + 273.15$.
4. Kelvin to Celsius: Conversion Formula: $C = K - 273.15$.

CHADWICK'S NEUTRON HYPOTHESIS: Physics: *Sir James Chadwick* (1891–1974), England. Chadwick was awarded the 1935 Noble Prize for Physics.

Physicists believed there were only two elementary particles in atoms: the positive proton and the negative electron. However, this concept was not adequate to explain many physical phenomena. One idea was that the internal nucleus of a helium atom has four protons, as well as two internal neutralizing electrons, which could explain the 2+ charges for helium. But scientists could not identify any electrons originating from the helium nuclei when it was bombarded with radiation. Another dilemma was that nitrogen has a mass of 14 but an electrical charge of 7+. This was confusing because if the nitrogen nucleus has 14 protons, it would have a charge of 14+. One solution was to assign seven negative electrons to neutralize seven of the positive protons, meaning the nucleus of nitrogen now has twenty-one particles. Because this did not make sense to many scientists, it was determined that the measured "spin" of particles in the nitrogen nucleus could be only a whole number; thus there could not be twenty-one particles in the nucleus. Further work demonstrated that gamma rays did not eject the electrons from positive protons, nor did the radiation eject the protons. Therefore, it was hypothesized that some new particle(s) with no electrical charge and very little mass must exist in the nuclei of atoms. Some years later this "ghost particle" was found and named the *neutrino* (*see also* Fermi; Pauli; Steinberger).

Chadwick's hypothesis: *Particles with the same mass as protons, but with no electrical charge, were one of the particles ejected from helium nuclei by radiation.*

> There was some opposition to Chadwick's construction of a cyclotron type of instrument by the British-based New Zealand physicist Baron Rutherford who thought it might lead to uncontrolled release of energy. This argument led Chadwick in 1935 to move from the Cavendish Laboratory in London to the physics department at Liverpool University. While chair of the physics department at Liverpool, he built the first cyclotron in Great Britain that produced results that supported the claims of Otto Frisch that a sustainable chain reaction is possible by the fission of a few pounds of Uranium-235, and Rudolph Peierls' concept that it was possible to separate fissionable Uranium-235 from a much larger quantity of Uranium-238. Frisch and Peierls believed that the atomic bomb was possible by releasing enormous amounts of energy by the process of the fission of nuclei. At the outbreak of World War II Chadwick moved to the United States as head of the British group working on the Manhattan Project that developed the first sustainable nuclear chain reaction. In 1945 Sir James Chadwick was knighted for his contributions in the field of physics and his service during World War II.

In the early twentieth century Walther Bothe bombarded beryllium nuclei with alpha particles (helium nuclei) and detected some unidentified **particle** radiation. Chadwick continued this work and called these nuclear particles *neutrons* because they carried no electrical charge but had a similar mass as protons. This discovery assisted in understanding the atomic number and the atomic mass of elements and provided an excellent tool for further investigations of the structure of atoms. It was this discovery for which Chadwick was awarded the 1935 Nobel Prize in Physics. Neutron bombardment is used today to produce **radioisotopes** for medical purposes. Chadwick also was the developer of the first **cyclotron**, or "atom smasher," in England. Some physicists consider Chadwick's neutron discovery the beginning of the nuclear age because it led to understanding the physics necessary to develop practical uses for nuclear fission such as nuclear energy used in electric power plants and the atomic fission bomb and nuclear **fusion** hydrogen bomb.

See also Anderson (Carl); Fermi; Heisenberg; Pauli; Soddy

CHAMBERS' THEORY OF THE ORIGIN OF LIFE: Biology: *Robert Chambers* (1802–1871), Scotland.

If the solar system can be formed by inorganic physical processes without the assistance of a supreme being, then it follows that organic plants and animals can develop by a similar physical system.

In other words, if the formation of an inorganic world (the universe) is possible without intervention of a supreme entity, then an organic world is also possible.

Robert Chambers was self-educated by reading the *Encyclopedia Britannica* and other science publications. He published several books while in partnership with his brother, William, after they established their own publishing house, which is still in business in Edinburgh, Scotland. He wrote extensively on the origin and development of the solar system and proposed that the organic world operated on the same principles as the inorganic world. He wrote that there was a progression from lower animals to higher life forms with no input by a god. The force responsible for life was based on his concept of chemical/electrical interactions.

A contemporary of Charles Darwin, Chambers based his theory for the origin of life on his acceptance of Pierre Laplace's nebula hypothesis for the origin and stability of the solar system. Many scientists accepted as a logical explanation Chambers' concept that the same laws of physics and chemistry that created the universe could create organisms. But theologians and the general public never accepted it because the theory denied credit for the origin of life to a supreme being (deism) or personal god (theism). Chambers proposed several other theories related to the embryonic development of species, which engendered tremendous negative reactions from the public.

See also Baer; Haeckel; Laplace; Ponnamperuma; Swammerdam

CHANDRASEKHAR LIMIT: Astronomy: *Subrahmanyan Chandrasekhar* (1910–1995), India.

The Chandrasekhar limit is a physical constant that states *if white dwarf stars exceed a mass of 1.4 greater than the sun's mass, they will no longer be self-supporting. As their mass increases past this limit, internal pressure will not be balanced by the outward release of pressure generated by atoms losing electrons to form ions. Thus, white dwarfs exceeding this limit will "explode."*

Interestingly, no white dwarfs have been found with a mass greater than the 1.4 mass limit. To date, most white dwarfs that have been discovered average about 0.6 the mass of our sun. It is speculated that the **supernova** observed and reported by Tycho Brahe in 1572 and another in 1604 by Johannes Kepler may each have been the collapse of a white dwarf star that pulled off mass from nearby red supergiants at a rate greater than could be eliminated by radiation from the white dwarf. This new mass was attracted internally by the gravitational pull of the white dwarf, causing an increase in its mass exceeding the 1.4 mass limit. Thus, a giant nuclear explosion occurred, creating a supernova.

Chandrasekhar also believed that as stars exhaust their nuclear fuel they begin to collapse by internal gravitational attraction, which ceases when a balance is established by the outward pressure of ionized gases. This gas is a high concentration of electrons, leaving behind ions, which become very dense. One thimbleful would weigh several tons on Earth due to the collapsed star's density.

It is estimated that most white dwarfs are "leftovers" of mass pulled from more massive stars. It is also speculated that in the long-distant future of the universe, all stars will become white dwarfs, and then we may have a static, or unchanging universe (thermodynamic equilibrium) because no other bodies would be available from which they could gather in extra mass and exceed the Chandrasekhar limit.

See also Brahe; Kepler

CHANG'S THEORIES AND CONCEPTS: Mathematics and Astronomy: *Cheng Chang* (c.78–142), China.

Chang's concept of pi: *Pi is the square root of 10, which equals 3.1622.*

This was one of the most original and most accurate methods of determining the value of pi up to this time, with the exception of Archimedes' geometric method of "perfect exhaustion." Many early mathematicians were intrigued by the relationship of the diameter of a circle to its circumference. This led to the idea of forming a series of geometric squares or polygons within the bounds and adjacent to the outer circumference of a circle and then calculating the known parameters of the squares or polygons (*see also* Archimedes).

Chang's concept of the universe: *The universe is not a hemisphere rising over Earth, but rather the universe consists of a large sphere with Earth at the center, similar to the yolk in the center of an egg.*

The early Greeks, including Aristotle, believed the universe consisted of a series of crystal hemispheres suspended above Earth. Chang's idea was the first to consider a spherical universe. Further, he developed an instrument to measure the major circles of the celestial bodies and demonstrated how they intersected at various points on Earth. He also developed an instrument that used flowing water to measure the movement of the stars (*see also* Aristotle).

Chang's concept of earthquakes: *Earthquakes are caused by dragons fighting in the center of the earth.*

From this idea Chang developed an instrument shaped in the likeness of several heads of dragons. A ball was held inside the mouth of each of the heads. When an earthquake occurred, a pendulum device would expel a ball from one of the mouths of one of the dragon heads, which would then determine the direction of the earthquake. Although Chang's "seismograph" was inaccurate, it was developed and used in the Far East for over seventeen hundred years before a more accurate one was designed in the West that used the Richter scale.

See also Richter

CHANG'S THEORY OF CAPACITATION: Biology: *Min Chueh Chang* (1909–1991), United States.

> *The male sperm must spend some time traveling in the reproductive organs of the female before it can fertilize the egg.*

Chang called this time factor, whereby sperm become more potent while in the female reproductive tract, the *capacitation factor*. He also believed that male seminal fluid has a "decapacitation" substance that prevents sperm from fertilizing the egg. Although no such factor or substance has yet been identified, it is assumed it is this decapacitation factor that prevents other sperm from uniting with an egg once one sperm has penetrated the ovum.

Using rabbits, Chang conducted much of the research with in vitro fertilization, where the ova (egg) is fertilized by a male sperm outside the body and then transplanted into the female rabbit's uterus. His research pioneered the way for others to perfect human in vitro fertilization.

Chang's work also provided much of the knowledge used to set up experiments by American biologist and researcher Gregory Pincus (1903–1967) to demonstrate that injections of progesterone into rabbits would prevent contraception. This was pioneering work for several modern methods of contraception.

CHAPMAN–ENSKOG KINETIC THEORY OF GASES: Mathematics/Physics: *Sydney Chapman* (1888–1970), England.

In the nineteenth century James Clerk Maxwell and Ludwig Boltzmann developed the general mathematical concept that describes the properties of gases as partially determined by the molecular motion of the gas particles, assuming the gas molecules follow classical mechanics. This is referred to as the *Maxwell–Boltzmann distribution* that gives gas particles a specific momentum.

In 1911 Chapman developed the next logical step in the theory of gas particles. He performed the mathematics necessary to prove the kinetic theory of gases.

The Chapman–Enskog theory provides a complete treatment and solution to the mathematics of the Maxwell–Boltzmann equation by using approximations to determine the average path of gas particles.

This joint theory is also known as the *Enskog theory* because the Swedish mathematical physicist David Enskog (1884–1947) also worked independently on this theory.

Chapman also used mathematics to predict the thermal diffusion of gases and the electron density at different levels of the upper atmosphere. In addition, he determined the detailed variations in Earth's magnetic field and related this to the length of the moon's day. (Because the moon keeps the same side pointed toward Earth, it rotates once every 27.3 days.) He demonstrated not only that the moon causes tidal effects on Earth's water and land, but there is a much weaker tidal effect on Earth's atmosphere. Chapman's work enabled other scientists to measure more accurately the kinetic motion of molecules as related to heat, and to understand better the ideal gas law, thermodynamics, and **geomagnetism**.

See also Boltzmann; Boyle; Gilbert; Ideal Gas Law; Maxwell

CHARGAFF'S HYPOTHESIS FOR THE COMPOSITION OF DNA: Biology: *Erwin Chargaff* (1905–2002), United States.

> *In the DNA molecule, the number of adenine (A) nucleotide units always equals the number of thymine (T) nucleotides, and the number of guanine (G) units always equals the number of cytosine (C) units.*

Chargaff's earlier work used paper chromatography and spectroscopy to study the composition of DNA in different species. He found that within one species, the DNA was always the same, but there was a difference in DNA composition between species. He believed there must be as many forms of DNA as there are species, even though much of the DNA was similar for all species. (It has been determined about 98+% of human and chimpanzee DNA is identical.) At this point, he realized that a pattern of consistency of AT nucleotide pairs and GC nucleotide pairs appeared in nucleic acid molecules. Although Chargaff did not follow up on this discovery, it enabled Crick and Watson to arrive at the placement of the AT and GC pairs inside the helix structure of the DNA molecule (*see* Figure C5 under Crick). Scientists are continuing to explore the many possible benefits of this discovery for the betterment of humankind.

See also Crick; Franklin (Rosalind); Watson

CHARLES' LAW: Chemistry: *Jacques Alexandre Cesar Charles* (1746–1823), France.

> *When the pressure remains constant, the volume of a gas is directly proportional to its temperature.* $V/T = \text{constant} = n/P$. *It might also be stated as: The volume of a*

fixed amount of gas that is at a constant pressure is inversely proportional to its temperature, VnP/T, where the nP represents the constant pressure on the fixed volume of gas.

Jacques Charles, originally an employee of a French government agency, became a physics professor at the Conservatoire des Arts et Métiers in Paris. He established the direct relationship between the temperature and the volume of gases. Aware of the flights of hot air balloons, Charles realized that as the air in the balloon became hotter, it expanded and thus became lighter due to a decrease in its density and an increase in its volume. As long as the fire at the bottom opening of the balloon heated the air, the air would expand and become lighter than the air outside of the balloon. When the air cooled, it decreased in volume and became heavier, causing the balloon to descend. He was also aware of the work of Henry Cavendish, a British chemist, who produced hydrogen gas that was much lighter than hot air and thus more buoyant. Even better, hydrogen did not lose as much of its buoyancy as did hot air as it cooled. On August 27, 1783, Jacques Charles, and his brother, Robert, made the first flight in a hydrogen-filled balloon. On a later flight, they reached an unprecedented altitude of just under 2 miles. This flight made him famous with the public and royalty. The danger of using hydrogen for lighter-than-air ships was recognized even in these early days, and its use for this type of airship was halted after the 1937 explosion of the *Hindenburg* zeppelin while mooring in New Jersey.

Charles' gas law led to further experimentation by Gay-Lussac, Dalton, and other scientists interested in the nature of matter and resulted in what is known as the "ideal gas law," a combination of several gas laws, including Boyle's Law.

See also Avogadro; Boyle; Dalton; Galileo; Gay-Lussac; Ideal Gas Law

CHARNEY'S THEORETICAL METEOROLOGY: Physics and Meteorology: *Jule Gregory Charney* (1917–1981), United States.

Charney proposed two important theories related to meteorology:

1. *Weather can be understood by the use of computer models and mathematics.*
2. *Biogeographical feedback mechanisms are responsible for desertification of landmasses.*

Jule Charney received his PhD from the University of California at Los Angeles in 1946. He then became a member of the Advance Studies Group at Princeton, New Jersey, and in 1956 became head of meteorology at Massachusetts Institute of Technology (MIT) until his retirement in 1977.

While they were at the Institute of Advanced Studies in the 1940s Charney, along with John Von Neumann, developed a large computer that required an extensive air-conditioned room in which it needed to be housed due to the tremendous heat its many vacuum tubes generated. They decided that it would be a logical use of the new Electronic Numerical Integrator And Calculator (ENIAC) computer, which was built during the period 1942 to 1945, to set up calculations that could predict weather for short periods of time. In 1949 they succeeded in more accurately predicting the weather for a four-day period. This exceeded the reliability of any other method of predicting weather over short periods of time. Today, computers can reasonably and accurately predict local and regional weather for about one week. The record of using computers, even supercomputers, to predict long-term climate changes is not nearly as successful. One reason: even the best computer/climate models cannot accurately predict long-term (or even short-term) climate changes on a regional or worldwide basis due to the many interacting variables that affect long-term cyclic climate factors, such as sunspots, temperature/pressure fluctuations of the sun, ocean, and atmospheric anomalies, and Earth's slight pertubations (wobble on its axis), among other factors. Short-term computer/mathematical weather predictions are possible, but they are not always accurate partly because weather (as well as long-term climate) does not behave as the computer models predict.

While at MIT, Charney proposed his theory of desertification in his 1974 *Theory of Biogeographical Feedback.* This concept considers the biological, geological, and geographical factors for a particular region. Because severe droughts have persisted on the African continent, particularly in the central and northern regions, Charney's theory considered the following: overgrazing and the cutting down of forest timber for firewood led to an increase in **albedo** of the land (the reflection of sunlight from the land into the atmosphere). This results in a cooler land surface and a warmer lower atmosphere, resulting in less cloud formation and therefore less rain. His theory that the less plant cover of land surfaces the greater the albedo is not completely accepted, particularly as the only or main causation of desertification.

CHARPAK'S CONCEPT OF TRACKING PARTICLES: Physics: *Georges Charpak* (1924–), France. Georges Charpak, the Polish-born French physicist was awarded the 1992 Nobel Prize for Physics for his work in inventing a device known as the "drift chamber" for detecting subatomic particles that provided a means of analyzing these particles by using computers.

> *Because only one particle in a billion interacts when exploring the deep parts of matter, the single event is not easily discernable, and photographic techniques are inadequate to the task. Therefore, a high-speed electronic device might be used to detect a greater number of events.*

The "cloud chamber," for which C.T.R. Wilson received the 1927 Nobel Prize for Physics, preceded the development of the "drift chamber" for detecting and analyzing subatomic particles. As charged particles passed through the cloud, they formed a "track" through the cloud in the closed chamber. Using the cloud chamber, Carl D. Anderson discovered the first antiparticle, known as the positron (a positively charged electron). He received the 1936 Nobel Prize for Physics for this endeavor.

Over the years other means of detecting subatomic particles and cosmic radiation used special photographic emulsions. The classic "bubble chamber" was the next device capable of detecting the movement and interactions of charged subatomic particles. Donald A. Glaser invented the bubble chamber in 1952 for which he received the Nobel Prize for Physics in 1960. This chamber was filled with overheated fluid; and as the charged particles passed through the liquid, they caused bubbles to form as the liquid boiled along the particle's tracks. These tracks could only be photographed at the rate of one per second. There were several problems with all of these instruments. One was that they only recorded a few particle "events" per second. Another was that they did not provide complete information about the particle. And another was that they do not lend the results to computer analysis.

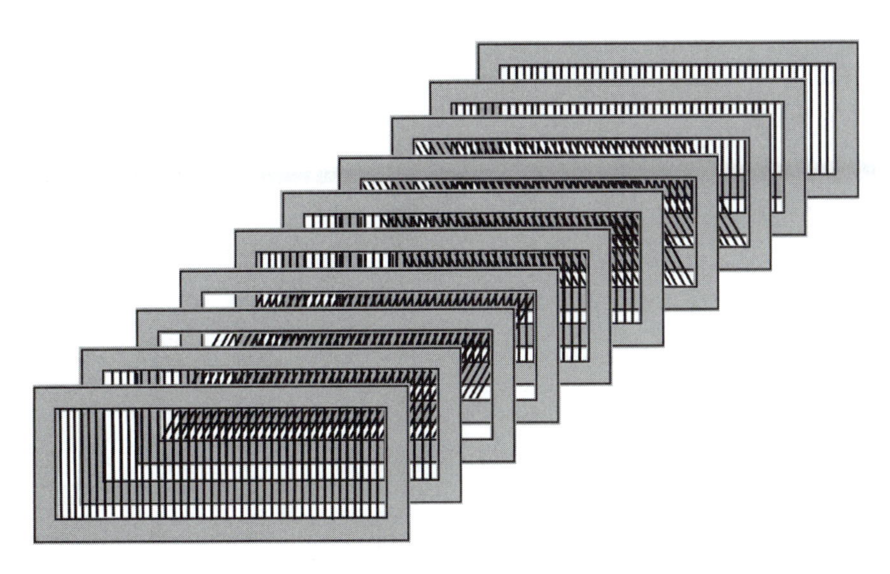

Figure C2. This multi-wire device can detect one million subparticle impulses a second that can then be analyzed by computers.

A number of new devices have been developed that actually track the subatomic particles' paths across a computer screen in three dimensions. These include the multiwire chamber that can track one million subatomic particles per second for computer analysis. Another was the wire spark chamber. During the 1970s Georges Charpak was doing his research at Conseil Européen pour la Recherche Nucléaire (CERN; also known as the European Organization for Nuclear Research) when he invented the improved "drift chamber" that helped revolutionize high-energy and subatomic particle physics. Conseil Européen pour la Recherche Nucléaire is located near Geneva, Switzerland. It is the world's largest subatomic particle physics laboratory with the world's largest particle accelerator.

As evidence of the importance of the invention of these detection devices is the number of Nobel Prizes for Physics that have been awarded to physicists and the discoveries made possible by their use. They have made great advances in related fields. For example, two important discoveries are 1) the J/psi particle in 1974 by Samuel Ting and Burton Richter, and 2) the W and Z subatomic particles by Carlo Rubbia and Simon Van der Meer. Not only does the drift chamber have application in physics, but it also contributed to research in the fields of biology and medicine as well as other industries.

See also C. Anderson; Glaser; Rubbia; Ting; Van der Meer; C. T. R. Wilson

CHARPENTIER'S GLACIER THEORY: Geology: *Jean de Charpentier* (1786–1885), Switzerland.

Glaciation is the agent responsible for the movement of boulders of one composition to an area where the boulders and rocks of a different composition are found.

Early geologists were puzzled for many years by the presence of large boulders that did not seem to belong in the areas in which they were found. Some thought that they were carried to their new locations by icebergs, but there was no evidence that icebergs ever existed in these areas. Charles Lyell contended these boulders were brought to their locations by enormous floods; however, there existed no other evidence of any such giant water flow that would have been needed to move these huge rocks. Others claimed that the boulders were really meteorites from outer space. But no firm evidence supported this notion.

Charpentier's **glaciation** theory was correct but not well accepted by most scientists, except Louis Agassiz, who believed the idea viable and published his glacier theory "Studies on Glaciers" in 1840 before Charpentier published his paper titled "Essays on Glaciers" in 1841.

See also Agassiz; Lyell

CHEVREUL'S THEORY OF FATTY ACIDS: Chemistry: *Michel Eugene Chevreul* (1786–1889), France.

> *When treated with acids or alkali, animal fats break down to produce glycerol and fatty acids.*

Chevreul experimented with saponification, the process of producing soaps by treating animal fats with alkalis. By using alcohols to crystallize the product, he identified several fatty acids, including oleic acid, stearic acid, butyric acid, capric acid, and valeric acid, which are used in organic chemistry. He published his first paper titled "Chemical Researches on Animal Fats" in 1823. Chevreul and Joseph-Louis Gay-Lussac patented the process to make candles from the stearic fatty acid. Until this time tallow candles rendered from animal fat were used. They had an unpleasant odor, burned poorly, and were messy. The new fatty acid candles were harder, burned brighter, and were less odiferous. They were a better product and made a fortune for Chevreul and Gay-Lussac. Chevreul also experimented with and produced other fats including lanolin, cholesterol, and spermaceti (a waxy substance obtained from the head of the sperm whale). Later in life he worked with dyes and produced coloring from logwood, yellow oak, and indigo. He became professor of chemistry, and later director at the Musée d'Histoire Naturelle (Museum of Natural History) in France in the 1880s. He had a long and productive life of 103 years.

See also Gay-Lussac

CHU'S HYPOTHESIS FOR "HIGH TEMPERATURE" SUPERCONDUCTIVITY: Physics: *Paul Ching-wu Chu* (1941–), United States.

> *The combination of the proper amounts of yttrium, barium, and copper can be used under pressure to reach a critical temperature above that of liquid nitrogen.*

The Swiss physicist Karl Alexander Müller (1927–) developed material that achieved **superconductivity**, at an unprecedented high temperature (35K or -238°C). This temperature is very low compared to ordinary temperatures, and liquid helium is required to maintain a system at this temperature. Using Müller's results, Chu's goal was to make superconductivity practical. To achieve this, it was necessary to make a material that would be superconductive at a temperature at which a material could be cooled by liquid nitrogen. Liquid helium has a lower boiling temperature than liquid nitrogen but is too expensive to use on a regular basis. Nitrogen becomes a liquid at about 77 kelvin (K) (−195.5°C). Chu tried several combinations of metals as superconductors and finally developed a mixture that was stable. It was a ceramic-type substance consisting of the elements lanthanum, copper oxide, and barium that would become superconductive of electricity at a temperature of about 93 K. He experimented with various ratios of these three elements and then substituted yttrium for the lanthanum. In 1987 he used a mixture of $Y_{1.2}Ba0_{.8}CuO_4$ and was successful in achieving superconductivity at about 93 K. Because there is very little or no resistance to the flow of electricity through superconductors at these low temperatures, the potential for research possibilities is limitless. Superconductivity at even higher temperatures may lead to superefficient magnets, improved electromagnetic devices, such as computed axial tomography (CAT or CT) scanning equipment and magnetic resonance imaging (MRI), as well as other electrical equipment. Practical superconductivity will lead to the development of super-fast trains and less expensive transmission of electricity.

See also Nernst; Simon

CLARKE'S SUPERGENE THEORY: Biology: *Sir Cyril Astley Clarke* (1907–2000), England.

> *Supergenes are groups of closely linked genes that act as an individual unit and carry a single controlling characteristic.*

Clarke was interested in the phenomenon of butterflies' inheriting particular wing patterns referred to as *mimicry*—that is, one species (mostly insect species) exhibits a color, body structure, or behavior of another species that acts as camouflage. For example, the "eye" spots on the wings of butterflies that mimic another species protect the butterfly from predators. He found that male swallowtail butterflies carry supergenes as recessive. However, the characteristics are expressed as patterns for the females. It is now known that many inherited human traits and characteristics are controlled by supergenes. Clarke recognized a similarity between the supergenes that resulted in butterfly wing patterns and the blood antigen of rhesus monkeys, referred to as the Rh-factor. An Rh-negative mother and Rh-positive father may produce an Rh-positive child, which can lead to the development of Rh-antibodies in the mother if the child's blood leaks through the placenta. This can cause the destruction of red blood cells in future Rh-positive children born to the mother. Clark's wife suggested that he inject the Rh-negative mothers with Rh-antibodies because this was what destroyed the red blood cells of the fetus. This proposed solution to the problem proved ingenious because after injecting antibodies into the mother's blood the incompatible Rh-positive factor of the mother would be destroyed before the mother's own blood made antibodies, thus preventing her blood from destroying the red blood cells of her unborn baby. Clark tested prospective mothers and injected those with the Rh-negative blood factor with Rh-antibodies, thus counteracting the effects on future Rh-positive children. This testing and these procedures are now used by most obstetricians and hospitals and have saved many lives of newborn children.

CLAUDE'S CONCEPT FOR PRODUCING LIQUID AIR: Chemistry: *Georges Claude* (1870–1960), France.

> *When air is compressed and the heat generated by the increase in molecular activity is removed, a point will be reached at which the major gases of air will liquefy.*

Claude was successful in applying this principle on an industrial scale, forming the worldwide Air Liquid Company. The gases that make up air can be fractionally separated by this process, producing oxygen, nitrogen, carbon dioxide, argon, neon, and other noble gases. Because each of these gases becomes liquid at a specific temperature, the reverse also occurs. As the temperature of liquid air containing all of these gases in liquid form increases, each specific gas "boils" off at its specific evaporation temperature, where it can be isolated and collected as a pure gas. Claude was the inventor of neon lighting (glass tubes containing neon gas at less than normal pressure). When an electric discharge is sent through the gas, it glows with the familiar red light.

CLAUSIUS' LAWS AND THEORY OF THERMODYNAMICS: Physics: *Rudolf Julius Emmanuel Clausius* (1822–1888), Germany.

Clausius' law: *Heat does not flow spontaneously from a colder body to a hotter body.* In an early form, the law was stated as: *It is impossible by a cyclic process to transfer heat from a colder to a warmer reservoir without changes in other close-by bodies.*

The conservation of energy is a fundamental law of physics and is often called the first law of thermodynamics, which states that the total energy of a closed system is conserved. Clausius' law is an early statement of the second law of thermodynamics, and it led to the concept of entropy.

Clausius' theory of entropy: *In an isolated system, the increase in entropy exceeds the ratio of heat input to its absolute temperature for any irreversible process.*

The three classifications assigned to thermodynamics processes are *natural, unnatural,* and *reversible.* The *natural* process proceeds only in a direction of equilibrium. It is a reversible process only if additional energy (heat) enters the isolated system. A hot body transfers some of its heat (hotness) to a cold body until both are at the same temperature as their surroundings—thus equilibrium. The *unnatural* process never occurs over extended time because the unnatural process would require moving away from equilibrium (reverse the arrow of time). And *reversible* systems are idealized in the sense they are always arriving at different states or stages of equilibrium (e.g., growing living organisms) until their death. In a perfect closed system, Clausius' ratio (entropy) would always remain constant. However in real life, every process occurring in nature is irreversible and directional (the arrow of time is pointed in only one direction) toward disorder. Thus, there is always an increase in entropy, which sooner or later leads to complete disorder and randomness and a static (unchanging) universe (i.e., equilibrium).

More recently, concepts of entropy have been expanded to the analysis of information theory.

Consequence of Clausius' second law of thermodynamics: *The amount of energy in the universe is constant while the entropy of the universe is always increasing toward a maximum. At some future point, this maximum disorder will result in the unavailability of useful energy.*

The second law of thermodynamics is also known as "time's arrow" because it only progresses in one direction—toward disorganization. For any irreversible process, entropy will be increased. In other words, a greater state of disorganization or randomness will exist until equilibrium is reached. The only way entropy or an entire system can be decreased is by extracting energy from the system. The entropy of a part of the system may decrease if the entropy of the rest of the system increases enough. For example, the sun continually supplies energy to Earth. Otherwise, total entropy (disorder) would soon occur on Earth. This also applies to small temporary systems, such as some chemical reactions, crystallization, and growth of living organisms. The sun provides most of the energy used on Earth (with the exception of radioactive minerals). Energy is pumped into all living systems through plants as food, resulting in highly ordered complex molecules and growth in living organisms. Thus, the entropy of some parts of the universe (but not the entire universe) is decreased as organisms grow and become more organized through the input of energy. However, the entropy of the universe is increased by the radiation of the sun and other stars. Upon death, the second law of thermodynamics proceeds, resulting in the disorganization of the organism and its complex molecules. Without Earth receiving the sun's energy, entropy will increase and become all-encompassing, leading to the death of all living organisms.

Clausius summarized the first and second laws of thermodynamics as follows: "The energy of the universe is constant, and the entropy of the universe will always tend toward a maximum."

See also Boltzmann; Carnot; Fourier; Heisenberg; Kelvin; Maxwell; Rumford; Simon

COCKCROFT–WALTON ARTIFICIAL NUCLEAR REACTION: Physics: *Sir John Douglas Cockcroft* (1897–1967) and *Ernest Thomas Sinton Walton* (1903–1995), England. They were joint recipients of the 1951 Nobel Prize for Physics.

> *The transmutation reaction progressed as follows:* $_3Li\text{-}7 + {}_1H\text{-}1 \rightarrow {}_2He\text{-}4 + {}_2He\text{-}4 + 17.2$ *MeV: where* $_3Li\text{-}7$ *is a heavy isotope of lithium,* $_1H\text{-}1$ *is a proton (hydrogen nucleus),* $_2He\text{-}4$ *are alpha particles (helium nuclei), and 17.2 MeV is in millions of electron volts of energy.*

This first artificial nuclear reaction was the lithium reaction that occurred in 1932 and was made possible by Cockcroft's development of a proton accelerator and E.T.S. Walton's invention of a voltage multiplier that increased the speed of the proton "bullets." This was the beginning of the nuclear era. The lithium nuclei (the target) were bombarded by high-energy hydrogen nuclei (protons), resulting in the production of two

Before joining the British Army in 1915 Cockcroft studied mathematics at Manchester University in England. After the war he studied electrical engineering and later attended Cambridge University where he graduated with a degree in mathematics. Following graduation he joined Rutherford's group at the Cavendish Laboratory. This is where he met E.T.S. Walton who had constructed a device that multiplied voltages that could be used to accelerate positive protons. They were the first to successfully produce an artificial nuclear reaction by bombarding lithium nuclei with protons. In 1940 Cockcroft traveled to the United States to join the Tizard Mission that was charged with the exchange of science and technology between Canada, England, and the United States (It was named after Henry Tizard [1885–1959] the head of Britain's Aeronautical Research Committee who headed the mission.) He also assisted in developing radar. In 1946 he returned to Great Britain to head up the new British Atomic Energy Research Establishment at Harwell, England. In 1948 he was knighted as Sir John Douglas Cockcroft. Later in 1959 he was appointed master of Churchill College in Cambridge that emphasized science and technology. Cockcroft combined his research and administration skills, as he became a leading statesman of science.

alpha particles (helium nuclei), plus a small amount of energy. There were no cyclotrons or particle accelerators available so they used particles (protons) from a natural radioactive source. This provided the information and knowledge needed to continue the development of powerful cyclotrons and particle accelerators that produced nuclear fission (transmutation) reactions, giving rise to the use of nuclear energy for the production of electricity, radioisotopes, and the atomic (fission) bomb.

See also Chadwick; Fermi; Hahn; Meitner; Szilard; Walton

COHN'S BACTERIA AND CELL THEORIES: Biology: *Ferdinand Cohn* (1828–1898), Germany.

Cohn's infectious disease theory: *Microscopic bacteria are simple organisms that can cause diseases in other plants as well as animals.*

Cohn's study of microscopic organisms led him to develop the first classification of bacteria titled "Researches on Bacteria" that he published in his journal in 1872. His classification system is basically still used today. He experimented with boiled solutions of bacteria and suggested that some bacteria can develop resistance to external environmental influences, including heat-resistant spores that led to his research on the formation of heat-resistant spores formed by *Bacilus subtilis*. Later, similar bacteria and other forms of life were found living around the steam vents on the bottom of the oceans where the temperatures were much too high to support normal surface life. Based on his research, and unlike many biologists of his day, Cohn did not believe in spontaneous generation, a theory that life can start in rotting garbage (*see also* Redi).

Cohn's theory of protoplasm: *The protoplasm found in plant and animal cells is essentially of the same composition.*

Protoplasm is the colloidal substance composed of mostly complex protein molecules found in all living cells. In green plants the protoplasm contains chlorophyll, which in the presence of sunlight manufactures complex organic compounds (mostly carbohydrates from carbon dioxide and the hydrogen from water), while liberating oxygen mainly from the water as a waste product. All animal and plant cells require the food produced by this process, called *photosynthesis*.

Cohn was a popular lecturer who presented biology in an interesting manner to a large and appreciative audience. He not only published his own research and lectures but also supervised the publication of Robert Koch's research on the life cycle of the anthrax bacillus. Others biologists including Louis Pasteur used Cohn's research and publications.

See also Koch; Pasteur; Redi

COMPTON'S WAVE–PARTICLE HYPOTHESES: Physics: *Arthur Holly Compton* (1892–1962), United States. Arthur Compton and Charles T.R. Wilson jointly received the 1927 Nobel Prize for Physics.

> Compton effect: *When X-rays bombard elements, such as carbon, the resulting radiation is scattered (reflected) with a wavelength that increases with the angle of scattering.*

For this effect to occur, the X-rays must behave as particles (photons) that during the collision transfer their energy to the electrons of the carbon. This reaction indicates that X-rays behave as particles as well as waves. Compton used Charles Wilson's "cloud chamber" to assist in detecting, tracking, and identifying these particles. The

Compton effect was the first experimental evidence of the dual nature of electromagnetic radiation, such as radio waves, light, ultraviolet radiation, microwaves, X-rays, and gamma rays that exhibit characteristics of waves and particles.

Compton's hypothesis for cosmic rays as particles: *If cosmic rays have characteristics similar to charged particles, then there should be a variation in their distribution by latitude caused by Earth's magnetic field.*

If such an effect by the magnetic field could be detected, then it could be concluded that cosmic radiation consists of charged particles and is not pure electromagnetic radiation. This hypothesis was later proved correct in 1938 by measuring the distribution of cosmic rays from outer space as they were affected by Earth's magnetism, which proved that their concentrations were different at different latitudes of Earth's surface.

See also C. Anderson; Hess; Rutherford; C. T. R. Wilson

CONWAY'S GAME OF LIFE THEORY: Mathematics: John Horton Conway (1937–), England/United States.

Conway's "game of life" is: *An infinite grid of cells, either alive or dead, that interact with eight surrounding neighbor cells that are located either above, below, or diagonally from adjacent cells. As time progresses, each of the following rules occur:*

1. *A live cell with fewer than two neighbors dies.*
2. *Any cell that is surrounded by three or more neighboring cells dies.*
3. *Any cell with two or three neighbors lives for the next generation.*
4. *Any dead cell with just three neighbors will come to life.*

The first pattern is just for the first generation of the system. For the second pattern, the rules apply to all (every) cells in the first generation as births and deaths occur at the same time. Repeat the rules to continue for future generations.

This game of life is the result of Conway's interest in John von Neumann's hypothetical machine that could replicate itself. In 1940 von Neumann found a mathematical model for such a machine with rules following a Cartesian grid. Conway simplified von Neumann's ideas for a self-replicating machine and devised the "game theory of life." His "game" was published in Martin Gardner's

John Conway was born in Liverpool, England, the day after Christmas in 1937. Although his family was not poor, they experienced difficult times during World War II in England. Interested in mathematics from the age of four, he was first in his class in mathematics in his elementary school and excelled in the subject in high school. He also developed an interest in astronomy that he still follows. He entered Gonville and Caius College at Cambridge where he was awarded a BA in mathematics in 1959. While at Cambridge, he began his research in number theory and games, as he liked to play games in the college's game room. He was awarded his PhD in mathematics in 1964 and was assigned as lecturer in pure mathematics at Cambridge. Following this period his interests covered many areas in the field of mathematics that led to his famous cellular automata theory (game of life), group theory, surreal numbers, knot theory, quadratic forms, coding theory, fractals, and tilings. John Conway became a fellow of the Royal Society of London in 1983. He received many awards including the Berwick Prize and the Poly Prize from the London Mathematical Society, the Esser Nemmers Prize in Mathematics from Northwestern University, the Leroy P. Steele Prize for Mathematics by the American Mathematical Society, the Joseph Priestley Award from Dickinson College located at Carlisle, Pennsylvania in 2000, and was given an Honorary Doctorate in Science by the University of Liverpool in the year 2001. In addition he has written several books on his original theories. He has taught at Princeton University in New Jersey since 1986.

column "Mathematical Games" in the October 1970 issue of *Scientific American* magazine. The game became very popular with professional and amateur mathematicians and also made Conway popular.

Conway also invented a new system of numbers and a theory of knots that identifies the differences in 801 different types of knots based on the number of "crossings" in the knot if the crossings are no higher than 11. He helped develop the field of "group theory" where groups, similar to prime numbers, cannot be broken down into smaller groups. Conway contributed to the identification of several types of groups leading to a total of twenty-six groups described in the fifteen thousand pages of the publication explaining the "enormous theorem."

COPERNICUS' COSMOLOGY THEORIES: Astronomy: *Nicolaus Copernicus* (1473–1543), Poland.

Copernicus' heliocentric theory of the universe: *All the spheres (planets and moons) revolve about the sun as their midpoint; therefore, the sun is the center of the universe.*

At the time of Copernicus' pronouncement, the concept of a sun-centered universe was not really new. Aristarchus of Samos (c.320–250 BCE) reportedly stated that the sun and stars were motionless and the planets, including Earth, revolved in perfect circles around the sun. However, no attention was paid to his theory for almost two thousand years because the Earth-centered universe, as postulated by Aristotle and Ptolemy, was the accepted truth—that is, until the Copernican **heliocentric** model was proposed. Copernicus was influenced by the Pythagorean Philolaus (c.480–400 BCE) who theorized that the planets, including Earth, moved around a central fire. Philolaus believed that we could not see this fire because we lived on the side of Earth that was always turned away from it. Copernicus' model engendered much controversy among the clergy, as well as most scientists, until others were able to study and understand his new model, which also explained planetary motion (see Figure C3).

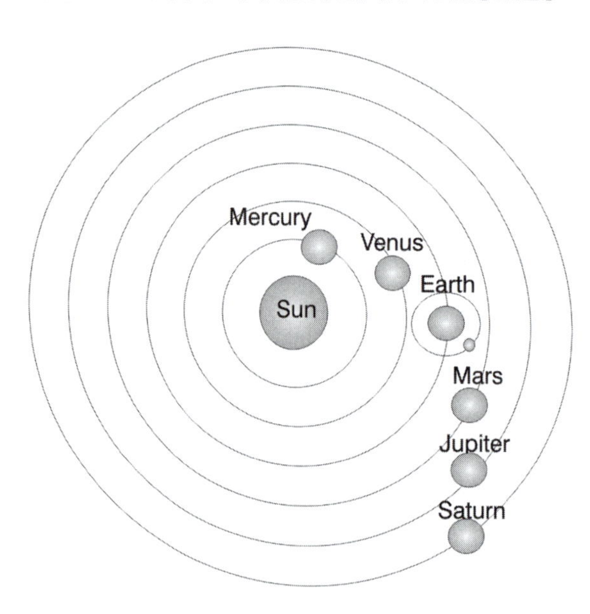

Figure C3. An artist's conception of Copernicus' heliocentric universe where all planets including Earth, revolve around the sun. His rationale was that if the sun is the midpoint of the planets' orbits, then the sun must be at the center.

Copernicus' theory of planetary positions: *Superior planets are those whose orbits are larger than Earth's and therefore, are farther from the sun than is Earth. Inferior planets are those whose orbits are smaller than Earth's and are closer to the sun.*

As a result of this theory, several other concepts to explain planetary motion were proposed. When an inferior planet is between Earth and the sun, it is in line of sight with the sun and is said to be in "inferior conjunction." When the planet passes on the far side of the Sun, away from Earth, the planet is in opposition and is said to be in "superior conjunction." Further observations of planetary motion led Copernicus to distinguish between the planet's ***sidereal period***—its actual period of revolution around the sun—and the *synodic period*—the time period of two successive conjunctions of a planet as seen from Earth. Therefore, the synodic period is what is directly viewed from Earth, as it is the time required for the planet to progress from one 180-degree opposition to the opposition of the next cycle, as related to Earth.

Copernicus was also aware his model required the inferior planets to move faster in their smaller orbits, while the speed of the superior planets was slower as they revolved around the sun. This information later became important for Kepler in the development of his laws of planetary motion.

Copernicus' theory of planetary distance from the sun: *Because the planets travel in perfect circles, they can be viewed as a direct line of sight to the sun, as well as at a 90-degree right angle, which forms a right triangle with the sun. The distance from the sun to the planets can be determined by geometry and trigonometry.*

At this time in history it was believed that all celestial motion, including planetary motion, progressed in perfect circles. Also, there were only six known planets in Copernicus' day. Copernicus' calculations for the distance of planets to the sun were excellent, considering that he believed their orbits to be perfect circles. His figures are close to what we now know about elliptical orbits of planets and their distances from the sun. Current figures are based on Earth's distance to the Sun as being 1 unit. This distance is the standard astronomical unit (AU), which equals the mean distance of 92,956,000 miles from Earth to the Sun. AUs are used to calculate the distances for the other planets to the sun.

Planet	Copernicus' Data (AUs)	Modern Distance (AUs)
Mercury	0.38	0.39
Venus	0.72	0.72
Earth	1.00	1.00
Mars	1.52	1.52
Jupiter	5.22	5.21
Saturn	9.18	9.55

Copernicus' theory of planetary brightness: *All the planets travel in circles around the sun; thus they are at different distances from Earth. Therefore, as their distances from Earth differ, so does their brightness.*

These theories raised another question: If Earth actually revolved around the sun, why didn't the position of the stars change every six months of Earth's yearly orbit of the Sun? This question is based on **parallax**—the apparent change in position of an object when viewed from two different positions.

Copernicus' conclusion about the size of the universe: *Stars seem not to move and therefore must be located at tremendous distances in space. Thus, the universe is much larger than formerly believed.*

Copernicus based this theory on the concept of parallax, which is dependent on the distance between two geographic sites at which the stars are viewed and the distance separating the observer and the actual object. In other words, even when viewed from opposite positions every six months in Earth's orbit, the stars are so distant they seem to stay in one position. About two hundred years later, with the development of improved instruments, a slight parallax of the stars was measured. But for the ordinary viewer on Earth, this slight parallax displacement is not noticeable. Copernicus was one of the first to conceptualize a vast universe that for the next hundred years remained incomprehensible.

Copernicus' theory of epicycle motion of the planets: *The minor irregularities in the motion of planets revolving around the sun can be explained by the epicycle each planet traces as it progresses in its own orbit.*

Ptolemy's Earth-centered model of the universe required the extensive use of epicycles to explain planetary motion. Because Copernicus maintained that planets revolved around the sun in perfect circles, he too required the use of epicycles to explain the irregularities in their motion. Several centuries later, it was determined that not all of celestial motion is in circles, and planets, moons, and comets traveled in ellipses of one type or another. An epicycle may be thought of as the planet moving in its own series of small circles as it progresses around the circumference of its orbit.

Copernicus' theory of the Earth spinning on its axis: *The motions of Earth consist of two or more component motions. One motion is revolving; the other is rotating.*

Most philosophers and scientists in Copernicus' day rejected the concept of Earth's revolving around the sun. The second contention—that Earth spins on its axis as it revolves around the sun—was even more difficult to comprehend. Most people claimed that their common sense dictated it was not possible and offered these arguments against a rotating Earth: 1) If Earth spun on an axis, why didn't objects fly off into space? 2) It would be impossible for anything moveable to be firmly affixed to Earth. 3) Birds would have to fly faster in the direction of Earth's rotation just to stay in the same place. 4) If a person jumped up, he would come down in a different spot because Earth would have moved.

Copernicus not only lived during the Renaissance, he was a Renaissance man with interests in economics, law, medicine, mathematics, as well as astronomy. During this period the Roman Catholic Church needed better astronomical information to set correct times and dates for special holidays and ceremonies. They approached Nicholas Copernicus, a monk in the church, for assistance. His studies on their behalf led to his astronomical theories that were not always accepted by the Church. His model of the solar system did not become generally known until 1515 as he shared his ideas only with friends. His studies were published in *De Revolutionibus Orbium Celestium* in the same year he died (1543).

See also Brahe; Coriolis; Galileo; Kepler; Ptolemy

COREY'S THEORY OF RETROSYNTHETIC ANALYSIS: Chemistry: *Elias James Corey* (1928–), United States. Elias Corey was awarded the 1990 Nobel Prize in Chemistry.

> *Retrosynthetic analysis occurs when the whole target chemical compound (the compound to be studied) is divided into subunits for analysis and then synthetically recombined.*

This was a new approach to analyzing chemical substances, particularly complex organic molecules. Corey and his research teams reduced complex molecules to smaller and smaller "pieces" and then recombined these units to arrive at the original or an altered molecule. In this manner, they determined how to synthesize many organic compounds for medical use. An example is the synthesis of prostaglandin hormones useful in treating infertility and inducing labor. Using his retrosynthetic concept, Corey developed a new computer program that greatly assisted chemists in their analysis and synthesis of organic compounds.

Elias Corey received his undergraduate degree at the Massachusetts Institute of Technology in 1948 and his graduate degree in 1950, both in chemistry. He became interested in chemistry after auditing a course in organic chemistry. After receiving his doctorate, he was employed by the University of Illinois in Champaign-Urbana, and nine years later moved to Harvard as a professor of chemistry. He is best known for new and unique methods for synthesizing organic molecules, particularly his "retrosynthetic analysis" technique that examines parts of whole molecules. He published his methods in *The Logic of Chemical Synthesis* in 1989.

CORIOLIS' THEORY OF FORCES ACTING ON ROTATING SURFACES:

Physics: *Gustave-Gaspard Coriolis* (1792–1843), France.

An inertial force acts on rotating surfaces at right angles to the rotating Earth, causing a body to follow a curved path opposite the direction of the rotating Earth.

The Coriolis effect is greatest if an object is moving longitudinally on Earth—from either pole to the Equator along longitudinal lines. In the Northern Hemisphere the apparent motion, when viewed from the North Pole, is to the right; for the Southern Hemisphere, when viewed from the South Pole, it is to the left. It affects the oceans and atmosphere on Earth but is a much weaker force than gravity. Even so, over great distances it causes cyclones, which are low-pressure areas that can develop into hurricanes, and water whirlpools to circle counterclockwise in the Northern Hemisphere. Conversely, anticyclones may develop into typhoons, which rotate clockwise in the Southern Hemisphere. The Coriolis effect influences ocean currents, including El-Niño, as well as other local and worldwide weather and climate phenomena.

The magnitude of the Coriolis effect is the velocity of the object compared to Earth's angular velocity for the given latitude and is the reason rockets and spacecrafts are launched to the east. The Coriolis effect gives rockets an extra boost as Earth spins eastward. Also, missile-launching sites are usually located on coastal areas so that any defective rockets or missiles will fall on water rather than land. Although other scientists recognized the effect caused by Earth's rotation as an inertial force, it was Coriolis who worked out the mathematics for this force and was the first to publish his results. The mathematical parameter for the Coriolis force is twice the component of Earth's angular velocity around the local vertical as expressed in: $2\Omega \sin \Phi$, where Ω is the angular momentum of Earth and Φ is the latitude on Earth's surface. This complex force resulting from the rotation of Earth on its axis was not apparent in the days of Copernicus and Galileo because it is a force much too weak to have been recognized or measured in their times.

Although the Coriolis force *deflects* or moves air and water masses to the right in the Northern Hemisphere, the *rotation* of the large air masses (hurricanes) is to the left (*see* Figure C4).

There are several misconceptions about the Coriolis effect. One is that it will affect the direction of rotation of flow of relatively small volumes of water, such as the

The Coriolis effect on Earth may be summarized as follows

1. The Coriolis effect is an inertial force caused by Earth's rotation to the east. Its strength is altered with the degree of speed of Earth's air and water masses.
2. The Coriolis effect *increases* as one goes from the equator toward the polar regions. The polar regions are at *right angles* to the axis of Earth's rotation.
3. The Coriolis effect *decreases* as one nears the equator. At the equator Earth's surface is *parallel* to the axis of rotation.
4. The Coriolis effect causes air and water masses to be *deflected* and turn *right* in the Northern Hemisphere and to be *deflected* and turn *left* in the Southern Hemisphere. Note: the direction of *deflection* is not to be confused with the direction of *rotation* of the air and water masses.
5. The geotropic flow is a gradient flow where the Coriolis force balances the pressure of the horizontal force of the wind.

draining of water down a bathtub or sink drain or when flushing a toilet. The direction of rotation of water down a drain can be made to go in either direction by swirling the water around before the plug is pulled. The Coriolis effect is much too small to be effective at this level of magnitude. In addition, the shape of the tub, the nature of its surface, the water's disturbance as it enters the tub, sink, or toilet, and how the drain is opened have more effect on the direction of the whirlpool of draining water than the Coriolis effect. This becomes more understandable when considering that Earth spins on its axis every 24 hours and affects very large masses of air and the oceans, whereas the tiny amount of water going down the drain might be said to be "just a drop in the larger bucket."

When firing a rifle, the bullet is affected by gravity and follows a curved path, but the distance the bullet travels is too small to be affected by the Coriolis effect. However, when firing very long-range artillery shells, the calculation of the ballistics of the projectiles' flight must take into consideration the Coriolis effect. In World War I the Germans developed a huge gun that bombarded Paris from a distance of 120 kilometers. This distance required the Coriolis effect to be taken into consideration when aiming the gun.

Cyclones are low-pressure areas in the Northern Hemisphere that, when under the right conditions, form hurricanes. They generally travel in a northerly direction, and the further north, the stronger the Coriolis effect. Hurricanes seldom move toward the equator because the closer you get to the equator, the less effect of the Coriolis force. In fact, the Coriolis effect is zero at the equator.

Figure C4. The Coriolis Effect causes winds and hurricanes to rotate counterclockwise in the northern hemisphere, and clockwise in the southern hemisphere.

THE CORI THEORY OF CATALYTIC CONVERSION OF GLYCOGEN: Chemistry: *Carl Ferdinand Cori* (1896–1984) and *Gerty Theresa Radnitz Cori* (1896–1957), both from the United States, jointly received one-half of the 1947 Nobel Prize for Physiology or Medicine. The other half of the 1947 prize was awarded to Bernardo Houssay (1887–1971) for his work with the pituitary gland and sugar metabolism. The only other husband-and-wife teams of scientists to share a Nobel Prize besides the Coris were Marie and Pierre Curie in 1903 and the Joliot-Curies in 1935.

Coris' hypothesis for glucose conversion: *The complex carbohydrate glycogen stored in the liver and muscles is converted, as needed, to energy, in the form of glucose-6-phosphate by the enzyme phosphoglucomutase. The process is reversible.*

Carl and Gerty Cori collaborated on biochemical research projects dealing with the analysis of enzymes and glycogen (sugars and starch). They isolated a chemical from dissected frog muscle identified as glucose-1-phosphate, where a molecular ring containing six carbon atoms joined this complex molecule. This new compound was named the Cori-ester and was shown to convert to the more complex sugar form after it was injected into animal muscles. Their research led to a more complete understanding of the role of high blood sugars, diabetes, insulin, hormones, and the pituitary gland. Their research also opened the door for a more complete understanding of the important role phosphates play in producing the energy in animal cells.

COULOMB'S LAWS: Physics: *Charles Augustin de Coulomb* (1736–1806), France.

> Coulomb's fundamental law of electricity: *Two bodies charged with the same sort of electricity will repel each other in the inverse ratio of the square of the distance between the centers of the two bodies.*

Coulomb devised this law after developing an extremely sensitive torsion balance, which consisted of a thin silk thread supporting a wax-covered straw (thin, very light reed or grass) with a small pith ball suspended on one end. The straw was balanced and suspended in an enclosed jar to prevent air drafts from affecting the results. His torsion balance could measure a force of only 1/100,000 of a gram by gauging the twist of the thread. There were markings around the circumference of the jar so Coulomb could measure, in degrees, any changes in the ball's position. Coulomb then used static electricity to charge the pith ball on the straw and another pith ball outside the jar. He brought the outside charged ball close to the jar. The ball inside the jar was repelled. He measured the distance of its movement on the degree markings on the jar and compared it with the distance between the centers of the inner and outer balls. He discovered that one of the basic laws of science applied (the inverse square law). When moving the outer ball twice the distance away from the inner ball, the effect on the inner ball was not one-half its movement but one-fourth. In other words, the effect of the electrical charge decreased as to the square of the distance between the centers of the charged balls.

Although Coulomb was given credit for demonstrating the effects of the "inverse square law" he was not the first to do so. Several other scientists, including Joseph Priestley and Daniel Bernoulli, of the early eighteenth century tried to find evidence of the inverse square law but failed. In the 1760s Henry Cavendish demonstrated the effects of the inverse square law experimentally. Unfortunately, the results of his experiments were not published until 1879. The law is really an analog of Sir Isaac Newton's law of gravity that is $F = G[m_1 \times m_2]/d^2$, where the F is the force, G is gravity, m_1 and m_2 are the masses of the two bodies that are involved, and d^2 is the square of the distance between the masses. Coulomb's law for electrical charges between two bodies is similar: $F = k[_1 \times q_2]/d^2$, where k is the electrical charge, q_1 and q_2 are the magnitude of the charges of the two bodies, and d^2 is the square to the distance between to two bodies. Or it can be written as the force between two charged particles as: $F = qq'/kd^2$ where F is the force, q and q' are the two charged particles, k is the dielectric constant, and d^2 is the distance between the two particles. It should be mentioned that electric and magnetic forces are four million, trillion trillion trillion times stronger than the force of gravity. Gravity is one of the weakest forces known, but it exists on a cosmic scale.

Coulomb's theory of the relationship between electricity and magnetism: *Electricity and magnetism follow the same physical laws, including the inverse square law.*

Coulomb demonstrated the similarity of the attractive and repulsive forces for electricity and magnetism and concluded that both were similar physical phenomena that followed the same physical laws. (See Figure A1 under Ampère.) Coulomb also concluded that electrical charges follow the same inverse law as does gravitation. Even so, many scientists of his day rejected this theory. When the theory of electromagnetism was later refined and better understood, it became important for the development and use of electromagnetic devices, such as motors and generators.

Coulomb's law of electrical charge: *The attractive and repulsive forces for electricity are proportional to the products of the charge.*

This famous theory is now known as Coulomb's law, which states that a coulomb is the unit quantity of electricity carried by an electric current of 1 ampere in 1 second. The unit of electrical charge, the coulomb (C), is named after him. As a result, a much better understanding exists of how to quantify electricity as a measurable current forced through a conductor by a voltage differential.

See also Cavendish; Faraday; Maxwell

COUPER'S THEORY FOR THE STRUCTURE OF CARBON COMPOUNDS:

Chemistry: *Archibald Scott Couper* (1831–1892), Scotland.

> *Carbon atoms have the unique ability to bond their valence electrons together to structure both chains and branches of chains to form carbon compounds (mostly organic).*

Couper knew carbon must have four valence electrons because it could form inorganic compounds such as carbon dioxide, CO_2, where oxygen has a valance of 2. Thus, it takes only one carbon atom with a valence of 4 to join with two oxygen atoms to form the molecule O=C=O (Note: the two = represents two valence electrons joining the single carbon atom with two oxygen atoms). Further research indicated that carbon atoms have the unique ability to bond with other carbon atoms to form chains and branches (*see* Figure V3.) Couper was the first to use notations such as =C=C=C= for carbon atoms to illustrate his theory, advancing the concept of isomers, which are compounds with the same molecular formulas but are structured differently. Isomers have different chemical and physical characteristics due to different arrangements of the elements making up the molecules. The greatest variety of isomers is found among the various hydrocarbons and the many complex organic compounds. A typical example is the hydrocarbon isomer of C_4H_{10} that can be structured as a chain of 4 carbon atoms with 10 attached hydrogen atoms, or a different structure of a branched group of 4 carbon atoms, with 10 hydrogen atoms attached to the carbon atoms.

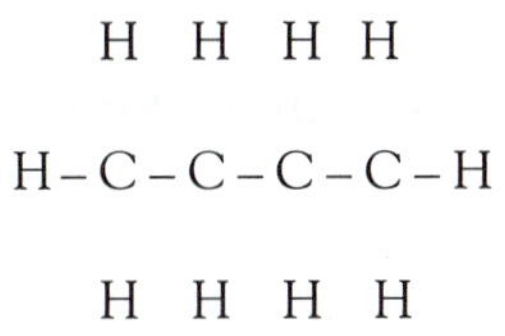

```
                                  H H H
                                   \ /
   H   H   H   H               H    C    H
                                \       /
H - C - C - C - C - H        H - C - C - C - H
                                /       \
   H   H   H   H               H    H    H

      butane                   2-methyl-propane
```

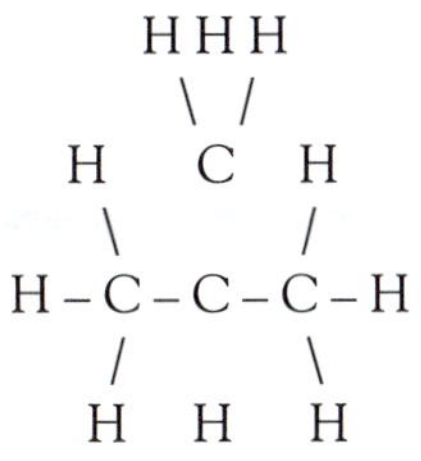

Their physical and chemical characteristics are quite different. The entire chemistry of carbon isomers, while complicated, has pioneered many new medicines and useful products.

Couper wrote an important paper titled "On a New Chemical Theory" in 1858 that explained his concepts of chemical compounds, but publication was delayed. In the meantime Fredrich Kekule arrived at similar isomeric straight chain and branching carbon molecular structures. However, after a dream of a snake eating its tail (see Figure K1), Kekule came up with a carbon ring structure that solved the problem of how some carbon compounds, including benzene, are formed in connecting rings. Even though Couper's designs for carbon compounds and isomers involved only straight and branching chain molecules, they anticipated and preceded Kekule's work.

See also Kekule; Van't Hoff

CRICK–WATSON THEORY OF DNA: Biology: *Francis Harry Compton Crick* (1916–2004), England. Francis Crick shared the 1962 Nobel Prize in Physiology or Medicine with James Watson and Maurice Wilkins.

> *DNA is a double helix joined by pairs of nucleotides of adenine + thymine (A+T) and guanine + cytosine (G+C), with the sugar-phosphate structure attached to the outsides of the helix strands.*

Early in Crick's career in England, he was joined by an American James Watson (1928–), who suggested that the first step in determining the structure of the basic molecule of life would be to learn more about its chemical nature, which they then pursued together. Crick and Watson were not the only scientists searching for the holy grail of life. Others, such as Maurice Wilkins (1916–2004) and his assistant, Rosalind Franklin, as well as the Nobel Laureate and Scottish biochemist Alexander Todd (1907–1997) in England, Erwin Chargaff, and Phoebus Levene in the United States, were conducting similar research.

There is a history of a number of scientists investigating the origins of replication of the DNA molecule. Frederick Griffith (1871–1941), a British medical officer, produced the first evidence of DNA when he experimentally identified the transformation of S and R strains of a bacterium. The Canadian-American physician and researcher Oswald Avery (1877–1955) experimentally proved that DNA was the transferring

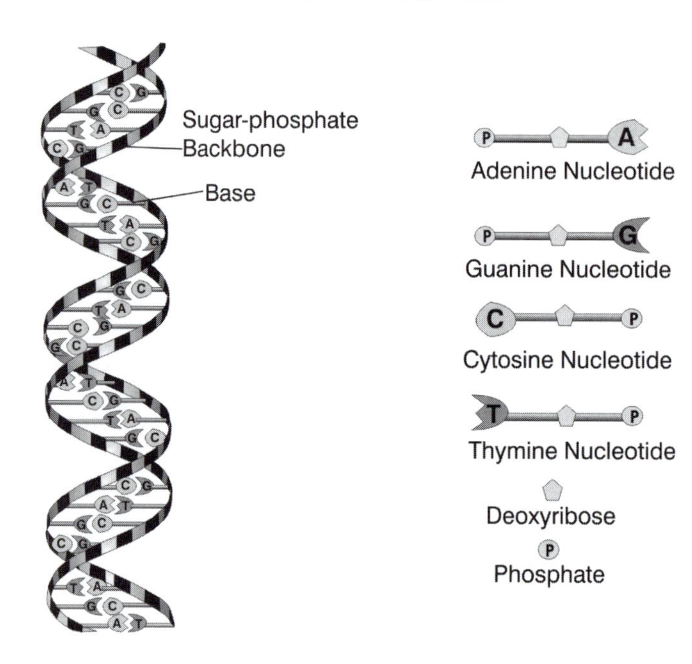

Figure C5. Crick and Watson determined the shape of the double helix shape for the DNA molecule from the crystal X-rays made by Rosalind Franklin.

agent for genetic information. And, Linus Pauling, who worked with the nature of chemical bonding, described the complex protein molecules (DNA) involved in the chromosomes of cells as being an alpha-helix structure, which came close to describing the double-helix structure proposed by Crick and Watson.

Nationalism, competition, jealousy, and secrecy, all of which are the antithesis of scientific investigations, were part of the search for the structure of DNA. Crick's work depended on information he obtained from crystallographic X-rays (X-ray photos) of DNA crystals to determine its structure (see Figure C5.) Franklin was very secretive about her work and refused to share her photographs or her crystallography techniques. Her supervisor, Wilkins, considered the study of DNA a joint project between Franklin and him. Franklin also withheld information about the placement of the sugar-phosphate backbone for the DNA molecule. Crick was acquainted previously with Wilkins, and as the story goes, there may have been a break in trust when Wilkins provided Crick with some of Franklin's vital information that enabled Crick to succeed with his project. In 1953 Crick and Watson completed their model based on information known at that time.

Francis Crick proposed a rather controversial hypothesis in his book *Astonishing Hypothesis: The Scientific Search for the Soul* in 1994. The hypothesis in essence states: *That a person's "you," your joys and your sorrows, your memories and your ambitions, your sense of personal identity and free will are in fact no more than the behavior of a vast assembly of nerve cells and their associated molecules.*

Humans have always wondered about their ability to be aware of themselves and attributed this phenomenon and their many personal thoughts, including beliefs in supernatural powers separate from their bodies. In the early seventeenth century René Descartes came up with the idea that the mind contains the essence of a human being and was distinct and separate from the brain and body. It was not until the discovery of the double helix of DNA that most scientists gave up the idea that consciousness, and so on, was too philosophical to study. Science experiments have indicated that all aspects of our subjective selves, our awareness, and all aspects of the mind are best explained by the behavior of over fifty billion nerve cells in the brain. Even today, not all scientists consider the mind, consciousness, and "self" as existing as a neurological phenomenon of the brain's multitude of neurons.

Crick and Watson shared the 1962 Nobel Prize for their discovery along with Wilkins, but Rosalind Franklin, being deceased, was not included. (Some say Franklin was not given adequate credit for her contributions to the final outcome.)

See also Chargaff; Franklin (Rosalind); Pauling; Watson

CROOKES' RADIATION THEORIES: Physics: *Sir William Crookes* (1832–1919), England.

Crookes' radiometer: *A closed glass container in which most of the air has been evacuated will continue to radiate heat.*

From his spectrographic work in identifying the element thallium, Crookes noticed that heat radiation caused unusual effects on the thallium gas while in the sealed glass container. He designed a device with four vanes. One side of each of the four vanes was painted black, and the other sides of the vanes were polished like a mirror. The vanes were balanced on a vertical pivot in a closed glass in which most of the air had been removed. When heat radiation (sunlight) struck the vanes inside the glass bulb, molecules on the dark hot side had greater momentum and thus pushed the vanes backward to a greater extent than did the molecules from the cooler shiny side. This led to further investigation of the effects of electricity on low-pressure gases. At the time, the radiometer demonstrated the essence of kinetic energy of gas molecules. Today, Crookes' radiometer is more like a toy used to demonstrate the effects of radiant heat on dark and shiny surfaces.

Crookes' cathode ray tube: *The air will glow when an electric current is passed through a closed glass tube containing low air pressure.*

To demonstrate that the glow in the Crookes tube and the slight fluorescence on the inner walls of the tube were due to electricity, Crookes placed a Maltese cross in the path of the rays. The form of the Maltese cross was used because its symmetrical design would produce a recognizable image as it interrupts the flow of cathode rays. At the point where the rays were blocked by the cross, a distinct shadow-like pattern appeared on the end of the glass tube.

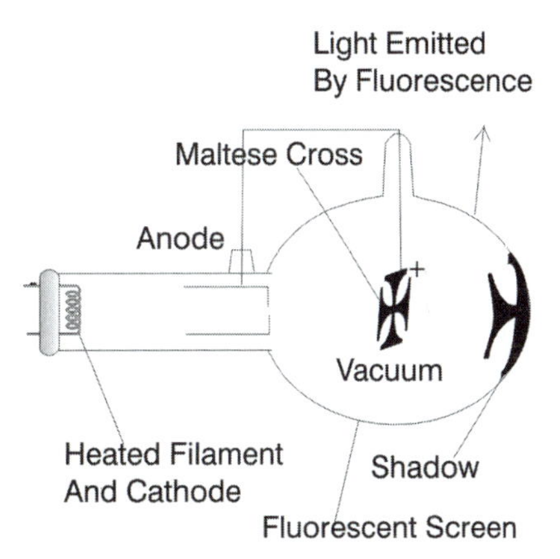

Figure C6. The symmetrical Maltese cross design was used to demonstrate that when a stream of electrons was sent to a target in the path of a fluorescent screen, the metal cross would block the screen. The shadow of the cross on the screen prevented the electrons from producing light on the fluorescent material on the screen. A similar device was used to demonstrate the nature of the electron; it has a negative charge and a magnet can deflect its path. Its mass is just 1/1840 the mass of the hydrogen atom.

Crookes also demonstrated that a magnet brought near the glass tube would deflect the cathode rays in a curved pattern that suggested the rays were composed of particles with an electric charge. He concluded it was impossible for electromagnetic radiation, such as light, to carry an electric charge and be deflected by a magnet. Therefore, the cathode rays must be charged particles. J. J. Thomson later demonstrated the cathode rays were really electrons. The shadow of the Maltese cross in Crookes' cathode ray tube might be considered the first TV picture because a similar process is used in modern television receivers.

See also Thomson

CRUTZEN'S THEORY OF OZONE DEPLETION: Chemistry: *Paul Crutzen* (1933–), Netherlands. Paul Crutzen, along with Mario Molina and F. Sherwood Rowland received the 1995 Nobel Prize for Chemistry. Crutzen demonstrated that nitrogen oxide accelerated the destruction of ozone (O_3) in the stratosphere, while Rowland and Molina, two American chemists, discovered in 1974 that CFC gas (chlorofluorocarbon) also causes ozone depletion.

Crutzen's theory states: *Nonreactive nitrogen dioxide* (N_2O) *gas is lighter than air, thus it rises in the atmosphere where solar energy splits it into two different reactive compounds* (*NO and* NO_2) *where they react as a catalyst with ozone gas* (O_3) *breaking it into* O_2 *and atomic oxygen, thus causing a depletion of the original ozone.*

These three scientists were not the first to arrive at the chemistry related to the ozone-oxygen cyclic reactions. In 1930 Sydney Chapman, a British astronomer and geophysicist who is best known for his work with the kinetic theory of gases (see Chapman), wrote an article that explained how ultraviolet radiation (UV) breaks stratospheric oxygen (O_2) into oxygen atoms (O) which then combine to form ozone (O_3), followed by the formation of two oxygen molecules ($2O_2$) when the oxygen atoms recombine with the ozone. At that time Chapman was unaware that there was a catalytic action that drove this sequence.

Paul Crutzen knew that nitrous oxide (N_2O), a stable gas produced by some soil bacteria (and later identified as a gas produced by automobile exhausts, jet airplanes, and fertilizers), could change into nitric oxide (NO) in the stratosphere that, in turn, acts as a catalyst and contributes to ozone depletion.

In 1974 Frank Sherwood Rowland and Mario J. Molina were the first to arrive at the idea that chlorofluorocarbons (CFCs) would have the same catalytic effect on the ozone molecule, as did nitrous oxide as proposed by Chapman. It was discovered that most of the CFCs produced since the 1930s were still in the atmosphere due to its high degree of stability. It was determined that UV light dissociates the CFCs and releases chlorine (Cl) in the ozone layer and is thus more effective in breaking down the ozone molecule than is nitrous oxide.

A number of other scientists became involved in the chemistry of ozone, but it was not until 1985 that Joe Farman, Brian Gardiner, and Jonathan Shanklin with the British Antarctic Society (BAS) published a paper in *Nature* that announced the discovery of a "hole" (more accurately a thinning area) in the ozone over Antarctica. Later a less extensive "hole" was discovered over the North Pole as well. Ozone is not really a "layer" but rather a diffuse screen of gases at an altitude of about 35 kilometers that partially blocks harmful ultraviolet radiation from reaching Earth's surface. Ozone is produced in the warmer stratosphere over the equator and drifts toward the poles. Crutzen showed that ozone is continually created and destroyed in the stratosphere. The concentration of ozone is also affected by natural processes such as volcanoes, atmospheric temperatures, and variations in solar activity so that, to some extent, it is replenished, but less so over the colder polar regions. The size of the "holes" over the poles varies with temperatures at different times of the year. A simplified depiction of the chemical reaction follows:

$$O + O_2 \rightarrow O_3$$
$$UV + O_3 \rightarrow O_2 + O$$

Ozone reacts to many things in the atmosphere such as nitrous oxide, several halogens (chlorine and bromine), methyl chloride, as well as hydrogen, where they capture an oxygen atom from the ozone molecule. An "X" is used as a generic chemical symbol to represent any substance that can react with ozone as follows:

$$X + O_3 \rightarrow XO + O_2$$

The great increase in the use of CFCs for refrigeration and pressure in spray cans over the years is held responsible as possibly the main cause of ozone depletion. In 1996 several nations began phasing out the production of CFCs, halogens, and related chemicals that seems to have affected the size of the ozone "hole." It might be noted that the "ozone hole" has recently been significantly decreased. This may also be a result of a "natural cycle" related to the formation and destruction of ozone, or it may be a result of several nations banning the use of CFCs.

See also Chapman; Rowland

CURIES' RADIATION THEORIES AND HYPOTHESES: Chemistry: *Marie Sklodowska Curie* (1867–1934), France. Marie and Pierre Curie shared the 1903 Nobel Prize for Physics with Antoine Henri Becquerel (1852–1908), who discovered spontaneous radioactivity. Madame Curie was also awarded the 1911 Nobel Prize for chemistry for her discovery of radium and polonium. She is one of only four people ever to receive two Nobel Prizes (the other three are L. Pauling, J. Bardeen, and F. Sanger).

Curies' radiation hypothesis: *Chemical reactions and mixtures of uranium with other substances do not affect the level of radiation. Only the quantity of uranium determines the level of radiation. Therefore, radioactivity must be a basic property of uranium.*

Pierre Curie (1859–1906) jointly received the 1903 Nobel Prize for Physics with his wife Marie, and Antoine Henri Becquerel. He determined that the slight deformation caused by the squeezing of opposite sides of certain types of ceramic crystals produces opposite electric charges on opposite faces of a crystal.

Pierre Curie was a well-known physicist who assisted his wife, Marie, in her research with radioactive elements. With his brother, Jacques Curie (1856–1941) they developed several techniques for detecting and measuring the strength of radiation. Their instrument, the electrometer, was sensitive enough to produce an electric current between two metal plates separated by the radioactive sample. They also discovered piezoelectricity, from the Greek work *piezo* meaning "to press." The piezo effect occurs when certain types of crystals are put under pressure. Pierre Curie and his brother also discovered it would work in the opposite manner, that is, by applying an electric charge to a crystal a change in the crystal's structure occurs. This discovery was incorporated in their electrometer used to measure minute electric currents, as well as radiation. The piezoelectric effect has found many useful applications including crystals in crystal microphones, sonar, ultrasound devices, phonograph needle pickup devices, radio transmitters, and as an analogous pendulum in timepieces. It is now found in many types of watches, clocks, and other devices where regular mechanical vibrations of a quartz crystal are used.

Pierre Curie also measured the amount of heat given off by radium. Each gram of radium gives off 140 calories of heat per hour, with a **half-life** of about sixteen hundred years. The Curies realized this amount of energy was beyond normal chemical reactions and must be from some other unknown part of the atom. Thus began the age of nuclear energy, even though the nuclei of atoms had yet to be discovered. Pierre Curie also discovered that permanent magnets lose their magnetism when heated to a specific temperature. This temperature point is now called the *Curie temperature* as a unit of measurement.

At the time the Curies worked with radiation, particularly radium, the extent of the dangers of radiation was unknown. It is assumed the Curies may have been the first humans to suffer from radiation sickness, but Pierre died after an accident with a horse and carriage. Marie died from illnesses related to radiation poisoning. The *curie,* the unit measurement for radioactivity, was named for Pierre Curie. Marie Curie's notebooks are still considered extremely radioactive.

Madame Curie separated chemicals from uranium minerals and found the ore pitchblende was more radioactive than uranium earth itself. Pitchblende is a heavy black ore containing a yellow compound that the German chemist Martin Heinrich Klaproth (1743–1817) thought was a new element. He named it *uranium* after the planet Uranus. The Curies brought many cartloads of pitchblende from northern Europe to her laboratory shed in France. (Pitchblende is also found in Colorado, Canada, and Zaire.) Over a period of months, Curie and her assistant chemically extracted this new element. She also theorized there must be more than one type of radioactive element in the ore, leading to a new hypothesis.

Curie's hypothesis for new radioactive elements: *Because the pitchblende ore contained substances with greater radioactivity than uranium, pitchblende must contain new radioactive elements.*

Curie continued to separate and test these new substances, which proved to be new elements. She named one polonium after her native country (Poland), and the other radium, for its high radioactivity. She discovered that the heavy metal thorium also exhibited radiation. Curie is credited with coining the word "*radioactivity*." She and her assistant used several chemical processes to separate the radium, which exists in very small amounts in pitchblende. After many months, she had produced only about 0.1 gram of radium chloride.

See also Becquerel

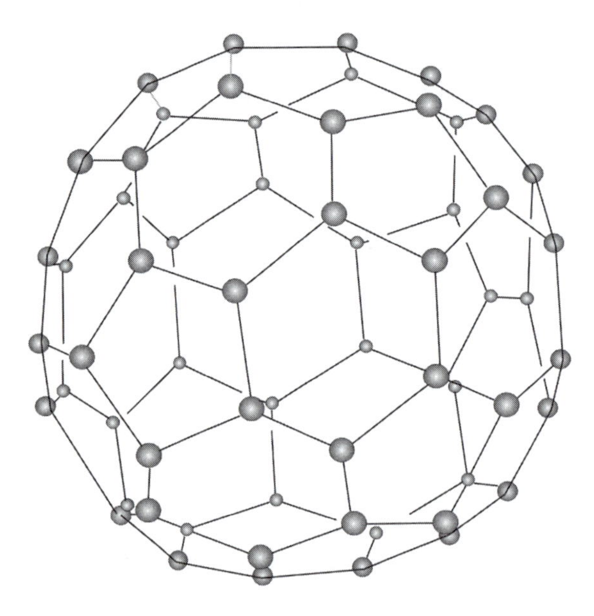

Figure C7. Artist's conception for the structure of the C_{60} atom discovered by Robert Curl and named *buckminsterfullerene* (nicknamed *Buckyballs*) by Harold Kroto. These compact masses of carbon are a third form of pure carbon, each composed of 60 carbon atoms into a soccer ball shape by 20 hexagons and 12 pentagons of bonded carbon atoms.

CURL'S HYPOTHESIS FOR A NEW FORM OF CARBON: Chemistry: *Robert F. Curl, Jr.* (1933–), United States. Robert F. Curl, Richard E. Smalley, and Sir Harold W. Kroto jointly received the 1996 Nobel Prize for Chemistry.

Curl's C_{60} hypothesis: *Vaporized carbon atoms in a vacuum can form single and double bonds, similar to aromatic carbon compounds, to produce a symmetrically closed shell with a surface consisting of multiple polygons.*

Curl and Smalley discovered the new complex carbon molecules, called fullerenes in 1985 while they were at Rice University in Houston, Texas, and Kroto was at the University of Sussex in England. The most common form is a group of 60 atoms shaped similar to a soccer ball, which is formed by 20 hexagons and 12 pentagons of bonded carbon atoms (*see* Figure C7). Due to the geodesic shape of the surfaces it was named *buckminsterfullerene*, after the architect R. Buckminster Fuller, who designed geodesic dome structures. However, it is usually referred to by its nickname, *Buckyballs*. Additional complex ball-shaped molecules of bonded carbon have been developed in addition to C_{60}. They include C_{69}, C_{70}, C_{76}, C_{78}, and C_{80}.

The discovery of fullerene Buckyball molecules has opened up new research vistas in the areas of superconductive materials, plastics, polymers, and medicines, as well as new theories to explain the beginning of the universe and the structure of stars.

See also Kroto

Robert F. Curl's life is an example of how parents and teachers can exert a positive influence on a young person's life. His father was a minister who traveled around to different churches in various towns in Texas. At the age of nine Curl's father became a church administrator and finally settled in one place. At that time, his parents gave him a chemistry set and a short time later Curl decided on a career as a chemist. He excelled in most subjects, but mainly in science. However, he was unable to take a class in chemistry until high school where his teacher recognized his outstanding interest and ability in this field. Curl was so far ahead of his class that his teacher assigned him special projects that he devoured. One of his early projects was a working model of a Cottrell precipitator, which is used in industrial smoke stacks to eliminate minute particles from the exhaust. He chose Rice University in Houston for its programs for students like himself, and his parents approved his choice because Rice did not charge tuition. He enjoyed his professors and did well in his first and second year and by his third year he was deeply involved in physical and organic chemistry, taught by some outstanding professors. He did his graduate work at the University of California at Berkeley with several outstanding professors. Curl said that his years at Berkeley were his most happy years. This is also when he met and married his wife. He did experimental as well as theoretical work on silicon oxides. Following Berkeley, Curl was appointed to a postdoctoral position at Harvard. Harvard was followed by an appointment as assistant professor at Rice University where he continued research in physical chemistry. Curl collaborated with several outstanding people who contributed to the discovery of fullerenes (Buckyballs). As a result of this research, Robert Curl, Rick Smalley, and Harry Kroto were awarded the 1996 Noble Prize for Chemistry.

CUVIER'S THEORIES OF ANATOMY AND TAXONOMY:

Biology: *Baron Georges Leopold Chrétien Frédéric Dagobert Cuvier* (1769–1832), France.

Cuvier's classification of animals: *Based on functional integration of animals there are four classes or "branches": 1) vertebrata (with backbones; e.g., mammals), 2) articulata (arthropods; e.g., insects), 3) mollusca (bilateral symmetrical invertebrates; e.g., clams), and 4) radiata (radially symmetrical sea animals; e.g., echinoderms).*

Cuvier demonstrated how the anatomy of different animals compared with one another and proposed these four **phyla** of animals in his classification scheme. Cuvier is often referred to as the founder of the science of comparative anatomy.

Cuvier's concept of form and function: *All organisms are integrated wholes into which parts are formed according to their functions.*

Cuvier was a firm believer that form follows function, not the reverse. Generally this means that the structure of tissue or an organ is based on what function the tissue or organ is required to execute. He believed that if a part of an animal changed, it would change the entire animal's form. He stated that all the parts of each animal are arranged to make it possible for the animal to be complete. Thus, he rejected organic evolution. The concept of form follows function is, to some extent, still debated in modern biology. The concept is also used in the field of architectural design, particularly by followers of the American architect Louis Henri Sullivan's (1856–1924) philosophy that a building's form should be designed to represent its intended function. In other words, for biology and architecture the structure should be built for the purposes of the organism it surrounds.

Cuvier's theory of evolution: *Organic evolution cannot exist because any change in an organism's structure would upset the balance of the whole organism, and thus it would be unable to survive in its environment.*

Cuvier rejected Darwin's theory of organic evolution based on natural selection that results in changes in species and the emergence of new species over long periods of time. Cuvier believed similarities between and among different organisms are due to common functions of their parts, not evolutionary changes. However, he did believe that catastrophic events occurred on Earth (he preferred the term "periodic revolutions"). Natural disasters such as floods, fire, and earthquakes caused massive extinction of animals and provide situations for the arrival of new species. Niles Eldredge and the late Stephen Jay Gould revived the catastrophic theory as *punctuated evolution*.

Cuvier's theory of fossils: *Fossils represent ancient species that became extinct due to period revolutions in their environment.*

In his study of fossils, Cuvier recognized some fossils were found deeper in the strata of rocks and earth and that the depth of the strata could determine their age. He used a similar classification system for fossils as his four phyla for living animals. Cuvier also recognized the detailed structure in some fossils, particularly the structure of wings. He was the first to identify the fossil of a flying reptile he named *pterodactyl*. Cuvier has been referred to as the founder of paleontology.

See also Buffon; Darwin; Eldredge–Gould; Gould; Lamarck; Wallace

D

DAGUERRE'S CONCEPT OF HOW TO "FREEZE" IMAGES MADE BY THE CAMERA OBSCURA: Chemistry: *Louis-Jacques-Mandé Daguerre* (1789–1851), France.

> *A light sensitive plate exposed to mercury vapors will change a latent image to a visible image.*

Daguerre was a French chemist as well as an artist who is sometimes credited with the discovery of photography, although several other competitors worked on the process contemporaneously. Daguerre and his assistants painted large scenes (14 × 22 meters) to use in dioramas for stage productions. The term "diorama" was coined by Daguerre in 1822. It refers to a rotating three-dimensional model or display of a landscape that depicts historical events, but with a faulty perspective. Changing the lighting as well as the scenes produced illusions of famous landmarks. These shows were famous in Paris and London and were in great demand. Initially, Daguerre traced images made by a camera obscura (a pin-hole camera that produced an upside-down mirror image of a scene) to copy for their dioramas.

Because this was time-consuming work, he theorized that there must be a method of "freezing" the camera's image, thus eliminating the laborious process of tracing. After some experimentation, Daguerre silver coated a copper plate. At the time, it was known that some salts of silver, such as silver iodide, were light sensitive. During one such experiment he left an exposed plate in a cabinet with a broken mercury thermometer. Later when he retrieved the plate, he observed that the latent image was now visible, but as a reversed, and more-or-less, negative image. Later experiments indicated that an image could be formed after just a few minutes exposure and that these images could be "fixed" in seawater. When left in daylight, the image faded, so Daguerre and his assistants attempted to secure the image in a "fixer" chemical known as hyposulfate

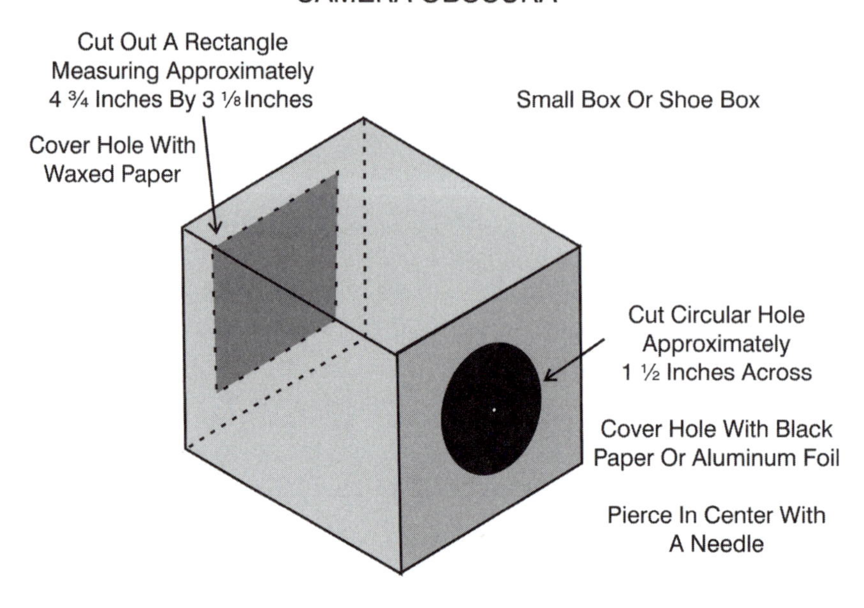

Figure D1. How to make a simple "box" camera.

of soda or "hypo." (The same form of hypo is still used today to dissolve the unexposed silver salts from the negative film or paper print so that it is no longer light sensitive.) The result was the "daguerreotype" positive image, which had to be enclosed in a glass frame with an inert gas such as nitrogen to prevent further fading. Some of these images still exist.

By the early 1800s daguerreotype photography had spread to the United States where it became popular with the masses as an inexpensive portraiture process. There were several disadvantages to daguerreotype photographs. They produced a positive and reversed image. It required a few minutes exposure. The image was not particularly permanent, nor could copies be made of the image. Even though Daguerre received patents for his process in England and Wales, the French government purchased the process from him and then gave it to the world without charge. At the same time William Henry Fox Talbot (1800–1877) of England (sometimes given credit as the discoverer of modern photography) was also working on a similar process for use as an entertainment medium. Other processes, such as the ambrotype, tintype, and collodion, of the mid-1800s were improvements over the daguerreotype. However, it was the introduction of the system that could produce a negative on a thin glass plate by the wet collodion process developed by Frederick Scott Archer (1813–1857) that was the beginning of the end of daguerreotypes. Archer's process produced a negative that, when developed and fixed, could be used to make positive prints that were superior to daguerreotypes. The improvement of this negative-to-positive process of photography proved the demise of daguerreotypes. Photography has progressed from the pinhole camera to the daguerreotype, to negative/positive films and paper, to cameras of today that produce digital images that require no negative film.

DALE'S THEORY OF VAGUS NERVE STIMULI: Biology: *Sir Henry Hallet Dale* (1875–1968), England. Sir Henry Hallet Dale and Otto Loewi shared the 1936 Nobel Prize for Physiology or Medicine.

Chemical and electrical stimuli are responsible for affecting nerve action.

While at the Wellcome Physiological Research Laboratories in England, Henry Dale discovered that the dangerous ergot fungus contained the chemical acetylcholine, a pharmacologically active extract that acts as a neurotransmitter. Later, along with the German American physician and pharmacological researcher Otto Loewi (1873–1961), Dale demonstrated that acetylcholine could stimulate and affect the parasympathetic nervous system that is responsible for controlling various organs. Acetylcholine is an **alkaloid** that poisons animal tissue. Eating spoiled grain that contains this fungus can result in a serious disease called *ergotism*. The symptoms are a burning sensation in the limbs that may lead to gangrene, hallucinations, and convulsions. It has been known to cause epidemics among poorer populations who eat rotting rye grain. Outbreaks of ergotism in the Middle Ages were called St. Anthony's fire. Along with the plague and scurvy, it caused psychic epidemics, with symptoms of dancing manias and mass madness where people claimed to be possessed by the devil, often ending in the killing of Jews, children, and women who were called "witches."

Dr. Loewi identified a chemical substance extracted from the vagus nerves of frogs that he called *vagusstoffe*. Dale recognized that it was similar to acetylcholine produced by ergot, which he associated as the same chemical resulting from the electrical discharge that stimulates the nervous system. Dale hypothesized that the electrical stimulation and acetylcholine were involved in controlling the heartbeat rate of humans and the nerve responses for other organs. This discovery that acetylcholine is a chemical released from **autonomic** nerve endings led to a better understanding of the electrochemical nature of the nervous system and the development of drugs similar to acetylcholine to control heart abnormalities.

Dale also explored the physiological affects of histamines and similar substances and their effects on human allergies. His work later led to the development of antihistamines used to relieve symptoms from hay fever and similar allergies.

D'ALEMBERT'S PRINCIPLE OF FLUID DYNAMICS: Mathematics and Physics: *Jean le Rond d'Alembert* (1717–1783), France.

The sum of the differences between the generalized forces acting on a solid or fluid system and the time derived from the motion of the system along an infinitesimal displacement of the system is zero.

D'Alembert's principle, a classical mechanics and dynamic systems concept, is somewhat difficult to understand without knowledge of mathematics and the concepts of "internal force" and "internal torque," both static and in motion (as in fluids). Any moving body is subjected to internal forces and torques caused by its rotation. This can be considered as something like an extension of Newton's third law of motion (where every action has an equal and opposite reaction) and Newton's law of angular acceleration that is only measured from the center of a mass. The difference is that for

D'Alembert's principle the inertial torque can act at any point within the mass, not just from the center as for Newton's law of angular acceleration. The sum of torques (including the inertial moment and force) as applied to any point is expressed in the equation $\Sigma T = 0$, where Σ is the sum, T is the torque, and 0 is zero.

D'Alembert was the illegitimate son of a rather well-known French army officer, the chevalier Louis-Camus Destouches. After his birth, his mother abandoned him at a church. He was later sent to an orphanage. Soon after, he was adopted by a workman and his wife who raised and educated him with funds secretly provided by D'Alembert's natural father. Educated by a religious group, he rebelled against an ecclesiastical career and chose instead a career in law, graduating as an advocate in 1738. He had broad interests and explored medicine and mathematics. In 1739 he pointed out mistakes in computations and equations he found in a popular mathematics publication. In 1740 he published his work in fluid mechanics that also explained his ideas on refraction, as well as his more famous D'Alembert's paradox that describes that the drag on a solid body that is within an incompressible fluid will be zero. He made one great error in the field of statistics when he published his argument that the probability of a tossed coin landing "heads" increases for every time that it comes up tails. His system led to a belief that when gambling one should bet less as one is winning and bet more if losing. Today, it is rather common knowledge that the statistics related to probability states that the odds of an honest coin coming up heads or tails is 50% for each side of the coin for each and every time the coin is tossed. This does not mean that the results might come up several heads or tails over several tosses—but rather each toss has a 50/50 chance of ending as heads or tails.

D'Alembert made several other contributions to mathematics, mainly writing most of the mathematics and scientific articles as editor of the huge twenty-eight-volume *Denis Diderot's Encyclopedia*. He also developed a theory of partial differential equations, a theory of winds, and the harmonics of vibrations. He also published eight books on his mathematical studies.

D'Alembert was in poor health later in life and died of a urinary disease. Because early in life he had rebelled against his religious education and background, he became a life-long nonbeliever and thus was buried in a common grave.

DALTON'S LAWS AND THEORIES: Chemistry: *John Dalton* (1766–1844), England.

Dalton's law of partial pressure: *At an initial temperature, the individual gases in a mixture of gases expand equally as they approach a higher temperature.*

Another way to say this is that all gases in a mixture of gases expand equally when subjected to equal heat. Because this relationship cannot be observed directly, it was established as a viable law by Dalton's observations and calculations dealing with his study of the atmosphere, humidity, dew point, and vapor pressure. This concept that all gases behave in a similar manner under similar temperatures led to other gas laws and Dalton's theories of the atom.

Dalton's atomic theory for elements: *1) The smallest particles of all matter are atoms; 2) Atoms are indivisible particles that cannot be either created or destroyed; 3) Atoms of the same element are the same; 4) Each element has its own type of atoms; 5) Atoms of one element cannot ever be changed into atoms of another element.*

Dalton's atomic theory was based on Democritus' philosophical concepts. A main difference was that Dalton was more empirical and documented his observations. He based his ideas about the atom on concepts developed by the "gas chemists," such as Avogadro, Boyle, Charles, and Gay-Lussac.

Dalton's theory for compounds: *1) Chemical reactions occur when atoms of different elements are separated or arranged in exact whole-number combinations, and 2) compounded atoms (molecules) are formed by the joined atoms of the elements that make up the compound.*

Dalton used his observations and measurements to assert his theory of compounds. Although molecules were not yet identified, his concepts of atoms combining by weight and whole numbers remain essentially correct.

Dalton's law of definite proportions: *A specific chemical compound always contains the same elements at the same fixed proportion by weight.*

Dalton rationalized these laws were based on his theories for elements and compounds and on what was known about atomic weights at the time. The law of definite proportions led to his law of multiple proportions.

Dalton's law of multiple proportions: *When two elements form more than one compound by combining in more than one proportion by weight, the weight of one element will be in simple, integer ratios to its weight when combined in a second compound.*

This means that atoms of one element can combine in different ratios, by weight, with atoms of another element. Dalton's laws were in essence correct. The problem he had at the time his laws were formulated was that accurate atomic weights of elements were not known, nor was the concept of **valence** for atoms forming molecules. Regardless, his insight enabled him to formulate two major laws of chemistry: the laws of definite and multiple proportions.

Dalton conceived these laws from his knowledge that oxygen and carbon can form two different compounds with different proportions of oxygen and carbon. For example, CO_2 (carbon dioxide, with a 2:1 ratio of oxygen) contains twice the amount of oxygen than CO (carbon monoxide). Dalton assumed the composition and ratio of elements in all compounds would be the simplest possible. This led to a mistake when he tried to apply his law to the compound water molecule. He assumed the ratio was 1:1 for hydrogen to oxygen (HO). This error occurred because at this time in history oxygen was given the atomic weight of 7, while hydrogen was given the arbitrary weight of 1, because it was the lightest of the elements. Once water molecules were separated by electrolysis, it became obvious there was twice as much hydrogen gas (by volume) derived than oxygen gas. Therefore, the water molecule had to be composed of two molecules of hydrogen (by volume not weight) to one molecule of oxygen (2H + O) or (H_2O).

Dalton's atomic theory was a combination of old and new ways of looking at the chemical nature of the world. His concept of atomic weights and the combination of atoms by whole numbers laid the foundation for further research into the makeup of matter. His experimental approach to studying chemicals and systems provided important information for future scientists. Also, his empirical and experimental techniques and the habit of recording his results showed others how to proceed in a rational manner. Dalton used some old symbols for chemicals known from the days of ancient Greece, and he added some of his own which were later replaced by the symbols now used for the elements (*see* Figure D2).

Not all of Dalton's laws were well received until other chemists rediscovered Avogadro's theories dealing with particle relationships of gases, and Dmitri Mendeleev's Periodic Table based on the atomic weights of the then-known elements was accepted.

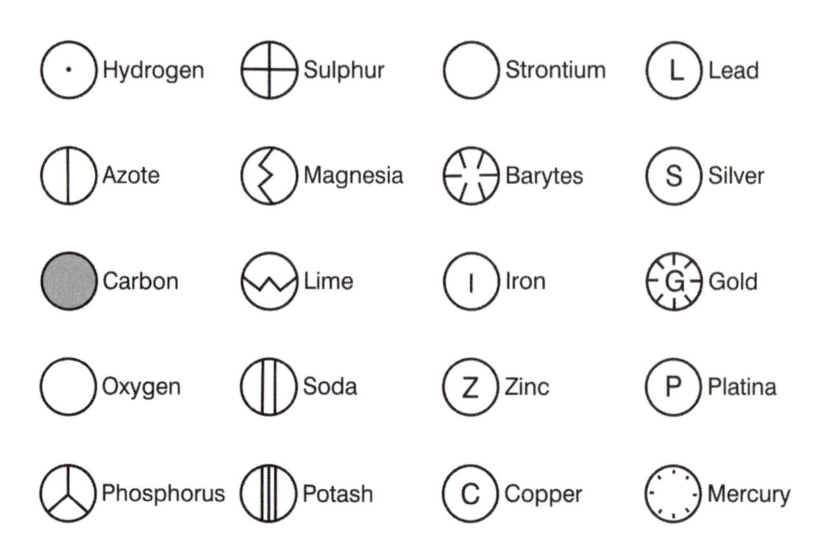

Figure D2. Ancient chemical symbols used by John Dalton.

(The Periodic Table of Elements was later revised according to the atomic *numbers* rather than the atomic *weights* of the elements.) Dalton's laws have been refined and improved over the years, but his work formed the central basis for modern chemistry. He is considered one of the fathers of modern chemistry.

See also Atomism Theories; Avogadro; Cannizzaro; Democritus; Dumas; Lavoisier; Thomson

DANA'S THEORY OF GEOSYNCLINE: Geology: *James Dwight Dana* (1813–1895), American.

> *A geosyncline is a gradual deepening of the ocean's basins that formed troughs or dips filled with sediments that were then compressed and folded to form mountain chains as the earth cooled and contracted.*

There are two kinds of geosynclines: 1) *miogeosynclines* (meaning "somewhat" like a geosyncline) are formed in shallow water at the edges of continents and are formed by sediments of sandstones, shales, and limestone that increase the thickness of continents. And 2) *eugeosynclines* (meaning "real" geosynclines) are rock formations in deeper ocean environments caused by submarine lavas from volcanoes erupting on the seafloor, and the formation of rocks such as slates, tuffs, cherts, greywackes, as well as igneous rocks formed from what are known as "plutons."

Many features of modern geology and mineralogy are built on Dana's scientific theories and concepts that are, for the most part, obsolete. These up-dated theories include midocean rifts and plate tectonics, revised concepts for the building of mountain ranges, volcanism, and similar geologic theories and concepts.

Dana was born into a middle-class family in the early 1800s. His father owned a hardware store in New York State. At a young age he had the "collecting bug" and

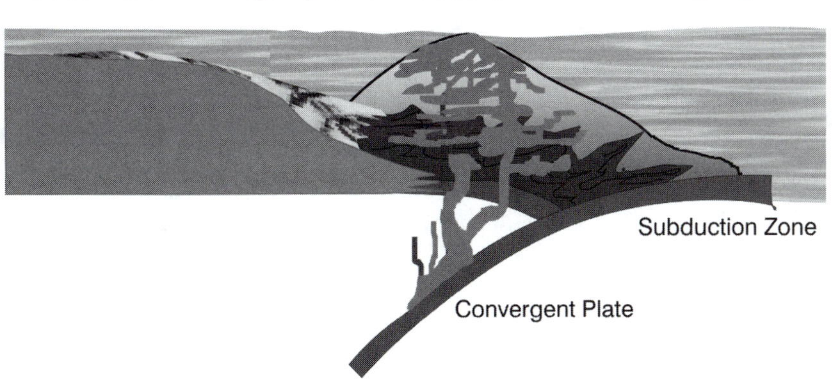

Figure D3. Examples of the two kinds of geosynclines: *Miogeosynclines* (means "like" a real geosyncline) and *Eugeosynclines* (meaning a real geosyncline).

spent much of his free time gathering rocks, insects, and plants. He attended Yale and studied under his future father-in-law Benjamin Stillman. At the age of 25 he was fortunate to be hired as one of several scientists on an expedition of six ships to explore and chart islands in the Pacific under Captain Charles Wilkes. At that time, the U.S. government needed to explore the Pacific for possible way stations for American merchant ships just as they needed to explore Antarctica for the whaling ships that ventured to this remote continent. This is one of the first examples of the government supporting a scientific venture. The expedition lasted five years and made many stops at the seemingly countless islands of the Pacific, including a visit to an active volcano of Kilauea in Hawaii. Soon after his return he married his former professor's daughter, the nineteen-year-old Henrietta Stillman.

Dana learned much on this voyage that he expanded and incorporated into the many scientific papers and books that he wrote throughout the rest of his life. His best known are *System of Mineralogy* (1837), *Manual of Mineralogy* (1848), *Manual of Geology* (1862), *Manual of Mineralogy and Lithology* (1887), *Corals and Coral Islands* (1872), and *Characteristics of Volcanoes* (1890). Several of these publications became well-known textbooks. One interesting publication in 1849 focused on the geology of the Sacramento Valley in northern California and the Umpqua River in Oregon where he mentioned the possibility of gold in these regions. It is said that this publication is partly responsible for the great gold rush of that period of history.

DANIELL'S CONCEPT OF THE ELECTRO-CHEMICAL CELL: Chemistry: *John Frederic Daniell* (1790–1845), England.

> *A two-fluid battery (cells) will produce a more reliable constant source of electricity over a longer period of time than will a single-fluid battery (cells).*

John Daniell was not the first to try to artificially produce electricity by the "wet method." Luigi Galvani thought he discovered "flowing" electricity when he hooked up a dissected frog's spinal cord and legs with a copper hook that suspended the end of one of the legs to an iron railing. He mistakenly thought he had discovered "animal electricity" because the frog's legs moved. Later, Alessandro Volta was unable to find any electricity in the legs of frogs and then went on to disprove Galvani's claim of discovering "current" electricity. Volta dipped a disk of copper and a disk of zinc into a container of salt solution that acted as an electrolyte. He reasoned that by placing alternate disks of the two metals separated by cardboard that was soaked in salt solution and then attaching a wire on each end, he could increase the flow of an electric current. This was the first battery composed of several "cells." The term "volts" refers to the "electrical potential" or "the pressure behind the electric current (amps)." The major problem with the Volta cell/battery was that the electrolyte emitted hydrogen gas that collected at the copper disk (the positive pole) and soon formed a screen-like barrier that prevented the flow of current. Improvements were made including the use of dilute sulfuric acid (H_2SO_4) as the electrolyte. However, after one use the Volta battery had to be dismantled to stop its chemical reaction.

The Daniell cells were a great improvement in the development of portable sources of small amounts of electricity. They were not only more reliable but safer. Daniell cells are also referred to as "gravity cells" and a "crowfoot cell" when a particular shape of electrodes are used. The Daniell cell's electrolytes are zinc sulfate ($ZnSO_4$) and copper sulfate ($CuSO_4$). The chemical reaction that takes place produces about 1.1 volts of electricity. There are various designs for gravity-type Daniell cells, but basically the cell is composed of a central zinc metal cathode that is placed into a porous pot containing a zinc sulfate

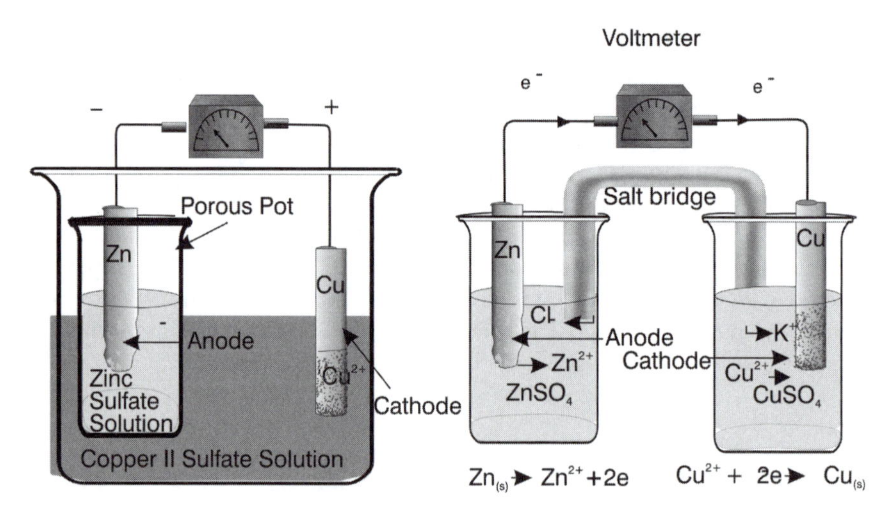

Figure D4. The Daniell cells were an improvement over voltaic cells by delivering a constant flow of current by separating the electrolytes in porous pots, thus preventing bubbles of hydrogen from forming on the electrodes.

solution. This porous pot is then placed into a solution of copper sulfate that is inside a copper container that acts as the anode. The purpose of the porous pot is for it to act as a barrier to separate the two solutions to prevent depolarization that would be caused by bubbles of hydrogen gas collecting at the anode. Unlike the Volta cell, the Daniell gravity cell can produce a continuous flow of electricity. Another version is known as the crowfoot cell that uses the difference in specific gravity (density) of the two solutions with the zinc sulfate layer over the copper sulfate solution. Each separated layer of electrolyte had its own electrode shaped like a crow's foot suspended in the solutions. Several versions of the Daniell cell were developed before the days of the "dry cell" (*see* Figure D4 for two modern versions of Daniell cells).

Although best known for his improved "wet" cell to produce electricity, John Daniell also had a career in meteorology. He invented the dew-point hygrometer known by his name, a pyrometer, and a water barometer. He also devised a way to produce gas that could be used for lighting purposes from a mixture of resin with turpentine. He wrote several articles on meteorology, climate, and chemical philosophy before his untimely death from apoplexy (stroke) at a meeting of the Royal Society in London in 1845.

See also Galvani; Volta

DARLINGTON'S THEORY OF CELL NUCLEAR DIVISIONS: Biology (Genetics): *Cyril Dean Darlington* (1903–1981), England.

> *Chromosomes in the first stages of normal cell division are divided by the process of mitosis, whereas the chromosomes are undivided in early stages of meiosis prior to gamete formation in the sex cells of male and females.*

Mitosis is the process of cell division in **soma** (body) cells whereas meiosis is the halving of the number of male and female chromosomes in their sex cells before recombining in the **zygote**. This concept was important for the understanding of how chromosomes "split" in the nuclei of cells.

Darlington is sometimes referred to as the man who discovered chromosomes. One of his contributions to the scientific theory of evolution was his discovery of exactly how chromosomes "cross over" during fertilization of the ovum (egg) cells by the sperm cells during meiosis.

A stern schoolteacher father and a firm bitter mother raised Darlington in a small town in Lancashire, England. He did not enjoy school, but as a six-foot three-inch young adult, he applied to an agricultural college, was turned down, but did receive a position as an unpaid technician. After the publication of his first scientific paper, he was hired by the John Innes Horticultural Institution of Merton, England. His mentors, the British geneticist William Bateson (1861–1926) and the British cytologist Frank Newton (d.1927) died within a year of each, and J.B.S. Haldane became head of the Innes Institute. Soon after, Haldane and Darlington became friends. As a consequence, Darlington became less hostile to authority and more self-confident in the academic environment and soon made his contributions to genetics. His first book the controversial *Recent Advances in Cytology* was published in 1932. It was not well accepted at first but later became known as containing the definitive science that described the mechanism of evolution observable at the chromosomal level. In 1937 he became director of the Innes Institute, and also a Fellow of the Royal Society.

Near the end of his career he cultivated an interest in the differences between races and how genetics might be applied to human history. He became involved in science and politics and condemned Russia for its practice of Lysenkoism over Mendelian genetics. During the last twenty-five years of his life he published a trilogy on genetics and humans, *Genetics and Man* (1953), *The Evolution of Man and Society*"(1969), and *The Little Universe of Man* (1978).

See also Haldane; Lysenko; Mendel

DARWIN'S THEORY OF EVOLUTION BY NATURAL SELECTION: Biology: *Charles Robert Darwin* (1809–1882), England.

> *Environmental pressures on organisms, such as climate and availability of natural resources such as food, act to select, by natural processes, those individuals better adapted to survive and who thus will pass viable traits related to survival to subsequent generations.*

In other words, natural selection is the force responsible for the development of advantageous traits in animals and thus plays the major role in the development of species. Modern concepts of evolution are as follows:

1. Different species of plants and animals vary in form and behavior. This variation is a basis for inherited characteristics.
2. Most species of plants and animals produce many more offspring than their environment can support.
3. Selected individuals of a species are better adapted to their environmental conditions than others and thus survive ("survival of the fittest"). Therefore, those individuals with the most favorable genes for survival will be the most "fit" for reproduction and pass these "survival" genes to offspring.
4. Thus, the genes for the most favorable characteristics are passed to future generations.
5. The "naturally selected" offspring will lead to new species by mutations and genetic changes.

Darwin was not the first to theorize about the nature of animals and how humans and animals exhibit some similar characteristics of structure, for example, four limbs, head end, tail end, head with mouth and eyes, similar internal organs, and so forth. Aristotle based his observations of plants and animals on specific anatomical differences and obvious characteristics and from these observations placed groups with common features on ascending steps of his "scala naturae" or "ladder of life" with man being on the top step (*see* Aristotle). Plato's concept of life was that each living organism possessed something that was unique to that type of organism. To Plato these "essences," although abstract, were what made all the differences in the types of organisms that were separate from other types of living things. This concept did not leave much room for the development of slow variations (evolution) or rapid changes (catastrophism) in organisms. During the Dark and Middle Ages the Roman Catholic Church based differences in living organisms on a hierarchy of God's creation where humans were below the angels while other creatures were below humans. Some adherents of religion believed this then, as well as today, as a part of God's grand design. Carolus Linneaus designed a classification or taxonomic system of organisms that grouped them by major characteristics into categories. In a sense, this was an elaboration of Aristotle's "ladder of life." Linnaeus used scientific names in a hierarchical system ranging from most general to the specific as phylum, class, order, family, genus, and species. Things began to change at the end of the Renaissance and the beginning of the period of Enlightenment with the study of Earth's geology. The age of Earth was greatly extended from the Biblical estimation of a few thousand years to a more realistic age of several billion years as determined by scientific studies. This extended period provided adequate time for slow variations within species to appear and even new species to emerge. Nicolaus Steno was the first person to realize that in a sequence of sedimentary rocks the oldest strata lay at the bottom and the younger strata are at the top. Scientists soon realized that fossils found in various strata of Earth's crust lived at different periods and that changes occurred over many years, as indicated by the depth of the layers of rock in the earth.

Darwin received a letter from Alfred Russel Wallace that outlined the same theory of natural selection as Darwin was developing. Because Darwin formulated the theory first, he is credited, along with Wallace, with the concepts of organic evolution. Darwin's book *On the Origin of Species by Means of Natural Selection* (1859) caused much debate in the general public and among religious groups. Although Darwin didn't emphasize the evolution of humans from lower forms of animals, it was the natural conclusion to be drawn from the theory. Because the concept of **genetics** and heredity was unknown to Darwin, he relied on the mistaken Lamarckian idea of inheriting acquired characteristics to explain the transfer of characteristics from parents to offspring. Many years before Darwin's time, Gregor Mendel had proposed the general concept of inherited characteristics. It was only after Darwin's death that Mendel's work was rediscovered and applied to organic evolution. Since then, the role of random genetic mutations (and RNA and DNA molecules) is better understood as to how these inheritable changes can equip living organisms to survive in their environments and thus produce more offspring with similar traits.

Even today there are substantial numbers of people who believe the development of species required some type of "intelligent design" by a supernatural designer because they believe that organic life is just too complex to have been formed and altered by the processes of evolution. To help counteract this antiscientific screed the National Academies of Science published *In Science and Creationism: A View from the National Academy of Science* in 1999 that outlined the following evidences for evolution:

1. *Paleontology*: Ancient bones and other fossils provide clues about what organisms of the past looked like and when they lived.
2. *Anatomy*: The structures of different organisms are based on similar plans, with differences reflecting their requirements for survival in different surroundings.
3. *Biogeography*: The myriad kinds of organisms on Earth and their many specialized niches are best explained by natural selection for useful adaptations.
4. *Embryology*: The embryos of very different animals, such as chickens and humans, look remarkably similar at certain stages in their development. Common ancestry seems to explain the similarities best.
5. *Molecular biology*: DNA is very similar in all organisms, especially in those that are closely related. The same twenty amino acids are used in building all kinds of

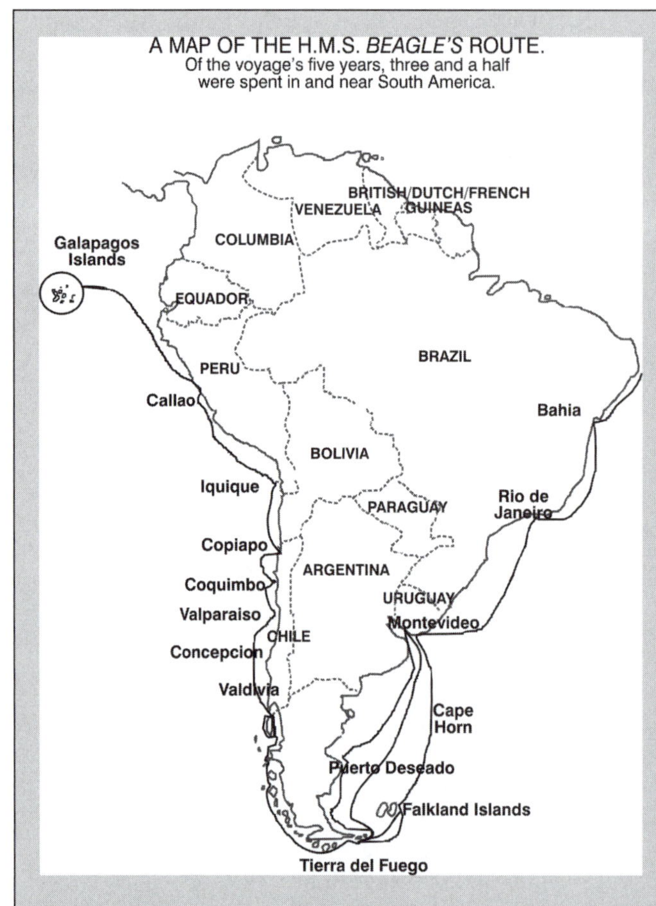

Figure D5. A map of Darwin's trip on the *H.M.S. Beagle.*

Charles Darwin, familiar with the principles of population proposed by the English demographer and political economist Thomas Malthus, also recognized that these theories are applicable to humans. Darwin was also knowledgeable of the book *Principles of Geology* by the English geologist Sir Charles Lyell who questioned the then-held concept of "catastrophic evolution" that proposed that animal species were created separately and were unchangeable forever. Lyell, as a geologist, was aware that Earth's surface changes constantly as natural forces act on it over long periods of time. Darwin's observation on the voyage aboard the *H.M S. Beagle* seemed to agree with Lyell's idea of "general uniformatarianism" where present geological events occur similar to past geological events.

Darwin based his new theory of natural selection on his years of observing the lands he visited and the great variations in geology. He associated this information with the great variety of plants and animals he collected and studied on this voyage. He spent five years, much longer than he anticipated, aboard the *Beagle* as he visited islands and coastal areas of South America. He recognized that the environment could affect the selection of individuals within species and could, over long periods of time, alter these same species, including the appearance of new characteristics and species. Darwin used the term "descent with modifications" instead of "biological evolution" as we think of it today.

life. Some stretches of DNA represent "excess baggage" carried from ancestors but not used in their descendants.

See also Dawkins; Dobzhansky; Elton; Lamarck; Linneaus; Lysenko; Mendel; Steno; Wallace

DAVISSON'S THEORY OF DIFFRACTION OF ELECTRONS: Physics: *Clinton Joseph Davisson* (1881–1958), United States. Davisson shared the 1937 Nobel Prize for Physics with George P. Thomson.

The angle of reflection of electrons from the surface of a crystal surface depends on the crystal's orientation. (In other words, electrons can be diffracted similar to light waves, i.e., photons.)

This theory was confirmed in 1927 by an experiment called the Davisson–Germer Experiment. Davisson in cooperation with the American physicist Lester Germer (1896–1971), who was a colleague at Bell Labs, built a vacuum box containing a heated filament that produced electrons that were accelerated at high voltage. They focused this electron light beam of known energy (momentum) at an angle to the surface of polished nickel metal. They were then able to determine that the reflected (or diffracted) angle of the electron beam from the nickel surface was the same as an electron's wavelength. This experiment was somewhat accidental and due, in part, to a patent suit involving Western Electric Company Laboratory, the former name of Bell Telephone Laboratory, and Irving Langmuir of the General Electric Company. An accident in Davisson's lab in 1925 resulted in the serendipitous result of light particles scattering at a particular angle. Davisson and Germer then enacted the famous 1927 experiment to confirm Davisson's earlier discovery that was used to confirm Louis de Broglie's 1924 experiment on wave/particle duality.

> Clinton Davisson was born in 1881 in Bloomington, Illinois. In 1902 after graduating from the local high school he enrolled at the University of Chicago. Due to the lack of funds he left the university and sought employment in various jobs until returning to Chicago to receive a BS degree in 1908. He entered graduate school at Princeton University and received his PhD in 1911. He sought enlistment in the U.S. Army in 1917 but was turned down. Soon after that he sought employment at the Western Electric Company, that later became Bell Telephone Labs. He spent the remainder of his career at Bell Labs experimenting with electron diffraction and how to apply his theories of the electron to engineering. After retiring in 1946 he spent a few years as visiting professor of physics at the University of Virginia in Charlottesville, Virginia. In addition to his Nobel Prize, he received the following awards in his lifetime including The Comstock Prize by the National Academy of Sciences, the Elliott Cresson Medal awarded by The Franklin Institute, as well as medals from the London Royal Society, and the University of Chicago Alumni. Davisson also held honorary doctorates from several universities.

The wave characteristics of an electron is also known as the de Broglie wave and is expressed by the equation $\lambda = \hbar/p$ where λ is the electron's wavelength, $\hbar$ is Planck's constant, and p is the electron's momentum (energy). Later Albert Einstein confirmed the particle nature of light waves (photons). The particle–wave duality of electrons became a debate of exactly what is the nature of electromagnetism and the underlying ideas of quantum mechanics. Davisson, de Broglie, and others believed in the concept of determinism in physics and that there was and is still more to be learned about the fundamental nature of matter. Later, Einstein confirmed the particle nature of light waves (photons). The particle–wave duality of electrons became a debate of concerning the exact nature of electromagnetism and the underlying ideas of quantum mechanics.

DAVY'S CONCEPT THAT ELECTRIC CURRENT CAN BE USED TO SEPARATE ELEMENTS: Chemistry: *Sir Humphry Davy* (1778–1829), England.

The process of electrolysis can be used to separate alkaline earth metals from their mineral ores.

Humphry Davy was familiar with the experiments that produced electricity by electrochemical experiments conducted in the past, especially Galvani's frog tissue, the Voltaic pile, and Daniell's gravity cell. It was well known in the early 1800s that water could be separated into hydrogen and oxygen gases by using Volta's pile, and that some solutions of salts could be decomposed in a similar manner. Davy's concept was based on the idea that electrical forces held together the different elements found in chemical compounds. In 1806 Davy gave a lecture on this topic titled "On Some Chemical Agencies of Electricity." Using this concept he theorized that passing a current through molten oxides of metals would decompose these substances. His first success was when he passed a current through a mixture of molten potash and sodium chloride (salt). When he placed several small pieces of metallic potassium metal in water, a gas was produced. The gas from these particles ignited with a bluish flame as the potassium earth metal reacted with the water. The gas was hydrogen that was ignited by the heat of the reaction of the potassium with the water ($H_2O + 2K \rightarrow K_2O + H_2\uparrow$). From there Davy went on to discover magnesium, strontium, and boron.

Several chemists of the day considered oxygen to be a major ingredient of all acids. Davy experimented with what was known as "oxymuriatic acid" which was thought to be composed of oxygen. Davy created a reaction of chlorine with ammonia that produced only muriatic acid (hydrochloric acid) and nitrogen gas, but not oxygen ($3Cl_2 + 2NH_3 \rightarrow 6HCl + N_2$). Later, he also proved that muriatic acid (also known as marine acid) was only composed of the elements chlorine and hydrogen, with no oxygen.

Davy taught himself chemistry and became an expert in applying the scientific method when performing many experiments. He was considered an excellent "qualitative" chemist but lacked the training as a "quantitative" chemist, meaning he was an excellent experimenter with some good ideas, but a bit unreliable with his measurements. He experimented with oxides of nitrogen as well as other gases. His data on the oxides of nitrogen (NO_2; NO, and N_2O) was used to confirm John Dalton's atomic theory that atoms combine in whole numbers when forming compound molecules. He was famous for his lectures and demonstrations that became popular social events. The demonstrations where he and other chemists demonstrated the effects of nitrous oxide on members of the audience became very popular. Davy himself had inhaled nitrous oxide gas (known as laughing gas) and became intoxicated. Davy became temporarily blinded when one of his experiments exploded. He hired an assistant, Michael Faraday, who had attended most of Davy's lecturers and showed an interest in Davy's research. As Davy's health failed from inhaling the gases he produced, Faraday became a valued assistant and friend. Faraday went on to become a successful experimenter in his own right. Davy was responsible for developing many scientific ideas and devices. One of his major accomplishments was the invention of the miners' safety lamp that could be used in mines without causing explosions by igniting mixtures of coal gas and air. He demonstrated that when two pieces of ice or any substance with a low melting point were rubbed together, they would melt without the addition of heat. This led to a better understanding of heat that eventually became known as the atomic kinetic theory rather than the then accepted "caloric theory" of heat. He also developed a method for using a gel of silver nitrate on a sheet of glass to produce a picture as a negative image. This was before tintypes were popular (*see* Daguerre). Davy developed a new chemistry approach to the study and practice of agriculture that was based on experimentation. He applied his knowledge of chemistry to develop an improved method for tanning leather and used his knowledge of electrochemistry to protect copper-bottomed ships by placing zinc plates on the copper. Toward the end of his life Davy became jealous of Faraday and twice opposed his membership in the Royal Society. However, in 1824 Faraday was finally made a Fellow of the Royal Society. Davy, in poor health from breathing his experimental gases, retired to Rome. He died of a heart attack in Geneva, Switzerland, in 1829.

DAWKINS' THEORY OF EVOLUTION: Biology: *Richard Dawkins* (1941–), England.

> *Hierarchical reductionalism occurs in genes and the DNA molecules, which are the basic units of natural selection responsible for the evolution of organisms.*

Richard Dawkins applied knowledge of genes and heredity to Darwin's theory of organic evolution. The genetic and molecular materials in the DNA base pairs of **nucleotides** are the fundamental units of natural selection. Dawkins refers to them as the "replicators," while the entire organism is the "vehicle" containing the genetic DNA "replicators." In his book *The Selfish Gene* (1976), Dawkins described his theory by stating that only the genes and molecules of DNA are important for natural selection to maintain the species. The individual organism is just a means of maintaining and replicating the DNA. How successful the species is depends on how well the replicating factors build the vehicles (bodies of plants and animals) that "store" and "reproduce" the DNA genetic material through natural selection. Dawkins expands his theory to include a form of sociobiology or "social Darwinism" he refers to as "kin selection" rather than individual selection as being responsible for the behavior that results in passing on the organisms' genes. He claims that kin selection explains many social and altruistic behaviors of some organisms. To expand and define this concept, he coined the term "**meme**" as the unit for cultural or social inheritance, with memes responsible for the evolution of ideas through natural (human) selection. Memes are also regulated by evolutionary processes in the sense that families, tribes, and social, and cultural groups create human environments that evolve with their culture. Memes, as units of cultural inheritance, evolve just as genetic material, through the process of natural selection. Dawkins also contends that current living organisms, including humans, are random "accidents." Dawkins' basic idea of evolution states that by following a few rules of physics and starting at very simple points (energy, **amino acids**, self-replicating organic molecules, etc.), life can evolve. Thus, under natural conditions, a variety of complex organisms and their cultures can evolve, but not necessarily in any one given direction. Dawkins believes no supreme being is required to start or direct the process.

There has been, and still is, much controversy over Dawkins' concept of hierarchical reductionism as applied to evolution. It is usually used by physicists to explain the structure and behavior of atoms in terms of elementary particles, molecules in terms of atoms, and so on, up the ladder to the structure and behavior of living cells as related to their component atoms and molecules. Sociobiology continues the hierarchical model to include not only the structure but also the behavior of humans and what species might follow humans based on the most elementary of quanta of energy and matter. Note that it is system-involving feedback. In other words, hierarchical reductionism also states that small, individual parts made up of differentiated cells and tissues evolve into an entire organism whose structure as well as behavior (culture, society, psychology) is expressed in a hierarchical sense in terms of the most basic particles such as **quarks** and **leptons**, **superstrings**, **membranes**, and finally energy.

In addition to *The Selfish Gene* (1976) Dawkins has published *The Extended Phenotype* (1982), *The Blind Watchmaker* (1986), *River out of Eden* (1995), and a number of scientific papers. More recently he wrote *The God Delusion* (2006). He is the professor for the understanding of science at Oxford University.

See also Darwin; De Vries; Dobzhansky; Wallace

DE BEER'S GERM-LAYER THEORY: Biology (zoology): *Sir Gavin Rylands De Beer* (1899–1972), England.

> *The development of animal cartilage and bone cells originates in the ectoderm of animal embryos.*

Up to this time, the accepted germ-layer theory stated that bone and cartilage cells were formed in the mesoderm (the middle layer of tissue) rather than the ectoderm, the outer layer of embryonic tissue. De Beer's theory contributed to the knowledge of how the vertebrae are developed in reptiles, birds, and mammals. Recent research used genetically engineered cells implanted in chicken embryos to produce a protein that determines the bone structure of a bird's wing. This led to the knowledge that similar genes shape the human arm, as well as the general skeletal structure and organs of all animals, including humans. De Beer also demolished Haeckel's law of recapitulation (also known as the biogenic law), which states that ontogeny (embryo development) recapitulates the phylogeny (evolutionary history) for each individual. In other words, each embryo undergoes all the stages of development that resemble all the stages of ancestral evolution of that organism's species. The law of recapitulation is an oversimplification of embryology as well as evolution and is no longer considered viable. Instead, De Beer framed his argument on the concept of *pedomorphosis*, which is the evolutionary retention of some youthful characteristics by adults. He also proposed the concept of *gerontomorphosis* where juvenile tissues are somewhat undifferentiated and are able to undergo further evolution, while more highly specialized tissues are not as likely to change through evolution. De Beer also came up with an explanation called *clandestine evolution* based on evidence of sudden evolutionary changes in fossil records, rather than gradual transformation as proposed in Darwin's concept of evolution.

De Beer used his study of the fossils of *Archaeopteryx*, the earliest-known prehistoric bird, to account for the obvious gaps in evolutionary changes in animals. According to De Beer, these "gaps" are due to the nonsurvival soft tissues in the fossils of animals. He referred to this as "piecemeal evolution changes" that explained the similarities between structures such as teeth, and feathers for reptiles and birds. He expressed his ideas by the statement that *each ontogeny is a fresh creation to which the ancestors contribute only the internal factors by means of heredity*.

Between 1924 and 1972 De Beer authored 17 books ranging from growth, embryology, zoology, evolution and Darwin, and unsolved problems in homology. After his retirement, De Beer moved to Switzerland where he published his work on Darwin and completed his massive *Atlas of Evolution*.

See also Darwin; Haeckel; Linnaeus; Wallace

DE BROGLIE'S WAVE THEORY OF MATTER: Physics: *Prince Louis Victor Pierre Raymond de Broglie* (1892–1987), France. De Broglie was awarded the 1929 Nobel Prize for Physics for his discovery that electrons exhibit wave characteristics.

> *A particle of matter with momentum (mass × velocity) behaves like a wave when the wavelength is expressed as* $\lambda = \hbar/p$. *(*λ *= wavelength,* $\hbar$ *= Planck's constant, and p is the particle's momentum.)*

In the *macro* world (large masses), when a body, such as an automobile is moving (the p in the formula), its energy or momentum is very great. Therefore, because Planck's constant ($\hbar$) is a very small number, and thus the value of the wavelength (λ) is so small, the wavelength behavior of the automobile cannot be discerned. However, in the nano or submicro world (very small particles of mass), such as electrons and protons, the particle will have little momentum (due to miniscule mass but great velocities), and therefore the particle's wavelength is easily detected and measured.

The effects of interference demonstrate the evidence for this theory (*see* Figure Y1 under Young). If a beam of submicroscopic particles is divided into two parts as it passes through two slits in a screen (diffraction), the small particles with minimal mass will arrive at different points on a target screen. The results can be measured, and the results are the same as they are for a similar experiment done with light photons and waves. The characteristics of constructive interference resulting from the split screen for the particles of matter are the same as the characteristics of wave motion. Two other scientists experimentally determined the wave-like behavior of minute particles. George Thomson and Clinton J. Davisson independently discovered the same principle. De Broglie's wave theory of matter supported and helped Erwin Schrödinger explain the theories of relativity, as well as wave and quantum mechanics.

See also Bohr; Born; Davisson; Einstein; Heisenberg; Schrödinger; Young

DEBYE–HÜCKEL THEORY OF ELECTROLYTES: Chemistry: *Peter Joseph William Debye* (1884–1966), Netherlands/United States, and *Erich Armand Arthur Joseph Hückel* (1896–1980), Germany.

> *In concentrated solutions, as well as dilute solutions, ions of one charge will attract other ions of opposite charge.*

Up to this time, the Arrhenius theory of ionic conductivity was correct only for very dilute solutions (*see* Figure A7 under Arrhenius). This theory, formulated in 1923, initiated the use of **electrolysis** for the separation of ions in very concentrated solutions (e.g., brine), for the extraction of sodium and chlorine, and led to the industrial production of gases, such as bromine, fluorine, and chlorine, as well as the extraction of some metals from their ores. Using X-ray diffraction, Debye and his assistant, Erich Hückel, determined the degree of the polarity of covalent bonds and the spatial structures of molecules, which disproved earlier theories of conductivity in strong electrolytes.

Peter Debye was an electrical engineer who received a PhD in 1910 in physics at Zurich University. This was followed by a series of administrative positions in universities and institutes in Germany and Switzerland. He moved to the United States in 1940 where he became a professor of chemistry at Cornell University in New York State. He became an American citizen in 1946. Throughout his career he was essentially a theoretician who considered varied problems related to the structure of matter, mainly how to apply experimental physics to solving problems of molecular structure. His early work involved the electron diffraction by gases, the formation of X-ray diffraction patterns from molecular substances, how to determine the degree of polarity and angles of covalent bonds in molecules, and how to use X-ray spectra to determine the spatial configuration of molecules. But his most famous work was the 1923 Debye-Hückel theory of electrolysis.

See also Arrhenius; Hückel

DEHMELT'S ELECTRON TRAP: Physics: *Hans George Dehmelt* (1922–), United States. Dehmelt shared the 1989 Nobel Prize for Physics with Wolfgang Paul and Norman Foster Ramsey.

> *By isolating an electron in an electromagnetic field, it is possible to suspend it, thus providing a means of continuously and accurately measuring its characteristics.*

Hans Dehmelt constructed what was then known as the *penning trap*, a combination of a strong magnetic field combined with an electric field. Both fields are contained in a vacuum inside a closed unit containing two negatively charged electrodes and one positive electrode that are used to isolate and suspend a single electron so it could be studied. He accomplished this by reducing the kinetic energy (motion) of the electron by cooling it, enabling him to measure the suspended single electron accurately. Dehmelt and his colleagues were also the first to isolate and detect individual protons, antiprotons, positrons (positive electrons), and ions of some metals. The penning trap could be used to analyze particles that had either negative or positive charges. When light was shone on a suspended single metal ion, it could be seen without the aid of instruments and appeared as a very small, bright, star-like light.

See also Ernst; Franck; Heisenberg; Hertz; Pauli; Ramsey; Rutherford; Stern; Thomson

Dehmelt and his assistants were the first to view what is known as the *quantum leap,* a very, very small "bit" of energy. This occurred when a single electron of a barium atom "absorbed" extra energy and was forced to a higher energy orbit. When the energized electron "jumped" back to its normal lower energy state (orbit), it emitted the tiny bit of energy that it had gained when forced to the higher energy orbit. This jump of the electron to a lower energy state is evident by the release of a quantum bit of energy that is detected as a photon (a particle of light). The media and others often refer incorrectly to a large increase in something as a "quantum leap," meaning something big. A quantum leap is really a very, very small amount of energy that becomes visible when an energized electron jumps to a lower energy orbit (*see* Figure D6). Dehmelt's work led to confirmation of quantum mechanics theory and a better understanding of the physics of subatomic particles.

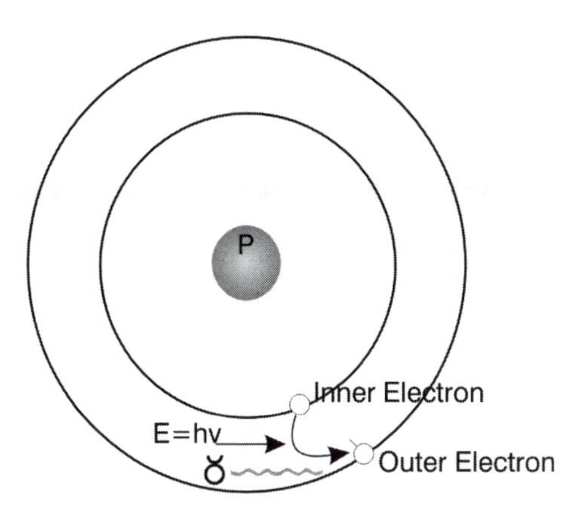

Figure D6. The quantum leap is based on Niels Bohr's idea of electrons being similar to planets orbiting the sun and with specific energy levels. The ''leap'' occurs when a electron, which cannot revolve around the nucleus in just any orbit (energy level), jumps from an inner orbit (higher energy level) to an outer, lower-energy-level-orbit, emitting a photon (a tiny packet/quanta of light). The energy level of the photon is equal to the difference in the energy levels of the two orbits of the electron, which is expressed as Planck's hypothesis $E = \hbar v$ where $\hbar$ is Planck's constant and v is the angular momentum of the electron.

DELBRÜCK'S AND LURIA'S PHAGE THEORY: Molecular Biology: *Max Delbrück* (1906–1981) and *Salvador Edward Luria* (1912–1991). Delbrück was born in Germany, Luria in Italy, but both moved to the United States. Along with Alfred Hershey (1908–1997) from the United States, they shared the 1969 Nobel Prize for Physiology or Medicine.

The phage theory states: *Bacteria develop resistance to phages by spontaneously mutating.*

Phage, a Greek word meaning "devour" or "eat," is the simplest genetic system known. Phages are simple viruses composed of plain strands of nucleic acid with a more complex "head" that contains DNA material. The phage that infects bacteria is referred to as *bacteriophage*.

Delbrück's and Luria's research sought to ascertain how phage could multiply so rapidly in bacteria—up to one hundred phage particles are produced in just a few minutes. Delbrück did the mathematical and statistical work on the problem, while Luria conducted the experiments. Along with Alfred Hershey, they investigated and determined the genome of phage virus. The **genome** is *the entire* DNA, including the DNA in the genes that are contained in the structure of an organism. Delbrück and Luria demonstrated that the phage virus inserts itself into the host bacteria and replaces the bacteria's DNA with the phage's own DNA, in essence cloning itself and resulting in mutation of the bacteria. Delbrück and Luria published "Mutations of Bacteria from Virus Sensitivity to Virus Resistance" in 1943 in which they speculated that bacteria could acquire resistance to lethal phages. They realized that the growth of bacteria would be different for each kind of bacteria within a colony of bacteria because in such numbers a resistant strain would be found. This led to their discovery that bacteria underwent mutations, just like other organisms. They also determined some genetically mutated bacteria develop a resistance to the destructive bacteriophage. The three Nobel Prize winners are credited with founding the field of molecular biology.

See also D'Herelle; Northrop

BACTERIOPHAGE

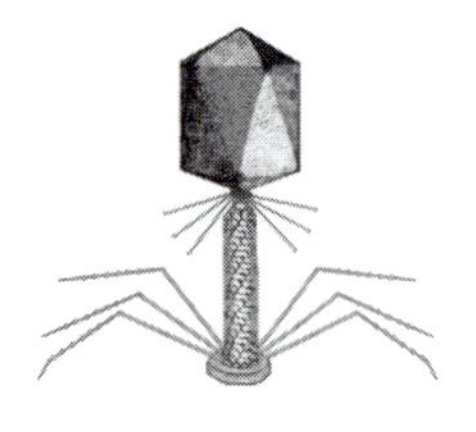

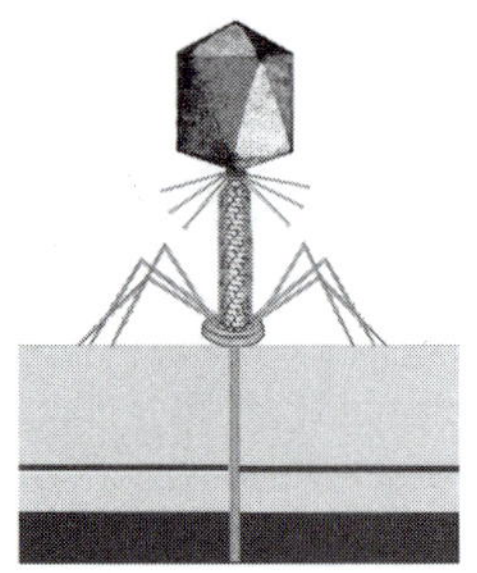

Figure D7. An artist's conception of a typical T4 bacteriophage virus that infects bacteria cells. It consists of two major parts, nucleic acid (RNA or DNA) in the "head" and a coating (capsid) made of protein to protect the nucleic acid. A specific bacteriophage infects only a limited range of bacteria species.

DEMOCRITUS' ATOMIC THEORY OF MATTER: Chemistry: *Democritus of Abdera* (c. 460–370 BCE), Greece. *see* Atomism theories.

DESCARTES' THEORIES AND PHILOSOPHY: Philosophy: *René du Perron Descartes* (1596–1650), France.

Descartes' theory of light and reflection: *White light is pristine light that can be reflected at an angle equal to the angle at which it strikes a mirrored surface.*

Descartes believed that pristine (white) light produced colored light only when there was a "spinning" sphere of light. He also stated that the angle of incidence is the angle formed between the perpendicular to the surface and the ray of light striking the surface. This was a simple way of explaining the law of reflection because the angle of incidence is equal to the angle of reflection, which was later developed and became known as Snell's law. Descartes' concept of light and matter was more metaphysical than empirical.

Descartes' concept of motion: *The force of motion is the product of mass times velocity.*

Today we refer to his force of motion as *momentum* (mass × velocity). This concept was accepted by Newton but opposed by Leibniz. Leibniz incorrectly assigned the force of a body as the product of mass times the velocity squared, which later developed into the idea of kinetic energy. This dispute led to conflicts in different schools of philosophy.

Descartes' mathematical concepts: *1) a systematic approach to analytic geometry (combining algebraic and geometric functions); 2) the invention of exponential notation; 3) a rule for determining the positive and negative roots for algebraic equations;* and *4) the development of equations to describe specific curved lines and curved surface area.* All were important to the development of calculus, a branch of mathematics that combines differential and integral functions used to determine the area within a curved surface.

Descartes was the first to relate motion to geometric fields. He saw a curve as being described by a moving point. As the moving point describes the curve, its distance from two fixed axes will vary according to that particular curve. This is known as the Cartesian coordinate system, which lends itself to the study of geometry by algebra as a means of interpreting graphs with the x (horizontal) and y (vertical) coordinates that compare two variables. Graphs have become a major geometric and analytic tool for all sciences. All of these concepts contributed to later developments by other mathematicians.

Descartes' philosophy and system of knowledge: *1) Nothing is true until a foundation (evidence) has been established for believing it to be true. 2) Start with* a priori *assumptions (first principles), proceeding by mathematics to deductions that use physics and mathematics. One cannot have complete knowledge of nature because there is always doubt; one is never certain of the nature of nature. In addition, reason deceives us. Therefore, a person can be certain only of doubting (not knowing for sure).* Note: These philosophical ideas of Descartes are also the basis of the modern scientific approach to research and knowledge.

These philosophical concepts led to his famous phrase, which summed up Descartes' total philosophy as "I think, therefore I am," which he asserted as justification of self and God. Descartes' greatest contributions were in mathematics and philosophy, and he believed there was great unity in the study of these two sciences. His application of algebraic methods to geometry was a major step in the progress of mathematics as well as other sciences.

See also Euclid; Leibniz; Newton

DE VRIES' "PANGENES" THEORY OF EVOLUTION: Biology: *Hugo De Vries* (1848–1935), Netherlands.

> *1) Organisms consist of large groupings of physical characteristics. 2) Each characteristic is attached to particles inside the nuclei of cells. 3) Although invisible, these hypothetical particles were referred to as "**pangenes**" by Darwin, which he theorized as units controlling heredity. De Vries equated them to the concept of chemical atoms. 4) Theoretically, when pangenes combine, they cause the appearance of unique characteristics for each species. 5) The more pronounced a particular character or feature of a species, the greater are the number of pangenes. 6) Pangenes may be dominant or latent (recessive and not visible in the organism). 7) During **meiosis**, pangenes may split, each causing new features or possibly new species. 8) When splitting or "mutating," pangenes are responsible for the formation of new pangenes, which produce new "mutants."*

Hugo De Vries rejected the idea that natural selection over long periods of time could produce new species and/or variations in existing species, as Darwin's theory of evolution proposed. De Vries believed new characteristics or species could come about only by genetic changes (mutations) in cell nuclei or organisms. His theories were not all correct, but they did explain that mutants have extra sets of **chromosomes** in their nuclei (triploids and tetraploids). Although the term "pangenes" is no longer accurate or used, De Vries' explanations give a reasonable account of how variations, later determined to be due to genetic mutations and natural selection, occur in the evolution of species.

See also Darwin; Dawkins; Dobzhansky; Lamarck; Mendel

DEWAR'S CONCEPT OF LIQUEFYING GASES: Physics and Chemistry: *Sir James Dewar* (1842–1923), England.

> *Increasing the pressure while reducing the temperature of a contained gas, and then rapidly releasing the pressure, will cause the gas to be liquefied.*

James Dewar was a chemist and physicist who worked in several fields but is best known for his work with low-temperature applications. In one sense, Dewar was more of a

James Dewar is known not only for his work on liquefying gases, but also for devising a special container for storing and transporting liquid gases. This container is known as the "Dewar flask" and is constructed as a metal or glass bottle-within-a-bottle that has a silver coating on the inside and outside.

The thin-walled flask is a double-layered jar with one container inside the other and a dead air space between them, usually with most of the air removed so that there is a vacuum seal between the two containers. The only opening is at the neck of the bottle with a plug device designed to release the pressure as the liquid gas slowly boils forming a gas under pressure. The reflective surface, particularly on the inside of the inner container is important to prevent heat radiation. When the Dewar flask is to be used to transport liquefied gases, the outside container is usually made of metal with a reflective surface.

The more common form of the Dewar flask is known as a Thermos bottle used to keep hot liquids hot and cold liquids cold. The name "Thermos" is a trademark for three independent companies, one in the United States, one in England, and one in Canada. German glassblowers manufactured small Dewar flasks and held a contest to name the device. The name "Thermos" won. It is derived from the Greek word *therme,* meaning "heat." Today, it is one of the most widely known and popular devices used from the Arctic and Antarctic regions to the equator.

scientific inventor who discovered some important and practical processes using known physical and chemical phenomena.

James Dewar was born in Scotland and educated at the University of Edinburgh in Scotland, where in 1869 he taught chemistry at that institution's Royal Veterinary College. He later became professor of experimental natural philosophy at the University of Cambridge, England, in 1875, and in 1877 he became a professor of chemistry at the Royal Institution in London where he did most of his experimental work. He used a known apparatus to successfully liquefy oxygen in commercial amounts but had no way to store oxygen in liquid form because it would quickly "boil" into gaseous oxygen. This problem led to his invention of the "Dewar flask" which is a specialized bottle to store liquefied gases (*see* the sidebar for details).

Some of Dewar's other accomplishments are:

1. Dewar is best known for his scientific work of the nature of matter at near absolute zero degrees temperature (0 kelvin [K] or −273°C). He was successful in lowering temperatures to 14 kelvin. It was some years later when helium gas was liquefied at 2.2 K that a lower temperature was achieved.
2. He developed the chemical formula for benzene (C_6H_6) through his long-time use of the spectroscope (*see* Kekule).
3. In 1889 Dewar in cooperation with the English chemist Sir Frederick Augustus Abel (1827–1902), invented smokeless gunpowder, now known as cordite.
4. He identified the specific heat of hydrogen gas and became the first person to liquefy hydrogen gas in 1898, and to produce solid hydrogen in 1899.
5. In 1905 he perfected a way to cool charcoal that can be used to create a high vacuum in an enclosed environment. This discovery has been useful in the study of atomic physics.

D'HERELLE'S BACTERIOLYTIC THEORY: Biology: *Felix d'Herelle* (1873–1949), Canada and France.

When bacteria are infected by viruses and the resulting fluids are filtered, the filtrate contains no live bacteria, but this filtered fluid can still infect bacteria. Because the infection agents are smaller than bacteria, they are thus "filterable viruses" that can still act as parasites on live bacteria.

After attending school in Paris, Felix d'Herelle returned to Canada, his birthplace, where he enrolled and graduated from the medical school of the University of Montreal. He started his work in bacteriology in Mexico in 1907 but returned to France in 1911 to accept a position at the Pasteur Institute. From there he moved to the Netherlands to the University of Leiden as a temporary professor, followed by a stint as a bacteriologist in Egypt, and ending up as the chairman of the Department of Bacteriology at Yale University where he remained until 1933.

Throughout his career d'Herelle traveled extensively to Argentina, Guatemala, Indochina, India, and in 1934 to Soviet Georgia. He wrote *The Bacteriophage and Phenomenon of Recovery* while working at the Tbilisi Institute. He dedicated his book to Joseph Stalin. However, due to Stalin's tyrannical rule, d'Herelle was forced to flee the country just as World War II began. He and his family returned to Paris. He was nominated for the Nobel Prize a total of 30 times but never won.

In the early 1900s while working in the Yucatan in Mexico Felix d'Herelle discovered the bacteriophage, which is a type of virus that infects and destroys bacteria (*Virus* is the Latin word for "poison." *See* Figure D7 under Delbrück). After filtering this virus from a mixture of infected bacteria, he discovered that the filtrate could still infect and kill bacteria. He referred to this filterable virus as a "bacteriolytic agent." He continued his work with cultures of dysentery bacilli and found that a filterable parasitic virus also kills these types of bacilli. In 1921 he wrote a monograph "The Bacteriophage, Its Role in Immunity" where he advocated using bacteriophages to treat diseases such as cholera and bubonic plague. Although he claimed some successes in India and Egypt, others using his techniques could not achieve the same level of successful treatments. His experiments and conclusions were not well accepted by those investigating bacterial-viral problems because it was not understood how, once filtered, the virus could still infect bacteria. The term "bacteriophages" and the importance of the filterable viruses in the field of bacteriology later became significant under Delbrück's phage group of researchers.

See also Delbrück; Northrop; Pasteur

DICKE'S THEORY OF THE BIG BANG: Physics: *Robert Henry Dicke* (1916–1997), United States.

The universe started about eighteen billion years ago (Note: the current estimate is thirteen to fifteen billion years as the age of the universe) as a tiny point of energy (singularity) "exploding" in an ever-increasing three-dimensional spread of energy at extremely high temperatures that soon formed elementary matter and radiation.

The big bang theory proposes that the formation of the universe originated with an infinitely dense pinpoint of compressed energy that, at the first fraction of a second of expansion, formed hydrogen and helium and, later, other matter which, under the influence of gravity, became stars and the planets and other objects found in the universe. This "explosion" was the source of all the energy and mass in the universe and was forceful enough to overcome the gravity of the expanding particles that are still expanding after billions of years.

The big bang theory is explained by considering a number of related physical phenomena:

1. A cataclysmic explosion occurred thirteen to sixteen billion years ago.
2. Within a few seconds the temperature reached about 10 billion °C.
3. This high temperature caused intense radiation to spread outward in all directions.
4. Soon after the initial explosion, particles of elementary matter were formed and radiated outward.
5. As the explosion expanded, it lost energy through both heat and radiation.
6. **Cosmology** confirms that this expansion is still taking place, and possibly accelerating.
7. Although it took place billions of years ago, remnants of this radiation should still be detectable.
8. If such residual radiation exists today, it should be detectable as "black body" radiation—radiation, such as X-rays from deep space, emitted from a black body at a fixed temperature.

These eight phenomena provide the basis for the origin of the big bang as a condensed, very small, extremely hot mass of energy. Several different wavelengths of radiation, in addition to visible light, have been detected. For example, radio waves, X-rays, and microwaves have been detected as originating from deep space and are used to further our knowledge of the cosmos and support of the big bang theory.

Calculations determined that the residual temperature from the big bang in the universe is about 3 kelvin (−270°C), and such radiation was later detected as background microwaves of about 7 cm. These factors led cosmologists to conclude the big bang theory is correct. The question remaining is: What was there before the big bang, and where did it all come from? This is more of a metaphysical question than a part of the theory that assumes infinity in time as well as space in both directions—before and after the event—just as it is possible to calculate infinity as negative and positive numbers.

See also Gamow; Hoyle; Weinberg

DIESEL'S CONCEPT OF AN INTERNAL COMBUSTION ENGINE: Physics (Engineering): *Rudolf Christian Carl Diesel* (1858–1913), Germany.

By using the compressed heat inside an engine's cylinder to ignite the fuel and drive the piston of an engine rather than using an external source of heat as in a steam engine, greater efficiency can be achieved.

Born in Paris, Rudolf Diesel and his parents, who were Bavarian immigrants, relocated to London during the Franco-Prussian war of 1870. Soon after, he moved back to Germany where he received a scholarship to the Munich Institute of Technology. Upon graduation, he worked in Switzerland and Paris as a thermal and refrigeration engineer. Diesel was not only an engineer who understood the theory of thermodynamics but was also a linguist who spoke three languages. He was also interested in the technology of the engine. This interest was partially responsible for his work to devise a more efficient source of power to reduce the size of engines and the cost of fuels, thus enabling the smaller independent craftsman to compete with larger corporations. After thirteen years of working on a design for a compression engine, he developed a model that consisted of a 10-foot-long single cylinder that had a rotating flywheel at one end. This was the first working model of a diesel compression engine. After receiving a patent in 1893 for his engine, he continued to make improvements to the engine and spent much of his time attempting to produce it commercially. It was known as the *diesel cycle compression engine* that could use almost any kind of combustible liquid as fuel, even powdered coal dust. In 1900 he demonstrated his engine at the World's Fair using peanut oil for fuel.

By the end of the nineteenth century Rudolf Diesel was a multimillionaire from selling franchises for the production and sales of his engine. It became widely used in all kinds of industries and production plants. As the diesel engine starts to run, it compresses the incoming air and atomizes fuel, raising the temperature in the cylinder to the point where the combustible mixture explodes, thus rapidly and forcibly expanding the hot gas driving the piston outward to suck in a new mixture of air and fuel. This is an excellent example of the *combined gas law* as well as the *law of thermodynamics*. There were two problems with the early diesel engine's use in automobiles. First was the need for a governor to control the speed of the engine's cycles of intake-compression-combustion-expansion-exhaust. The other factor was that most diesel fuels thicken when cold, and this higher viscosity does not enable fuel to flow freely to the engine. Heating the fuel tanks, lines, cylinder blocks, or reformatting the fuels to maintain fluidity at low temperatures has solved this problem. Many users of large diesel construction equipment in the Arctic allow the diesel engines to run, sometimes continuously, between periods of use.

The size and weight of the modern diesel engine has been reduced as the efficiency has been increased to about 80% (whereas the steam engine is only about 10% efficient). There are basically two kinds of diesel engines: the two-stroke and the four-stroke varieties. Most engines are of the four-stroke, six-cylinder-in-line type of engines. There are also four-cylinder and eight-cylinder versions. Greater efficiency is partly achieved due to the high internal temperature of about 700 to 900° Celsius that increases the compression rate to 25:1 (as compared to about a 10:1 ratio for a gasoline engine). Automobiles obtain about 40% more miles per gallon with cars using turbo-diesel engines. Modern diesel engines are also used in trains, buses, heavy equipment, and to power manufacturing industries as well as in automobiles and trucks. New fuel mixtures and fuel injection systems are being developed to help reduce the pollution created by the spark-ignited and pressure/heat-ignited fuels in internal combustion engines.

Later in life Rudolf Diesel experienced financial and severe health problems and suffered from headaches and depression. In 1913 while on the London to Antwerp mail ferry, and after finishing a pleasant meal with friends, he was found missing. Later his body was spotted in the English Channel by another ship but left at sea.

See also Boyle; Carnot; Calvin; Clausius; Ideal Gas Law; Maxwell

DIRAC'S RELATIVISTIC THEORIES: Physics: *Paul Adrien Maurice Dirac* (1902–1984), United States.

Dirac's quantum mechanics: *The physics of electromagnetic radiation and matter should interact on the micro subatomic, very small scale of nature, as well as at the macro, very large scale.*

For almost three centuries, fundamental Newtonian mechanistic laws for classical physics were accepted. These laws explained the stability of atoms and molecules but could not explain the small packets of energy emitted by excited atoms. Albert Einstein developed the special and general theories of relativity, and Edwin Schrödinger developed the wave equation representation of quantum mechanics. Enrico Fermi and Paul Dirac revised and improved the concept for quantum mechanics by incorporating Einstein's calculations for relativity into the equation and making some corrections in the energy levels of "spinning" atoms which resulted in the Fermi–Dirac statistics. This statistic was designed to satisfy the Pauli exclusion principle required to account for the behavior of particles with half-integer spins.

Dirac's theory of negative energy: *Energy exists below the ground state of positive energy.*

To explain his theory, Dirac claimed that areas below ground state (normal positive charge) were already filled with negative energy, and if a light photon tried to enter this area, it would become an observable electron, leaving behind a vacant "hole" representing a potential similar to a positive charge. This theory led to a new understanding about the nature of matter (*see* Dirac's theory of antimatter, which follows). This positive electron was named a positron (^{+}e) and was confirmed by Carl Anderson in 1932. It proved to be a new way to look at the universe, which now included antimatter somewhat as a mirror image of matter. If there were exactly the same number of electrons (^{-}e) as positrons (^{+}e) in the universe, they would collide and destroy each other, resulting in energy with no matter left over. The reason that matter exists in the universe and there are more electrons than positrons is that more electrons (photons/electromagnetic radiation) were produced at the beginning of time. This theory has led to speculation of a "sister" universe based on antimatter (*see also* Anderson).

Dirac's theory of antimatter: *1) Each negatively charged particle (called an electron) has a counterpart. 2) These counter particles are positively charged electrons called positrons. 3) A positron must always occur in conjunction with an electron (but not the other way around). 4) Their collision destroys both.* Thus, *5) electromagnetic waves (radiation) are released, producing more electrons than positrons.*

Dirac's theoretical positron was discovered later by Carl Anderson, who confirmed Dirac's concept of antimatter as applying to all matter. Dirac's theory was an explanation of the dual nature of light where in some instances light behaves as a wave (as indicated by interference and diffraction). In other cases, light resembles particles called **photons**, which are matter similar to electrons with measurable energy, as indicated by their frequency and momentum. Thus, the concept of particle–wave duality is dissimilar to the Newtonian classical mechanics concept of matter.

Dirac's theory of "large number coincidences": *There is a 1 to 10^{40} ratio that is constant in the universe. It is exhibited in various physical phenomena and represents a model of the universe.*

Dirac's mathematics revealed this phenomenon as a relationship between natural physical constants and the quantification or natural properties: 1) The ratio of gravitational attraction and electrostatic attraction between electrons and protons is 10^{40}, 2) Earth's radius is 10^{40} that of the electron's radius, and 3) the square root of the number of particles in the universe is 10^{40}. Dirac considered there was a universal relationship between the ratios of the radii of all objects (e.g., Earth) and all forces (e.g., gravity). He also theorized that as the universe continues to expand with age, this force ratio would not change, even when the distance between objects increases. Rather, the gravitational constant will change; thus the effects of gravity become less as time and space increase (expand). The theory that the gravitational constant is decreasing over time is no longer accepted. On the other hand, to explain the increased rate of expansion of the outermost galaxies in the universe, some cosmologists are suggesting that antigravity might exist at the horizon of the universe. This is not the same as Dirac's theory for the change in the gravitational constant, but both concepts end up with the same conclusion about the expanding universe.

See also Anderson (Carl); Born; Einstein; Fermi

DJERASSI'S THEORY FOR SYNTHETIC ORAL CONTRACEPTION: Chemistry: *Carl Djerassi* (1923–), United States.

> *Physical methods can be used to determine the structure of organic molecules leading to the translation of this knowledge into the production of synthetic organic steroids related to female sex hormones that control contraception.*

Carl Djerassi's father was a physician in Bulgaria and his mother was from Vienna, Austria. He and his parents moved to the United States during World War II in 1939 when Carl was a young teenager. Carl was an exceptional student who graduated from Kenyon College in Ohio with a BS degree at age 18 and received his PhD from the University of Wisconsin by the time he was 22 in 1945. This is the same year he became a U.S. citizen. He recalls that he admired Eleanor Roosevelt, the president's wife, and wrote her a letter stating that he needed a scholarship. Her secretary replied, and he did receive a scholarship. Following graduation he secured a job with CIBA Pharmaceuticals. Then in 1949 he joined Syntex located in Mexico City where he worked with a team that was attempting to identify and extract the steroid cortisone from plants. His team of chemists at Syntex was the first to successfully extract cortisone ($C_{21}H_{28}O_5$) from Mexican yams. Following this success, the team started work on synthesizing the steroid hormone progesterone, known as nature's contraceptive. The team did encounter one problem, however. Hormones produced from the urine, testicles, or ovaries of animals proved to be very expensive. In addition, if taken orally it would lose its effectiveness. In the 1950s Djerassi and his team were successful in the synthetic production of progesterone, which caused a huge drop in cost of the drug. To make this synthetic oral steroid more effective, Djerassi altered the structure of progesterone ($C_{21}H_{39}O_2$) by removing a methyl group (CH_3) from the formula. He also added a bit of the male hormone, testosterone, to the formula to make it more effective as a contraceptive. After some more manipulations of the formula, he received U.S. Department of Agriculture (USDA) approval in 1962 for this synthetic steroid to be used as a contraceptive. From this date the story of how the "pill" has affected lives all over the world by "liberating" women from unexpected and unplanned pregnancies is well known. In 1951 Dr. Djerassi resigned his post at Syntex but continued his career at several universities and was involved with other business interests. He developed a unique form of insect control by applying a biodegradable spray with a hormone that prevents young insects from maturing. Djerassi is a prolific writer and has published many scientific papers, five novels, including *Cantor's Dilemma,* some poetry, short stories, essays, two autobiographies, and a memoir.

DÖBEREINER'S LAW OF TRIADS: Chemistry: *Johann Wolfgang Döbereiner* (1780–1849), Germany.

When considering three elements with similar characteristics and with atomic weights within the same range, the central (middle) element will have an atomic weight that is the average of the total atomic weights of all three elements.

Johann Döbereiner was interested in chemical reactions that involved catalysts, which are chemicals that either speed up or slow down chemical reactions but are not changed or consumed in the reaction. One of his discoveries was that hydrogen gas would ignite spontaneously when passed over powdered platinum. Using this concept, he developed the Döbereiner lamp, which uses metal platinum as the catalyst. (The catalytic converter in the exhaust systems of modern automobiles uses platinum as the catalyst to convert toxic exhaust gases to less harmful compounds.) In his experiments involving catalytic actions, he observed how the atomic weights changed incrementally for elements with similar compositions. This led to his law of triads proposed in 1829. An example is the triad for the nonmetals chlorine (atomic weight 35.5), bromine (atomic weight 80), and iodine (atomic weight 127). When calculating their average atomic weight, $242.5 \div 3 = 80.8$ the approximate atomic weight of bromine, the middle element of the three examples, is attained. A second example is the triad for three metals: calcium, strontium, and barium. This triad's average atomic weight calculates as $265 \div 3 = 88.3$, which is the approximate weight of the middle element, strontium. This relationship of atomic weights to characteristics of similar elements was an important discovery that Dmitri Mendeleev used in designing his Periodic Table of the Chemical Elements (*see* Figure M5 under Mendeleev). Being aware of this phenomenon provided Mendeleev a means to leave vacant spaces in his table that could later be filled. He called yet-to-be-discovered elements *eka elements*. He predicted these eka elements would fit into specific blank spaces where the other two elements in the triad, with known atomic weights, were directly above and below the missing eka element. For example, in Group IVA (14) the yet-to-be-discovered element germanium was named eka-silicon to fill in the vacant space. His eka-predicted elements were very accurate as far as atomic weights were concerned, but they were not always correctly arranged according to specific characteristics of elements in groups. Later, his periodic table was corrected to

Johann Wolfgang Döbereiner's life and education is a good example of bringing yourself up by your "bootstraps." His family was not wealthy, and his father, a coachman, did not contribute to his early education. At fourteen years of age he was apprenticed to several apothecary (pharmacy) businesses where he developed an interest in chemistry which he pursued the rest of his life. Leopold Gmelin (1788–1853), a well-known German chemist, whom Döbereiner met in Strasbourg, encouraged him to continue his study of chemistry. Despite his limited finances, he tried starting several businesses. Although they failed, he was soon given the position as assistant professor of chemistry at the University of Jena, which is now located in Germany. His first success in the field of chemistry was his discovery that when hydrogen gas passes over a spongy-type of platinum, ignition would spontaneously take place. This phenomena of catalytic activity, which he pursued the rest of his life, led to the development of his first successful business venture—the manufacture of the Döbereiner lamp. His interest in chemistry led to his theory of the "law of triads" where the average atomic weight of three related elements will approximate the atomic weight of the middle element of the triad.

represent the elements arranged by their atomic numbers (protons) rather than atomic weights.

See also Dalton; Mendeleev; Newlands

DOBZHANSKY'S THEORY OF GENETIC DIVERSITY: Biology: *Theodosius Dobzhansky* (1900–1975), United States.

Populations with a high genetic load of debilitating genes confer an advantage to organisms by providing more versatility within changing environments.

Theodosius Dobzhansky believed **species** that have a wide variety of genes, even recessive dormant debilitating genes, will be more successful by providing the entire species with greater genetic diversity. This diversity is related to the evolution of race and species and provides more effective adaptation to changing environments. Historically, this meant that those species that survived over long periods of time were the ones with the greatest pool of genes. When these genes no longer provided an advantage in overcoming environmental conditions, natural selection contributed to their extinction over time. Dobzhansky made detailed studies of the fruit fly (*Drosophila*, that has yielded extensive genetic information) and proved his theory that there is considerable genetic variation within a population. Those species that had the greatest genetic variety survived, including lethal genes, which prove to be even more important for survival in changing environments. By all standards, the dinosaurs must have had a very wide and diverse genetic load because there were so many different types of dinosaurs that successfully survived for millions of years. Despite this, they became extinct about sixty-five million years ago. Humans arrived many millions of years after the dinosaurs' extinction. Prehuman types of beings existed about five million years ago, followed by prehistoric humans, and later modern humans (*Homo sapiens sapiens*), who have existed for fewer than one hundred thousand years. Over 98% of all plant and animal species that ever lived are now extinct due to many natural causes. How long present species of plants and animals can or will survive, including modern humans, may depend on the extent of their genetic load, which may mean that humans are at the end of their evolutionary evolvement as a species. Dobzhansky's theories were important in understanding the mathematical relationships of natural selection, as well as Mendelism (also known as Mendelianism).

Born in Russia, Dobzhansky received his BS from Kiev University in 1921 followed by a few years of teaching. He moved to the United States in 1927 on a fellowship at Columbia University in New York City. He soon accepted a position teaching genetics at the California Institute of Technology and became a citizen of the United States in 1937. He wrote several books. The best known is *Genetics and the Origin of Species* in 1937.

See also Darwin; Dawkins; De Vries; Mendel; Wallace

DOMAGK'S CONCEPT OF DYES AS AN ANTIBIOTIC: Chemistry: *Gerhard Domagk* (1895–1964), Germany. Gerhard Domagk received the 1939 Nobel Prize in Physiology or Medicine. Because of World War II, he was unable to receive the award until 1947.

By adding sulfonamide compounds to selected dyes, bacterial infections can be controlled.

For many years physicians sought drugs that could cure patients, particularly those with bacterial infections, such as gonorrhea, meningitis, pneumonia, childbed fever known as puerperal septic (a blood poison), urinary infections, and so forth. Over the centuries alchemists (*see* Paracelsus) treated patients with many toxic drugs including mercury, arsenic, lead, and gold. Their treatments were just as likely to kill the patient as cure their ailments, but some successes occurred. Once bacteria became known as the source of many diseases, the concept of using toxic substances to cure specific diseases became even more urgent. From the late 1930s into the early 1940s the production and use of sulfa drugs increased rapidly. It was used in 1936 to treat President Franklin Roosevelt's son for a streptococcal infection. Sulfa drugs were responsible for saving the lives of many servicemen wounded or suffering from dysentery during the early years of World War II. During the war Churchill was treated with sulfa drugs for pneumonia. By 1942 over ten million pounds of the drug were being produced. Some problems developed with the use of sulfa drugs partially due to their poor solubility, which means crystals of the drug could build up in the kidneys. In addition, some strains of bacteria developed a resistance to sulfa. By the end of World War II a new, more effective antibiotic, penicillin, became available. Today there are numerous varieties of antibiotics, as well as antiviral drugs, available to treat the seemingly growing number of known bacterial and viral diseases.

It is not known who first considered using new types of synthetic dyes to treat diseases, but Domagk based his ideas on research done by Paul Ehrlich and several other scientists. These scientists had some degree of success in using dyes to treat conditions caused by some large organisms, such as protozoa, but were unsuccessful in treating infections caused by small cocci (ball shaped), bacilli (rod shaped), and spirochete (spiral shaped) types of bacteria. After years of research, Paul Ehrlich, a German medical researcher, developed in 1909 an arsenic based compound that was known as "606" because it was the 606th trial for the drug he called *arsphenamine* that proved to be the first effective drug for the treatment of syphilis which is caused by a bacterium known as the spirochete. (Salvarsan, the trade name for arsphenamine, was first marketed in 1910.) There was still a need for an effective drug to use against many types of bacterial infections and diseases. In 1927 the large German chemical company I.G. Farben, for whom Domagk was the director of laboratory research, decided to explore the medical possibilities of the dyestuff it was producing at the time. Because most dyes strongly attach to the protein molecules of cotton and wool fibers (they become "fast"), Domagk theorized that some dyes might also attach themselves to the protein molecules that composed infectious bacteria. Domagk added a chemical called 4-sulfonamide-2′, 4′-diaminoazobenzene hydrochloride to an orange-red dye. He infected rats and rabbits with streptococci bacteria and treated them with the new dye. It did not kill the animals, thus it was nontoxic, while killing the bacteria. For the first time in history, a chemical was found that would combat bacterial infections in humans without poisoning the patient. The drug was first given the name *streptozoan* but was later changed to *prontosil,* the first sulfa drug. Although the drug was not used commercially for several years, it has been reported that Domagk used prontosil to cure his daughter of a streptococcal infection following a pinprick in her arm, saving her arm from almost certain amputation. Sulfonamide compounds soon proved effective in treating streptococcal diseases, such as gonorrhea and epidemic meningitis, as well as staphylococcal infections. They were extremely effective in treating erysipelas (an acute inflammatory disease of the skin and underlying tissue), urinary tract infections, and undulant fever due to bacilli. Sulfa drugs saved many lives during

World War II until about 1944 when penicillin, a superior antibiotic, became available

See also Ehrlich; Fleming (Alexander); Paracelsus

DOPPLER'S PRINCIPLE: Physics: *Christian Johann Doppler* (1803–1853), Austria.

Doppler arrived at his principle as it related to sound. In essence: *the Doppler effect applies to any source of sound (as well any source of electromagnetic frequencies, such as light, radio, and radar frequencies) moving away from or toward an observer that will change in frequency with reference to the observer*. Another way of stating the principle is: *The observed frequency of a wave depends on the velocity of the source relative to the observer, or the observer to the source of the sound or electromagnetic wave*.

Christian Doppler originally applied his effect as it relates to waves of air particles (sound). Later it was experimentally determined to apply as well as to electromagnetic waves such as light. Doppler arrived at his equation about frequency related to velocity of waves based on a unique experiment with sound. He reasoned that when the source of a sound is coming toward an observer, the sound waves that reach the ear are compressed and arrive at shorter intervals and thus at a higher pitch frequency. Although the opposite is also true—as the source of sound waves (or the observer) moves away from each other, the waves that reach the ear are spread out and arrive at longer intervals, therefore, at a lower (pitch) frequency. Doppler placed a group of trumpet players on an open train car and had them play loudly as the train moved away. As the train moved closer, he noted the change in the tone and pitch (frequency of the sound waves) of the trumpet notes. Almost everyone has experienced the Doppler effect. For example, when a train rapidly approaches, its whistle is shrill or high pitched due to the

The Doppler principle provided astronomers with a valuable tool. Armand Fizeau, who determined the speed of light with a unique instrument, also applied the Doppler effect to light waves. This principle when applied to light is sometimes referred to as the *Doppler–Fizeau shift*. The Doppler principle is much more important in the field of astronomy. It was first used to measure the rotation of the sun on its axis. As the sun rotates, the light spectrum on the side of the sun rotating toward Earth is slightly compressed, which makes the light appear bluer. Conversely, the light from the side of the sun rotating away from Earth spreads its spectrum, and thus the sun's light looks more reddish. This principle is also used to measure the motion of stars. If the star appears reddish, it may be receding from us, as its light spectrum spreads out to the red area of the electromagnetic spectrum. This is known as the *redshift* that is due to the shift toward the longer infrared frequency of light waves as stars move away from us. Conversely, if stars are approaching us, they appear bluer due to the compression of the electromagnetic spectrum toward the shorter-frequency blue area of the spectrum. (Basically, most stars are receding from us, but at different rates.) The Doppler principle enables astronomers to measure the distances of stars and galaxies and is used as one of the arguments for an ever-expanding universe.

During World War II, British engineers designed radar (**ra**dio **d**etection **a**nd **r**anging), which was based on the Doppler effect. It used a specific radio wavelength that could be bounced off a moving object. The returning wave was at a different wavelength from that of the transmitting wavelength. The different frequencies between the sent and the returning wave that was bounced off the target object was used to determine the object's position, altitude, and the rate it was approaching or receding from the radar operator who picked it up. It was also used to develop more accurate bombsights. Since that time, radar has become a valuable scientific tool for navigation, meteorology, and astronomy.

compressed sound waves, which increase the sound waves' frequency. The reverse takes place as the train recedes from the receiver because the sound waves are less close together, and thus at a lower pitch (frequency). Although Doppler predicted that a similar effect would apply to electromagnetic waves, he did not have much success using his principle with light waves. However, using his principle, other scientists demonstrated a color shift of light waves (frequencies) under the same conditions of motion as there was for sound waves.

See also Fitzeau; Olbers; Watson-Watt

DOUGLASS' THEORY OF DENDROCHRONOLOGY: Astronomy: *Andrew Ellicott Douglass* (1867–1962), United States.

> *Climate and environmental history can be determined by examining the formation of the rings in the cross sections of tree trunks.*

Andrew Douglass' first interest was in trying to decipher the eleven-year period of high sunspot activity to the period of low sunspot activity as measured on Earth. (The eleven-year high to the low in a cycle is just one-half of the cycle. A full cycle is from high to high.) The actual complete cycle is twenty-two years from one crest to the next crest in the complete cycle. Douglass initially tried to relate the sunspots' high-to-low part of the cycle to the distinct rings in trees that represent yearly growth. He also hypothesized that there might be a correlation between the rate of growth of other vegetation and Earth's climate and the sunspot cycle. He did not find a correlation between the two but determined the rings were a more interesting area for study as a means of dating the past.

Dendrochronology, Greek for "time-telling by trees," is defined as the study of the rings of growth in mature trees to verify historic climate, weather, temperature change, rainfall, insect populations, diseases, and so on. For instance, it has been determined there were periods of devastation of plant growth due to insect plagues and volcanic eruptions, as well as extended droughts, long before today's pollution problems. The time period during which these events occurred are easily ascertained by carefully examining and counting the rings and their sizes in the cross sections of older trees. Douglass' goal was not only to learn about prehistoric chronology of climate but also to use dendrochronology as a means of predicting future climatic changes, particularly global climate changes. There are some limitations to dendrochronology as a dating tool for historical conditions. Living trees have a definite age; thus you can go back only so far, whereas fossil tree rings can be read back to prehistoric times. Also, using tree rings to correlate global or even hemispheric climate changes has proved to be very inaccurate due to the lack of correlation between the growth rings of trees located in different parts of the same continent and the wide distribution of trees worldwide. More accurate methods for determining past and future climate changes are now available.

THE DRAKE EQUATION: Astronomy: *Frank Donald Drake* (1930–), United States.

The Drake Equation states that: $N = R^* \times f_p \times n_e \times f_l \times f_i \times f_c \times L$, where

N is the number of possible civilizations that might be found in the Milky Way galaxy that may, at any time, communicate with us;

R^* is the rate of star creation in our galaxy (estimated by the National Aerospace and Space Administration [NASA] at 6/yr);
f_p is the fraction of those new stars that may have planets (about 0.5);
n_e is the average number of planets (or satellites) that can support life (about 2);
f_l is the number of the above planets that actually develop life (1);
f_i is the fraction of those that will develop intelligent life (0.01);
f_c is the fraction of planets with intelligent life willing to communicate with us (0.01);
L is the expected lifetime of these civilizations (ten thousand years).

The Drake equation is also known at the *Green Bank equation* or the *Sagan equation* (after the popular astrophysicist Carl Sagan who, based on the huge numbers of star systems, assumed that statistically there must be life on planets similar to Earth in some of these systems). This equation is associated with the fields of astrosociobiology, xenobiology, and Search for Extraterrestrial Intelligence (SETI).

Interested in science from an early age, at eight years of age Frank Drake considered the possibility that life existed on other planets. (He never mentioned this idea because of the religious atmosphere at school and home.) After high school, he enrolled at Cornell University and studied astronomy. After a brief "hitch" in the U.S. Navy, he entered graduate school at Harvard University to study radio astronomy. He did his research in this field at the National Radio Astronomy Observatory (NRAO) in Green Bank, West Virginia, and later at the Jet Propulsion Laboratory in California. In 1960 Drake formed the project known as *Ozma*, which was the first radio search for extraterrestrial intelligence. After conducting extensive searches, no evidences of such signals were found. However, Drake refused to give up and was convinced that sooner or later either radio or light contacts were coming to us from outer space.

In 1961, in cooperation with J. Peter Pearman and the Space Science Board of the National Academy of Sciences, they organized a conference at Green Bank that became known as SETI where he first introduced his equation. The N in the equation is an example of a large range of outcomes estimated to be between 1 and 1,000,000. The implications of this are that intelligent life on Earth may be a one-time cosmic event or the universe actually holds many such sites of intelligent life. As Carl Sagan stated, there are billions, and billions of suns with planets out there and some of them statistically will support intelligent life.

Frank Drake remains involved in a SETI project known as Project Phoenix. He served as director of the Arecibo Observatory in Puerto Rico, now known as the National Astronomy and Ionosphere Center (NAIC), from the mid-1960s until 1984. From 1984 to 1988 he was professor of astronomy and astrophysics and Dean of Natural Sciences at the University of California, at Santa Cruz, California, where he is currently Professor Emeritus.

DRAPER'S RAY THEORY: Chemistry: *John William Draper* (1811–1882), United States.

> *Electromagnetic rays, absorbed by some chemical substances, can cause chemical changes in that substance. In addition, the rate of chemical change is proportional to the intensity of the radiation.*

The daguerreotype field of photography was based on exposing silver salts to light, which caused the image on the surface of a glass plate coated with silver salts to darken at the points where the greatest amount of light occurred. The problem was that the silver salt continued to darken as it became exposed to more light outside the camera. Draper solved this problem when, after exposure, he dissolved the unexposed silver from the glass plate with a solution of sodium thiosulfate, also known as "hypo." This chemical is still in use today to "fix" the photographic image by dissolving the unexposed silver salts in the emulsion coating the film and paper prints. Draper determined that electromagnetic rays (visible light, X-rays, etc.) cause a chemical reaction when absorbed by some light-sensitive chemicals (e.g., silver nitrate) and that the amount of light and time of exposure are proportional to the chemical changes. Draper was the first scientist to apply the new field of daguerre-type of photography to his work. He was the first to realize that not all electromagnetic rays (including light) were involved in chemical changes, but that only those rays that were absorbed would cause changes. He explained in a scientific paper that the amount of chemical change was related to the intensity of the radiation multiplied by the time of exposure to the radiation. He built cameras out of cigar boxes and perfected the process to the point that he was able to take short exposure photographs of the moon in 1840 and to take the first pictures through a microscope. He was also the first to record the solar spectrum photographically through a **prism** (*see* Figure F8 under Fraunhofer.) In addition, using his sister as a model, he was the first to take a successful, short-exposure portrait that is still in existence. Draper experimented with the size of the **aperture** for the lens, which, if enlarged, would reduce the time of exposure required to expose the image.

See also Daguerre

DULBECCO'S THEORY OF CANCER CELL TRANSFORMATION: Biology: *Renato Dulbecco* (1914–), United States. Renato Dulbecco shared the 1975 Nobel Prize in Physiology or Medicine with Howard Temin and David Baltimore.

> *Normal cells, when mixed with cancer-producing viruses in vitro, will kill some of the cells, while other cells are changed by the virus and will continue to grow and multiply as cancerous tumors.*

Renato Dulbecco refers to this theory as "protective infection." The significance of Dulbecco's concept is that it is possible to grow in the laboratory cells that have been infected by tumor-causing viruses. This simplifies the process of understanding the nature of malignancy and experimenting with possible treatments. It is easier to experiment with different chemical treatments using cancer cells **in vitro** than in the human body. His theory and laboratory techniques advanced cancer research.

In 1986 Dulbecco published a paper "A Turning Point in Science" that appeared in the magazine *Science*. Dulbecco pointed out that to learn more about the science of cancer it will be necessary to know more about the human cellular genome. This paper was published shortly after a group of scientists met in 1985 to discuss the possibility of sequencing the complete genome of humans. This was a huge task spurred on by this important paper that was completed in 2003. Dulbecco moved back to Italy in 1993 where he became director of the National Council of Research in Milan. In early 2006 he left Italy and retired, at the age of 92, to La Jolla, California. He continues to follow the latest advancements in cancer research, primarily at the Salk Institute located in La Jolla.

See also Baltimore; Gallo; Temin

DUMAS' SUBSTITUTION THEORY: Chemistry: *Jean Baptiste André Dumas* (1800–1884), France.

An atom or radical can be replaced by another of known quantity that will produce the same result.

Jean Dumas believed that organic chemistry was similar to inorganic chemistry as related to the formation of **radicals** of the same types. Organic chemistry involves the element carbon in the construction of large molecules found in living tissues, such as large protein molecular compounds. Inorganic chemistry involves the reactions between and among all types of elements that form inorganic (nonliving) compounds. Radicals might be thought of as molecules that contain an electrical charge (similar to inorganic ions) and can act as units when combining with other elements or compounds, regardless of whether they are "organic" or "inorganic." An example is the hydroxyl radical OH–, which is part of the water molecule with a negative charge and is thought to have some effect on the aging of cells. Dumas contended that the site of these radicals is where replacements of atoms of one type for another take place. He showed that several different compounds that were composed of the same atoms or radicals exhibited similar characteristics. His famous example demonstrated that trichloroacetic acid was a similar compound to acetic acid. Dumas contributed to the advancement of chemistry with his work with atomic weights based on whole numbers as multiples of hydrogen as 1, and thus carbon as 12. This led him to his theory of types, which today is referred to as *functional groups of elements*. Historically, the credit for the theory known as the *theory of types* was disputed between Dumas and Auguste Laurent.

See also Dalton; Laurent

DYSON'S THEORY OF QUANTUM ELECTRODYNAMICS: Physics: *Freeman John Dyson* (1923–), United States.

Quantum theory can explain the relationships and interactions between minute particles and electromagnetic radiation.

Quantum electrodynamics, known as QED, is a theory that is relativistic in the sense that the subatomic particles in a field are rapidly changing their values (sometimes referred to as the "jitters") due to their constantly changing movements at their small submicroscopic scale. This leads to problems of measurement of exactly where they are at any particular moment and exactly how they are moving. This involves many possibilities and requires the use of mathematical probabilities as the only way to arrive at a single outcome. In other words, there is "uncertainty" introduced when measuring a particle's position (location) and momentum (unending movement) when being observed. In the 1920s and early 1930s this problem led to the development of the theory of quantum mechanics that solved this problem by using mathematical probabilities to describe the realm of atoms and subatomic particles. By the late 1920s quantum mechanics that dealt with single particles was expanded to include quantum fields as well. By the 1940s this research led to combining particles and fields that is known as QED. Those involved in the research at that time were Richard Feynman, Freeman Dyson, Julian Schwinger, and Sin-Itiro Tomonaga. (All but Dyson received a Nobel Prize for this work.) QED was the first viable quantum description of a physical field that involved quantum particles that can calculate observable quantities at this physical scale.

Freeman Dyson combined several related theories into a general theory that described the interactions between waves and particles in terms of quantum concepts (*see* Planck for a description of *quantum*, which means "how much"). This single theory enabled scientists to better understand quantum electrodynamics based on the traditional ingredients of particles and fields that synthesize waves, particles, and the interactions of radiation with electrons and atoms. This theory is also known as the *quantum theory of light or radiation*.

See also Dehmelt; Einstein; Feynman; Planck; Schwinger

E

EDDINGTON'S THEORIES AND CONCEPTS: Astronomy: *Sir Arthur Stanley Eddington* (1882–1944), England.

Eddington's star equilibrium theory: *For a star's equilibrium to be maintained, the inward force of gravity must be balanced by the outward forces of pressure caused by both the gas and radiation produced by the star's fusion reaction.*

Sir Arthur Eddington also developed the system indicating that heat that was generated inside stars was not transmitted outwardly by convection—as heat could be on Earth—but rather by a form of radiation. This theory of equilibrium not only explained why stars do not usually explode, but also provided a much better understanding of the internal structure of stars. It was William Higgins who determined that the sun (also a star) as well as all stars, and presumably the entire universe, are composed of the same basic elements as is Earth, with carbon being the most important of the first elements formed whose atoms were larger than hydrogen and helium. Our sun consists of several layers that are not sharply divided. The inner layer or core of the sun is about 250,000 miles in diameter and consists of hydrogen undergoing fusion to form helium, resulting in great quantities of energy in the form of radiation. The core's temperature is about 27,000,000°F; its pressure is about 7,000,000,000,000 pounds per square inch. (The air pressure on the earth's surface is less than 15 pounds per square inch.) The next layer or "shell" is the convection zone that surrounds the core, which transmits the radiation generated in the core to the outer layers. The photosphere, which means, "sphere of light," is the surface layer where the radiation is converted to light and heat as we know it on Earth. It is about five hundred miles thick, with a temperature of almost 10,000°F. Light and heat as well as other forms of electromagnetic radiation are constantly sent outward in three dimensions from the convection zone to the photosphere. The *chromosphere*—the outer "shell" or layer—is about two thousand miles thick with a variable temperature ranging from about 8000 to 90,000°F. The outermost layer is called the *corona*, which is a low-density collection of ionized gases. This layer forms the "spikes" that shoot out from the sun's surface and can be seen during an eclipse. As Eddington's theory of equilibrium states, if this arrangement becomes imbalanced, a star could explode. Exploding stars, which are extremely bright, were recorded in ancient history. Today, a very bright star that lasts only a few days or a week is called a nova, or supernova (*see also* Higgins).

Eddington's theory of star mass–luminosity relationship: *The more massive (very dense) the star, the greater its luminosity.*

The brightness of a star is determined almost entirely by its density (mass per unit volume). It is a fundamental principle of astronomy that for stars of constant mass, their luminosity is also constant. The mass of a star is related to its density but not necessarily to its size. Therefore, a small, high-density star is also massive in the sense that it may contain more matter than a star that is larger but less dense. A small, dense star may be much brighter than a very large star of low density. Up to the time of this concept, only masses of binary stars could be directly calculated—that is, a pair of stars close enough in proximity that their mutual gravity causes them to rotate around a common, invisible center of gravity. This led to the theory that stars, even of different spectral classes, with the same masses also had the same luminosities. This relationship of mass to brightness was of great significance in not only determining the nature of stars but also their distance from Earth. Using this mass-luminosity theory, along with the gravity-equilibrium theory, Eddington calculated there was a limit to the size of stars which is about ten times the mass of our sun. Any star forty to fifty times the mass of the sun would be unstable due to the excessive internal radiation. Eddington's mathematical equations are considered fundamental laws of astronomy and provide a new way to look at the evolution of stars, including our sun.

Eddington–Adams confirmation of Einstein's special theory of relativity: *Albert Einstein predicted light from distant bright stars would be "bent" by the gravity of another star as it passed by that star.*

Eddington reported that during a total eclipse, the light from several bright stars was slightly bent as their light came past the sun. This demonstrated that light exists as quantum bits called photons as predicted by Einstein, as well as waves of electromagnetic radiation that are affected by the gravity of the sun (*see* Figure E1 under Einstein). Eddington's work confirmed Einstein's theory of special relativity. The American astronomer Walter Sydney Adams (1876–1956) further tested the theory by measuring the shift in the wavelength of light from Sirius B, a very dense white dwarf with strong gravity. Einstein predicted the light from a massive star would shift to the red end of the spectrum. Thus, as the light from a massive star was slowed due to that star's gravity, a reddish shift occurs in the star's light (not to be confused with the **Doppler effect** for the "redshift," which is based on the lengthening of the frequency for light waves from a star that is rapidly receding from us) (*see also* Doppler; Einstein).

Eddington's constants for matter: *The total number of protons and electrons in the universe is* 1.3×10^{79}, *and their total mass is* 1.08×10^{22} *masses of the sun.*

Eddington arrived at these figures after considering the concept of an ever-expanding universe with "curved" space as theorized by Einstein. This theory was based to some extent on the tremendous velocities of nebulae. Eddington extended this theory and combined it with the theory for the atomic structure of matter to calculate his constants by theory alone. These are considered fundamental constants of science, which are important for the concept of an expanding universe.

Eddington's physical theory of the big bang: *The universe started to expand in all directions when a small, very dense point of energy (and possibly matter) exploded with tremendous force.*

Eddington was not the first to come up with a "cosmic egg" concept of the origin of the universe. This "egg," about the size of a marble or less, was assumed to be extremely dense as it contained all the mass-energy in the universe. Eddington's contribution was in developing the mathematical equations to explain the physics of the expanding universe. The concept goes back thousands of years, but it was Willem de Sitter (1872–1934), the Dutch mathematician/physicist/astronomer, who first developed a viable cosmological model based on the theory of an expanding universe. In 1964 Robert Woodrow Wilson and Arno Penzias of the Bell Telephone Laboratories detected cosmic microwave remnants of a hot primeval fireball as evidence of the big bang. Although some of the details of the big bang are still elusive, the concept of an inflationary universe is now accepted by most astronomers. It theorizes that at the time of the bang, all the original particles and energy could defy the speed of light and expand at any speed. Several other scientists have contributed to the cosmic egg/big bang/inflationary universe concept. There are still questions as to the origin and state of the nascent universe: What existed before the big bang? Is the universe really expanding? If so, do we really know the rate of expansion? Will it continue to expand forever? Will it reach equilibrium, then contract and start all over again? Will it regenerate or is it continually generating new matter? Is it static?

See also Dicke; Gamow; Hale; Hubble; Lemaître

EDISON'S THEORY OF THERMIONIC EFFECT: Physics: *Thomas Alva Edison* (1847–1931), United States.

> *Thermions (negatively charged electrons) generated at the hot cathode filament (of the light bulb) will jump to a cooler wire some distance from the filament.*

This is commonly referred to as the "Edison effect" and is the only physical theory Edison developed. All of Edison's other accomplishments were inventions that led to the development of important industries.

In developing his light bulb, Edison followed the lead of Sir Joseph Wilson Swan (1828–1914), the English physicist and chemist, who developed the first light bulb and was the first to use a carbonized (charred) thread as a filament. Swan's bulb did not work very well because he could not produce a good vacuum inside the bulb, nor could he develop a battery to produce a strong enough current to cause the carbonized thread to incandesce (glow white hot and give off light). Just one year before Edison announced his invention, Swan refined his carbon filament incandescent bulb and demonstrated it to the public. Given this set of circumstances, Swan should really have been given credit for the invention because he demonstrated his light bulb before Edison perfected his model. Even so, Edison is generally credited with the invention, although sometimes Swan is listed as a coinventor of the incandescent light bulb.

Edison was the first to explain that electric current flows in only one direction—from the filament to the electrode, and not the other way around. He experimented with several hundred different types of materials to act as filaments as he developed his incandescent light bulb. In 1883 he inserted a metal wire next to the filament, but not connected to it or the source of electricity. His expectation was that such a "cold" piece of metal would reduce the amount of air, thus improving the vacuum and prolonging the life of his filaments. He noticed that some electrons (he called them *thermions* because the filament was hot) flowed across the space gap in the bulb to the cooler metal wire, producing a noticeable glow. He patented his "Edison effect" but did not exploit it. Later, this arrangement of the filament next to a metal grid proved to be a valuable design for the development of the electronic vacuum tubes used in radios and early television sets before the days of semiconductors and transistors. His two-filament tube was called a diode. Later others added a small grid of wires between Edison's hot negative filament and the cool positive metal plate. This design became known as a triode. Over the years the Edison effect has been redesigned using semiconducting metals (instead of hot glass tubes) to construct small triode-like chips used in the integrated circuits for computers and other electronic devices.

One of Thomas Edison's major contributions to research was his establishment of private, independent research laboratories. His first was in Menlo Park, New Jersey, that contained a large library and expensive equipment and was staffed by over twenty technicians, including mathematicians and physicists. This laboratory produced many inventions that led to new industries and products that are widely used to this day. One for which he was internationally famous was the phonograph. His original model was a cylindrical device that recorded sound with a needle that formed the sound's imprints on tinfoil. Edison is also credited with the invention of many other devices, including moving pictures, types of office equipment, the storage battery, and many more. He received over one thousand patents in his lifetime and is considered the icon of inventors (although Nikola Tesla's inventions are equally important in today's world). After coinventing the incandescent light with Swan, he raised enough money to form the Edison Electric Light Company that was limited to small sections of New York City because his system generated direct current unlike the alternating current system devised by Tesla that could send current over long distances. Tesla's system was the beginning of today's large-scale power stations and national electric distribution systems. In 1887 Edison moved his laboratory to a larger facility in West Orange, New Jersey, and later established another laboratory outside of Fort Myers, Florida.

See also Bardeen; Esaki; Tesla; Shockley

EHRLICH'S "DESIGNER" DRUG HYPOTHESIS: Chemistry: *Paul Ehrlich* (1854–1915), Germany. Paul Ehrlich was awarded the 1908 Nobel Prize for Physiology or Medicine.

Using the molecular structure of synthetic compounds, specific pharmaceutical drugs can be designed and produced to treat specific disease.

Aware of the aniline dyes (coal tar dyes) developed by the English chemist Sir William Henry Perkin (1838–1907) and how different dyes could stain animal fibers (e.g., wool and hair), Paul Ehrlich assumed these dyes could also differentially stain human tissue, cells, and components of cells. Using an aniline dye, scientists saw for the first time the chromosomes of cells, which Ehrlich called *colored bodies*. Other scientists identified specific germs that caused specific diseases. Ehrlich then hypothesized that because specific dyes will stain specific tissues selectively, it might be possible to design a chemical to attack specific types of germs that are also composed of specialized living material. This led to the theory that certain substances could act as "magic bullets" and attack specific disease-causing organisms while not attacking normal healthy cells in the body. He also formulated his side-chain theory of immunity, which led to the development of synthetic chemical compounds designed specifically to attack microorganisms, and thus enable the body to establish immunity to those specific microorganisms. Although many coal tar dyes can cause disease, including some cancers, the large dye molecules can be manipulated to attack specific types of bacteria, such as those that cause sleeping sickness and syphilis. However, drugs derived from coal tars were not effective for treating other diseases, including streptococci and cancer.

Ehrlich demonstrated that toxin–antitoxin reactions are not only a chemical reaction but, like most chemical reactions, these specific types of reactions can also be accelerated by heat or retarded by cold. He also was able to develop a standard for which antitoxins could be more accurately measured. His work in immunology provided Ehrlich the information necessary for him to arrive at his theory for side-chain immunity.

Paracelsus, the sixteenth-century alchemist, was known as the ancient founder of chemotherapy; Ehrlich is known as its modern founder.

See also Koch; Domagk; Elion; Paracelsus

EIGEN'S THEORY OF FAST IONIC REACTIONS: Chemistry: *Manfred Eigen* (1927–), Germany. Manfred Eigen shared the 1967 Nobel Prize for Chemistry with George Porter and Ronald Norrish.

> *Ionic solutions in equilibrium (same temperature and pressure) can be disarranged out of equilibrium by an electrical discharge or sudden change in pressure or temperature resulting, within a short time, in the establishment of a new equilibrium.*

Manfred Eigen used the "relaxation technique" along with ultrasound absorption spectroscopy to determine that this reaction occurred in 1 nanosecond (one-billionth of a second). Using this information and his techniques for measuring fast reactions, he ascertained how water molecules are formed from the H^+ (hydrogen ion) and the OH– (hydroxide ion) to form H_2O, or (H-O-H). He continued to use his "fast reaction" theory to explain complex reactions and characteristics of metal ions and, later, more complex organic biochemical reactions and nucleic acids. The theory of fast ion reaction is important to the understanding of chemical reactions in all types of living organisms. Understanding the steps involved in a series of mechanisms of fast reaction led to the development of the theory for the understanding of acid-base reactions.

Following his work with fast reactions Manfred Eigen turned his attention to problems dealing with biochemistry, particularly with the chemistry involved with the storage of information in the central nervous system.

EINSTEIN'S THEORIES, HYPOTHESES, AND CONCEPTS: Physics: *Albert Einstein* (1879–1955), United States. Albert Einstein was awarded the 1921 Nobel Prize in Physics.

Einstein's theory for Brownian motion: *The motion of tiny particles suspended in liquid is caused by the kinetic energy of the liquid's molecules.*

In 1827 the Scottish botanist Robert Brown (1773–1858), while using a microscope, observed that pollen grains suspended in water were in constant motion, which he believed was caused by some "life" in the pollen. He added minute particles of nonliving matter to water and observed the same motion. This phenomenon was not explained until the kinetic theory of molecular motion was discovered. Albert Einstein derived the first theoretical formula to explain why these small particles moved in a liquid when the particles themselves were not molecules. His equation was based on the concept that the average displacement of the particles is caused by the motion resulting from the kinetic energy of the molecules in the liquid. This resulted in a better understanding of the atomic and molecular activity of matter, and thus heat.

Einstein's theory of the nature of light: *Electromagnetic radiation propagated through space (vacuum) will act as particles (photons) as well as waves since such radiation is affected by electric and magnetic fields, and gravity.*

James Clerk Maxwell developed an equation stipulating that electromagnetic radiation can travel only as waves. This concept disturbed Einstein, as did the experiments by Philipp Lenard who had observed the photoelectric effect of ultraviolet light "kicking" electrons off the surface of some metals. It was determined that the number of electrons emitted from the metal was dependent on the strength (intensity) of the radiation. In addition, the energy of the electrons ejected was dependent on the frequency of the radiation. This did not jibe with classical physics. This dilemma was solved by Einstein's famous suggestion that electromagnetic radiation (light) flows not just in waves but also as discrete particles he called *photons*. Max Planck referred to these as *quanta* (very small bits). Using Planck's equation, $E = \hbar v$, where E stands for the energy of the radiation, $\hbar$ is Plank's constant, and v is the frequency, Einstein was able to account for the behavior of light as massless particles with momentum (photons) that have some characteristics of mass, for example, momentum (mass × velocity), as well as characteristics of waves. It resulted in Einstein being awarded the 1921 Nobel Prize for Physics.

Einstein's concept of mass: *The at-rest mass of an object will increase as its velocity approaches the speed of light.*

When a body with **mass** is not moving, it is at rest as far as the concept of inertia is concerned, meaning it is resistant to movement by a force. An analogy would be sluggishness, inertness, or languidness in a human being. Once an at-rest mass is in motion (i.e., **velocity**), it attains momentum (mass × velocity). When there is an increase in its velocity, there is also an increase in the body's mass. Thus, if a mass attained the speed of light, it would not only require all the energy in the universe to accomplish this, but also would equal all the mass in the entire universe. Therefore, it is impossible for anything with mass (except electromagnetic radiation, i.e., light) to attain the speed of light. This is one reason that light must be considered as being both a wave and a particle.

Newton's three laws that relate to mass and motion represent a classical, mechanistic concept of the universe. Newton's laws are deterministic based on the conservation of mass that states that matter cannot be created or destroyed. Although **weight** is proportional to mass, the *weight* of an object varies as to its position in reference to Earth (or other body with mass in the universe) and thus gravitational attraction, whereas the *mass* of an object is independent of gravity. The mass of an object (matter) is the same regardless of its location in the universe and thus is independent of gravity. At the same time, one might say that in deep outer space, mass has near zero weight.

Einstein's theories of relativity ultimately changed the Newtonian concepts of mass and motion. In modern physics, the mass of an object changes as its velocity changes, particularly as the velocity approaches the speed of light. This phenomenon is not noticeable on Earth because our everyday velocities are far less than that of light. For instance, the at-rest mass of an object will double when it attains a velocity of 160,000 miles per second. This is approaching the speed of light, which is 186,000 miles per second, and even a very small mass is incapable of attaining the speed of light. When masses with extremely high velocities interact, nuclear reactions can occur, where mass can be converted into energy—thus the famous Einstein equation, $E = mc^2$, where E is the energy, m is the mass of the object, and c^2 is the constant for the velocity of light squared.

Einstein's theory of special relativity: *1) Physical laws are the same in all inertial reference systems. 2) The speed of light in a vacuum is a universal constant. 3) Measurement of time and space are dependent on two different events occurring at the same time. 4) Space and time are affected by motion.*

An inertial reference system is a system of coordinates (anywhere in space) in which a body with mass moves at a constant velocity as long as no outside force is acting on it. From this concept, other components of the special theory of relativity follow.

Albert Einstein's special theory of relativity provides an accurate and consistent description of events as they take place in different inertial frames of reference in the physical world, with the provision that the changes in space and time can be measured. He developed the special theory of relativity to account for problems with the classical mechanistic system of physics. Many people had (and still do have) difficulty understanding his theory. In essence he is not describing the nature of matter or radiation, although he recognized their association. His theory describes the world or event, as it might look to two individuals in different frames of reference. For example, in classical Earth-bound physics, a person in a car going in one direction at 50 miles per hour (mph) meets a car approaching from the opposite direction going 100 mph. This describes how the speeds of the two cars are observed and judged (measured) independently by another person standing by the side of the road and not moving. But this does not apply to the drivers of two approaching cars. The driver in a car going 50 mph would perceive a car approaching him from the opposite direction at 100 mph, as going at 150 mph. Conversely, a driver who is going just 50 mph and is passed by another car going 100 mph in the same direction will perceive the passing car as going just 50 mph. This is just common sense and can be proved with classical equations of adding and subtracting velocities, which is known as Galilean transformations. However, this is not how it works with electromagnetic radiation waves such as the velocity of light. Einstein's theory states that the time between the two events (of the cars) is dependent on the motion of the cars. The special theory states that there is no absolute time or space.

According to experiments by Albert Michelson and Edward Morley, the speed of light is independent of the motion (velocity) of its source or the observer. For instance, if both cars are traveling at astronomical speeds in space and one car is going twice as fast as the other car and both turned on a spotlight toward the approaching car, the light would travel the same speed in both (either) directions. One driver would not perceive the light as coming toward him at a greater speed than would the other driver because they would judge the combined speeds of 50 and 100 mph of the two cars on earth. In contrast to Earth-bound car drivers, Einstein stated that despite how fast you are going, the speed of light will be constant for all frames of reference. The drivers of the two "space" cars, regardless of how fast they are going, will be in two different frames of reference of both time and space, but the speed of light will remain constant.

Thus, from their individual frames of reference (points of view), they will not be aware of "Earth-bound commonsense" differences in their speeds. No matter how fast you go, the speed of light will always be the same, even if you are speeding in the same direction as the light being propagated. The theory later included the concept of the three Euclidean coordinates of space—width (x), height, (y), and depth (z)—with the addition of the coordinate of *time* to arrive at a space-time continuum as developed by Hermann Minkowski.

There is much confusion about the word "relativity." In science, it is used as something "relative" to something else that can be measured mathematically or statistically. Specifically, Einstein's special theory of relativity is related to frames of reference as measured for the four coordinates of space and time (*see also* Minkowski).

Einstein's principle of gravity: *Gravity is the interaction of bodies equivalent to accelerating forces related to their influence on space-time. Gravity measurably affects the space-time continuum.*

There are two related concepts of gravity: the Newtonian classical concept and the Einsteinian concept related to his theories of relativity. Newton's law states that the gravitational attraction between two bodies is directly proportional to the product of the masses of the two bodies and inversely proportional to the square of the distance between them, as expressed in $F = G\ m_1m_2 \div d^2$. Following is an example that relates acceleration to the force of gravity on Earth. If you are in a train or car that is traveling on a perfectly smooth surface and cannot see out the windows, you cannot tell if you are going backward, forward, or not moving at all if the vehicle is traveling at a uniform speed. But if the vehicle accelerates or decelerates, your senses will react as if gravity is affecting you. A person also becomes aware of G forces (simulated gravity) when a car or airplane rapidly accelerates. To sum this up, classical physics stated that all observers, regardless of their positions in the universe, moving or stationary, could arrive at the same measurement of space and time.

Einstein's theories of relativity negate this concept because the measurements of space and time are dependent on the observers' *relative* motions regardless of their inertial frames of reference within space coordinates. Einstein combined the ideas of several other physicists and mathematicians that dealt with non-Euclidean geometry, the space-time continuum, and calculus to formulate his gravitational theory. In essence, Einstein's concept of gravity affected space and time, as in his theory of general relativity. Even so, Einstein's concept of gravity was not quite correct because he did not take into account the information developed by quantum theory for very small particles and their interactions, even though these subatomic particles are much too small (or even massless) to be affected by Earth's gravity. His concept dealt with the macro (very large) aspects of the universe. As with all other laws of physics, the laws concerning gravity are not exact. There still is room for statements that more precisely interpret the properties of nature. For Einstein, the interactions of bodies are really the influence of these bodies (mass) on the geometry of space-time.

For many decades, scientists have tried to explain gravitational waves in relation to the theories of relativity or some other principle. *How* gravity acts on bodies (mass) can be described, but exactly *what* gravity is or *why* it is has not been discerned. Another hypothesis is based on the theoretical particle called the **graviton**, proposed by quantum theory. Gravitons behave as if they have a zero electrical charge and zero mass. Although assumed to be similar to photons, they do have momentum (energy). The concepts of gravity waves and gravitons are still under investigation. Recently, there have been several experiments designed to detect the graviton. The space

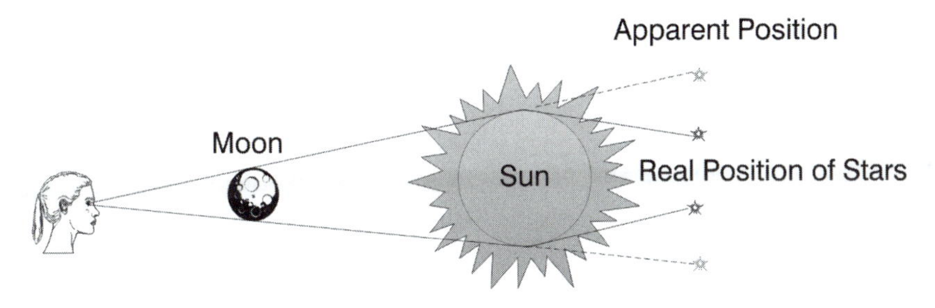

Figure E1. Einstein predicted that light from stars located behind the sun can be viewed during an total eclipse. As the star's light passes the sun, it is bent toward Earth by the sun's gravity, indicating that light has momentum (mass x velocity) and thus is affected by gravity leading to the theory that light is composed of minute packets (quanta) (i.e. photons). (The actual bending of the light is less than depicted in the diagram.)

satellite Gravity Probe B was launched in 2004 and lasted until 2005 when the data analysis started. It was designed to measure general relativity's predicted "frame-dragging" effect by measuring the space-time curvature near the Earth. Also, National Science Foundation (NSF) has funded two land-based installations designed to directly detect gravity waves on Earth. The Laser Interferometer Gravitational-Wave Observatory (LIGO) consists of two installations, one in Hanford, Washington, and the other in Livingston, Louisiana, that operate in unison as a single observatory. Other experiments to detect and measure gravity waves and gravitons are planned.

Einstein's theory of general relativity: *The interactions of mass (as related to gravitational force) are really the influence of bodies (masses) on the geometry of space and time. Space and time are affected by gravity.*

This theory is based on two main ideas: that the speed of light is a universal constant in all frames of reference and that gravitational fields are equivalent to acceleration for all frames of reference within the space-time continuum.

The first proofs of the general theory of relativity came from astronomy. It explained the previously unknown reason for the variations in the motions of the planets. The theory then was used to predict the bending of starlight as it passed massive bodies, such as the sun, and as it was detected during a total eclipse.

The theory also predicted that electromagnetic radiation in a strong gravitational field would shift the radiation to longer wavelengths. This was demonstrated by using the **Mössbauer effect**, which predicts the effects of a strong gravitational field on radiation. An experiment using a strong source of gamma radiation was set up just seventy-five feet above Earth to measure the gamma rays as they approached the surface of Earth. A minute lengthening of gamma rays (very short wavelength electromagnetic radiation) caused by the gravity of Earth was detected, thus confirming the theory.

Einstein's unified field theory: *A simple general law combining the four forces of nature (the electromagnetic force, gravitation, the strong force, and the weak force):*

1. *An electromagnetic force is exerted between electrical charges and magnetic fields.*
2. *Gravitational force is related to mass and acceleration and can affect electromagnetic radiation.*

3. *The* **gluon** *of the strong force holds the nuclei of atoms together. The positive protons and quarks in an atom's nucleus would repel each other and fly apart if it was not for the ''glueballs'' of the strong force that ''bind'' the quarks and protons.*
4. *The weak force is responsible for the slow nuclear processes that produce radiation, such as beta decay of the neutron to generate high-speed electrons.*

Albert Einstein spent the last thirty years of his life attempting to combine these four fundamental forces and the equations incorporating them into a general unified theory (GUT). He never achieved this goal. Einstein did not completely accept the new quantum theory, which did not lend itself to his concept for a unified field theory. One reason for not completely accepting quantum mechanics was because of the concept of indeterminacy where the exact position and momentum of a particle could not be determined at the same time. Thus, he considered that the "fuzziness" of matter was inconsistent with his concepts of relativity and the real world.

Toward the end of his life and later, physicists used **particle accelerators** to separate and identify numerous particles and forces from atoms and their nuclei, which made the unified field theory impossible. But the idea is not completely dead. Today there are several efforts to combine or find symmetry between various theories of matter and energy. Some examples follow:

- *The grand unification theory (GUT)*: an attempt to derive an equation to combine the strong and weak forces and explain how the particles of matter were dispersed from each other at the time of the big bang at speeds greater than the speed of light. The GUT has led to another concept referred to as the *inflationary universe*.
- *The theory of everything (TOE)*: an attempt to state that there is only one simple force and one ultimate particle in the universe. They have not been found.
- *The string theory*: a mathematical concept to explain everything in the universe with just one theory, based on the premise that all elementary subatomic particles are really strings that are single-dimensional loops, sometimes described as a doughnut folded over itself several times. Presumably the string theory has as few as ten or as many as twenty-six dimensions (not the three coordinates plus time with which we are familiar). Thus, it may not be related to the real universe but is an intriguing concept for mathematicians and theoretical physicists. When various mathematical equations and techniques are used to combine other mathematical equations into one final statement, the results always seem to come out as noise or lead to infinity.

Even before his work on relativity and gravity he published several important papers before and during 1905 that addressed areas of theoretical physics that changed many scientific concepts of physics.

In summary, these papers addressed the following topics:

1. One of his first papers dealt with the atomistic nature of the new science of thermodynamics. He considered the mechanical (Newtonian) view of the world intricately related to the second law of thermodynamics.
2. His next paper was the use of this concept to explain the Brownian motion of microscopic particles suspended in a fluid that led to the concept of the kinetic theory of motion and heat.

3. At the age of sixteen he wrote a paper explaining how time and motion are related to the observer if the speed of light is constant. This paper might be called the "seed" of his theories of relativity.
4. He published several papers that addressed why the Newtonian and Galilean laws of gravity, time, and space required an inertial frame of reference, as well as the compatibility of Maxwell's equations and relatively.
5. He proposed a quantum hypothesis, particularly for light (photons). Even though this led to the development of quantum mechanics, and so forth, he had some difficulty in trying to incorporate relativity, gravity, quantum theory, and other ideas into an overall grand unification theory (GUT). Scientists are still trying to arrive at a theory of everything (TOE).

As a young boy Albert Einstein was shy but curious even though he did not talk until he was three years of age. He did not do particularly well in school except when he was introduced to subjects dealing with nature. He had an ability to understand mathematical concepts and taught himself Euclidean geometry at twelve years of age. He left school at age fifteen but later entered a school in Switzerland. However, he disliked the methods of formal education and spent much of his school years studying physics on his own and playing his violin. He passed his graduation exams by studying his classmates' notes but was not recommended for graduate education. He spent two years tutoring students and later was hired as an examiner in the Swiss patent office in 1902. He was married in 1903, had two sons, was divorced, remarried, and reportedly had ten mistresses later in life. He received his doctorate from the University of Zurich in 1905. He believed it important to simplify and unify the system of theoretical physics.

See also Eddington; Galileo; Minkowski; Newton

EINTHOVEN'S THEORY THAT THE HEART GENERATES AN ELECTRIC CURRENT: Biology: *Willem Einthoven* (1860–1927), Netherlands. Einthoven was awarded the 1924 Nobel Prize for Physiology or Medicine.

> *An accurate measurement of the electrical responses of the human heart can be made by viewing the magnetic lines of force that are proportional to the strength of the current that is produced by the heart, and thus indicating specific conditions of the heart.*

Willem Einthoven was not the first to consider that the living tissue of animals produces an electrical current that can be measured. In the eighteenth century Luigi Galvani experimented by touching a frog's leg with a discharge from a static electric machine. He then tried clamping a dissected frog's spinal cord and leg with a brass hook onto an iron railing during an electrical storm. The frog's muscles twitched. His deduction, which proved incorrect, was that the electricity was produced by the frog's tissue. Alessandro Volta, a contemporary of Galvani believed Galvani's theory to be flawed and set out to test the hypothesis. Volta demonstrated that, in actuality, a slight current was generated when two dissimilar metals (brass and iron of Galvani's original experiment) were brought together under moist conditions irrespective of the frog's tissue. In the late 1800s Augustus Waller (1856–1922), the French-born cardiologist, was the first to detect, as well as the first to attempt to measure, the electric current generated by the heart. Waller used a special type of meter to measure millivolts produced by the beating heart. His device was complicated to use as well as inaccurate unless the electrodes were actually placed directly on the heart. Waller's device also required a series of mathematical calculations.

In 1901 Einthoven developed a series of prototypes that were called "string galvanometers." They were complicated, but they worked. His string devices used a thin conductive wire filament suspended between a strong electromagnet. When a small current was passed through the wire in the magnetic field, the wire filament would move slightly. By shining a light on the wire string a shadow would be cast that could make an image on a moving roll of photographic paper. His original string galvanometer weighed six hundred pounds, was water cooled to keep the powerful electromagnets cool, and required five people to operate it. Despite all these drawbacks, it was the first machine that could, with some degree of accuracy, measure the condition of the heart through the chest wall without attaching electrodes directly to the heart. Einthoven used his galvanometer to study and describe many cardiovascular disorders. The early string galvanometer was, over the years, improved to what we now call an electrocardiogram or electrocardiograph (ECG or EKG) machines that are relatively small, portable diagnostic devices that can accurately determine many possible conditions of the human heart.

See also Galvani; Volta

EKMAN'S HYPOTHESIS OF THE CORIOLIS EFFECT ON OCEAN CURRENTS: Geology (Oceanography): *Vagn Walfrid Ekman* (1874–1954), Sweden.

> *The effect of Earth's rotation as well as other factors on ocean currents is more evident at the poles than at the equator.*

In the 1890s the Norwegian Arctic explorer Fridtjof Nansen (1861–1930) made an important observation. While sailing in the Arctic regions, he noted that drift ice did not move in the direction of the wind but rather deviated 45 degrees to the right. Ekman used this information in a 1905 publication "On the Influences of the Earth's Rotation on Ocean Currents" in which he described the complex forces responsible for ocean currents. After observing ocean currents near the equator and the Arctic Circle, Ekman noted that near-surface water moves in the opposite direction to the motion of the surface water resulting in a movement of water at right angles to the wind directions. In the Arctic the motion of the ice floes were determined by the currents that in turn were produced by a complex interaction between the forces of the surface wind, the friction between the different layers of water in the oceans, and most important, the rotation of the earth. The results of these complex forces creating variations of water velocity with depth of forces became known as the *Ekman spiral*.

The Ekman Spiral is also a good example of the Coriolis force interacting with other forces. Sir George Biddell Airy based his theory of internal waves and *dead water* on the work of Vagn Walfrid Ekman. The region of *dead water* in the oceans is caused by a thin layer of freshwater that is produced by melting Arctic freshwater ice spreading over the sea that forms waves between the layers of fresh and saltwater that are at different concentrations, thus at different densities. This phenomenon has been known to slow down or even stop a sailing ship's progress. In 1903 Ekman developed what is known as the *Ekman current meter* to detect and measure the differences in ocean

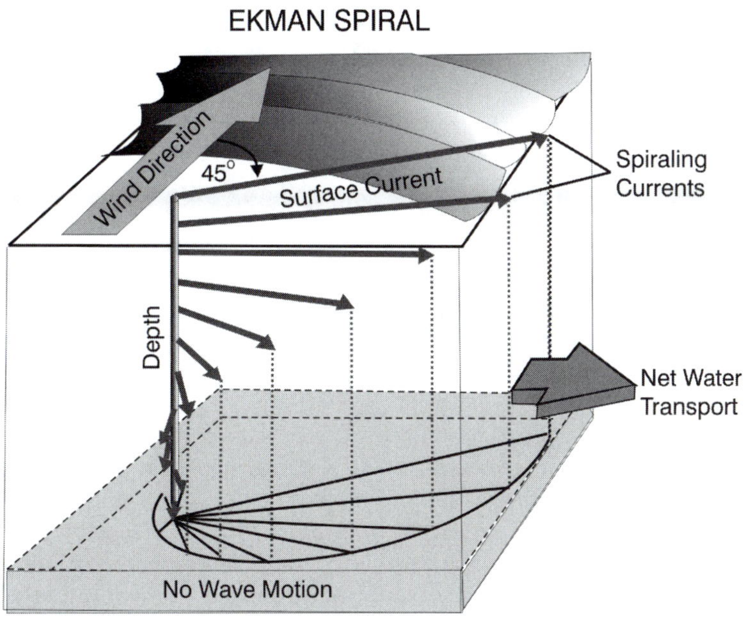

Figure E2. Several forces (surface winds, Coriolis Force, layers of water, etc.) create complex variations of water velocity and direction at various depths.

currents. It could also measure the strength and direction of ocean currents. Over the years it was improved. Today, high-precision instruments are used to measure various factors affecting ocean currents.

See also Airy; Coriolis

ELDREDGE–GOULD THEORY OF PUNCTUATED EVOLUTION: Biology: *Niles Eldredge* (1943–) and *Stephen Jay Gould* (1941–2002), United States.

Evolution of species and individuals by natural selection results from pressure brought about by relatively rapid changes in the environment.

The paleontologists Eldredge and Gould found 1) patterns in the fossil records indicate the abrupt appearance of new species, 2) evidence suggests the relatively stable **morphology** of present species, 3) there is wide distribution of transitional fossils, 4) there are apparent differences in the morphology between older "parent" species and current "daughter" species, and 5) distinct patterns can be found in extinct species. All of this evidence led to their theory of punctuated evolution.

Punctuated evolution is also referred to as *punctuated equilibria* or, at times, as *catastrophism*. Although still accepting many of the tenets of slow organic evolution, such as natural selection, their theory claims the evidence indicates that long periods of slow evolution were "punctuated" by very rapid environmental changes. Presumably, earthquakes, volcanoes, meteors or asteroid bombardments, or other catastrophes altered the atmosphere and thus the food supply and hastened other rapid changes in the environment. In other words, the abrupt appearance of a new species is the result of *ecological succession* and *dispersion*. One possible example is a 10-kilometer asteroid that impacted about sixty-five million years ago in Chicxulub, located in Mexico's Yucatan, which exploded dirt and other matter into the atmosphere, blocking sunlight. It is generally accepted as causing the elimination of most plant life on Earth. Once plant life was eliminated, so were the dinosaurs and many other animal species. Some primitive forms of life, such as bacteria, minute multicellular organisms, seeds, and small animals, survived and continued to evolve into new species. There are a number of large gaps in the fossil records of early organisms that indicate that evolution may not have been a continuous slow process, but new species were derived from the survivors of catastrophic ecological events. On the other hand, some of these gaps in the fossil records could be an indication that soft-bodied organisms were not fossilized during certain periods of Earth's development.

Today most scientists accept the concept that the continuing evolution of species is driven by natural selection. However, there is disagreement concerning the mechanisms of how and why new species appear, and how long it takes for genetic mutations and environmental changes to appear in organisms. One of the problems is "paleospecies" which makes a distinction between the classifications of ancient fossils as compared to the classification of the remains of species found in more recently observed populations. An incomplete fossil record may be created by several factors including geological and geographic changes, such as continental drift, mountain building, overlapping of stratified sedimentary layers of Earth, and other forces that have altered Earth's surface over time and amplifies the problem of paleospecies.

See also Buffon; Darwin; Gould; Lamarck; Wallace

ELION'S THEORY FOR CELL DIFFERENCES: Biochemistry: *Gertrude Belle Elion* (1918–1991), United States. Gertrude Elion shared the 1988 Nobel Prize in Physiology or Medicine with George Herbert Hitchings and James Whyte Black.

> *Nitrogen-based molecular compounds (purines) found in the nucleic acid of abnormal cells can be altered to cause death to the abnormal cells without causing death to normal cells, thus, a drug can be designed to kill the abnormal cells that cause normal cells to die.*

Gertrude B. Elion was born, raised, and educated in New York City during a period when few women joined the ranks of physicians yet alone research scientists. After receiving a degree in chemistry from Hunter College at the age of 19, she was unable to get into graduate school primarily because of financial considerations brought about by the Depression. However, she had a variety of jobs to pay for a master's degree in chemistry. She was the only female in the class of 1941 in New York University's master's degree program. During World War II when many men went off to war, a job opened at Burroughs Wellcome Laboratory working with Dr. George Hitchings (1905–1998) who allowed her to conduct her own basic research in virology and immunology. At this time she began her studies for a doctorate but never finished, which she always regretted.

While at Wellcome, she began studying nucleotides and purines and how they might be chemically altered and used as drugs for specific diseases. This was some years before James Watson and Francis Crick determined the structure of DNA.

The team of Elion and Hitchings developed several compounds designed to "cure" leukemia. They were tested on mice but proved to be too deadly for use on humans. By rearranging the structure of purine compounds, they developed the drug called 6-mercaptopurine (6-MP) that was approved for use with patients with terminal leukemia. It proved to be effective when used as part of a treatment with other drugs. Since 1988 about 80% of children suffering from acute leukemia can be considered as cured.

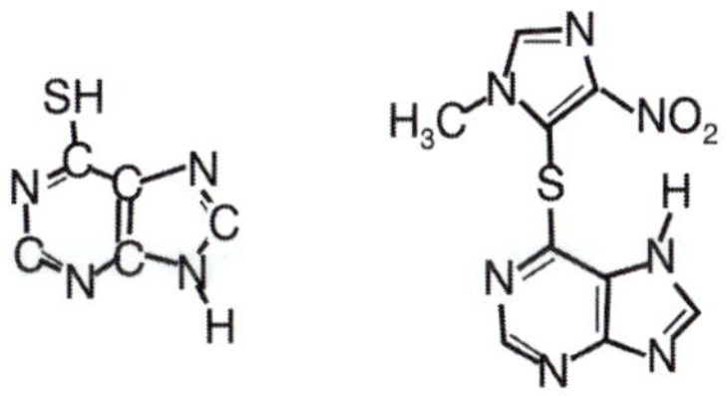

6-MP

6-mercaptopurine (6-MP) azathioprine

Figure E3. Elion and Hitchings altered a dangerous drug so it could be used to fight leukemia. They changed the structure of a purine compound to form a new drug called 6-MP that was more effective.

During the mid-twentieth century organ transplants were performed, but many of the transplanted organs were rejected by the recipient's own immune system. The immune system produces antibodies in white blood cells called lymphocytes that are designed to fight infections. Organs that were transplanted from one person to another were seen as "foreign" to the immune system (with the exception of organs from identical twins). Elion improved on the drug she called 6-MP so that it was more effective in controlling the production of the white blood cells. Thus, the immune system would not reject the implantation of the organ of one person to another. This made possible the thousands of organ implants performed today.

Elion, Hitchings, and the Scottish pharmacologist James Black (1924–) realized that a form of their designer drug 6-MP was altered in the body by an enzyme that also was responsible for the formation of uric acid. When the body produces an excess of uric acid, it results in a painful condition known as gout. A compound called allopurinol would block the production of uric acid and proved very effective for the control of excess uric acid and thus the relief of gout. Today, with modern medications gout is one of the easiest human conditions to treat successfully in a short period of time. Research at the Burroughs Wellcome laboratory continued to develop other designer drugs. Some were variations of the original 6-MP version designed specifically to control high blood pressure (Propranolol), gastric ulcers (Tagamet), and AIDS.

Upon retirement, Gertrude Elion worked for the World Heath Organization. Although she never received her PhD, she did receive many honors and several honorary doctorate degrees.

ELTON'S THEORY OF ANIMAL ECOLOGY: Biology: *Charles Sutherland Elton* (1900–1991), England.

As more species arrive in a given area, space and resources become a limiting factor, restricting the habitat and resulting in the extinction of some species while other species adapt to their limited (changed) environment.

Charles Elton, an early student of the science of **ecology**, performed many of his animal ecology studies on Bear Island off the coast of Norway. Elton named his concept "packing," evidenced by the island's limited number of existing species, which resulted from specialized evolution within the island's limited environment. He surmised that a species' place in the ecological environment is directly related to the availability of food as well as its predators. This idea led Elton to advance for the first time the concept of an interactive food chain, called a *food cycle* and often referred to as the *food web.* Bear Island's geology was suited to a limited number of plant species, which became food for the island's birds. The birds then became food for a limited number of the island's mammal species, which completed the island's community of species, and food chain. Because the mammals and most other species of organisms on the island could not migrate to escape any limitations of the packing imposed by environmental conditions, they would be subject to different evolutionary pressures than would exist if the animals were located in much larger, diverse environments. Animals living in an environment with a widespread limited number of species can "practice" environmental selection by migrating to change their habitat. Elton's theory of how a limited environment affects the types and distribution of plant and animal species is considered an important contribution to the theory of evolution. However, some question its applicability to humans despite humans' ability to change their environment to make it more suitable for habitation.

See also Darwin; De Vries; Dobzhansky; Haeckel; Wallace

ENDERS' THEORY FOR CULTIVATION OF VIRUSES: Microbiology: *John Franklin Enders* (1897–1985), United States. John Enders, Frederick C. Robbins, and Thomas H. Weller shared the 1954 Nobel Prize for Physiology or Medicine.

> *Viruses can be cultivated in test tubes by using selected live cells from animals with the addition of the penicillin antibiotic that will prevent bacterial growth in the culture, thus producing a safe and effective vaccine.*

John Enders, the son of a wealthy banker, attended Yale and Harvard Universities. While in college, he had difficulty in deciding on a career. In 1917 he left college to become a flight instructor during World War I. He later became a real estate agent and then entered Harvard to study languages. While at Harvard, he became interested in biology and changed his major to microbiology after entering Harvard Medical School.

Prior to 1948 whole live chick embryos, as well as the nerve cell tissues of live monkeys, were experimentally used to develop treatments for human viral diseases. However, the results were unsatisfactory. It was inconvenient and cumbersome to use live chick embryos, and live monkey nerve tissue created problems of rejection in the human body. John Enders established an Infectious Disease Laboratory in 1946 at The Boston Children's Medical Center and was the first to use just the living cells and not the entire living embryos in the development of viral cultures. In 1948, along with two other Americans, Frederick C. Robbins (1916–2003), a pediatrician and virologist, and Thomas H. Weller (1915–), also a virologist, he cultured the mumps virus using chick embryo cells mixed with ox serum and some added penicillin. He hypothesized that by adding penicillin to the viral culture, it would result in an attenuated virus, and thus a more effective treatment for several viral diseases would be possible. This was a continuation of the work on the mumps virus that he, along with several others, had begun in 1941. A similar improved technique was used to cultivate the poliomyelitis virus in 1949 and later in 1950 the measles virus. Enders, Robbins, and Weller received the Nobel Price in 1954 for their work on the poliomyelitis virus, an especially crippling and deadly disease of young children. Through their research, large quantities of the polio virus were produced and analyzed. Eventually this led to the mass manufacture and distribution of Sabin and Salk polio vaccines. It should be noted that the time period between the successful cultivation of a viral vaccine and its actual introduction into the general population is significant and not without setbacks. A successful polio vaccination program was not established until 1963; measles and mumps vaccine programs were developed in the late 1960s. Research on improvements to all these vaccines continued for sometime even after their widespread use.

See also Robbins; Sabin

EÖTVÖS' RULE (LAW): Physics: *Baron Roland von Eötvös* (1848–1919), Hungary.

There is a measurable relationship that relates surface tension, temperature, density, and the molecular mass of a liquid.

This "law" explains why surface tension causes water to rise on the inside walls of a capillary tube (a capillary tube is a glass tube with a very small diameter hole through its length). This action is due to the rounded shape of drops of water that causes them to slightly rise on and cling to the inside surface of the glass tube forming a concave surface on the water. The same force applies to the shape of the surface water in a drinking glass, but it is not as noticeable. Evötös devised a unique technique and device called the *reflection method* that made it possible to compare and measure the surface tension of various liquids. Using this sensitive reflection device, he was able to determine the surface tensions of several liquids and thus calculate their molecular weights. This relationship became known as Eötvös' rule that, in essence, states: *The rate of change of molar surface energy with temperature is a constant for all liquids*. This rule is so fundamental to all liquids that it is compared to the universal gas law.

During the last decades of his life, Eötvös conducted detailed studies of gravity, magnetism, and geophysics in general. In 1888 he developed an instrument, called the "torsion balance," designed to measure and to determine any difference between inertial and gravitational forces. It was constructed as a bar with two weights made of different materials attached at the end of the bar by a thin torsion fiber (this is similar to Cavendish's experimental instrument used to measure gravity). Eötvös decided to use this instrument on the smooth surface of a large frozen lake so as to minimize topological features that might interfere with sensitive measurements on land. By 1903 he had collected measurements from forty different locations. Later, his torsion balance data from the lake measurements were used to determine that the lake's axis was parallel to a tectonic line. This was the first time such geological data was established by use of a torsion balance and is now considered the birth of the field of geophysics research.

Eötvös also developed an instrument called the "gravity compensator," which was somewhat similar to a *variometer* that is curved with deflectors at both ends. The instrument could be rotated around a horizontal axis with the deflectors oriented in a vertical position to measure a zero balance. By changing the position of the deflectors, gravity could be measured with some accuracy. He also was concerned with the two forces that affect mass—the inertial force that is apparent when a force accelerates a mass, and the gravitational force that the mass experiences due to attraction to another mass. He determined that the difference between inertial to gravitational mass is very small, something like 1 to 20,000,000.

The *Eötvös effect* was experimentally demonstrated by considering the motion of boats in the water, as well as the effects of gravity on the boats. This effect was based on his experiments that indicated the weight of a moving body on Earth's surface is dependent on its direction and speed. In other words, the weight of the boat on Earth's surface changed as a result of the direction and speed at which they were proceeding. This experiment was conducted by observing one boat heading east and one west. This Eötvös effect is important in the study of gravity.

Eötvös was also interested in measuring magnetism. Using his torsion balance he determined the **declination** and **inclination** of Earth's magnetic field at various points on Earth. Evötvös attempted, however unsuccessfully, to determine Earth's magnetic inclination in past eons of time. Today it is known that the geo and magnetic poles have shifted since the beginning of Earth's existence.

See also Cavendish

ERASISTRATUS' THEORY OF ANATOMY AND PHYSIOLOGY: Biology: *Erasistratus of Chios* (c.304–250 BCE), Greece.

> *A three-way network of veins, arteries, and nerves supplies every organ and other tissues of the body with vital fluids; and the brain is composed of two sections with many folds and convolutions.*

Erasistratus of Chios (or Kos), which is an island off the coast of Greece in the Aegean Sea, founded an independent medical school for the study of anatomy and physiology of the human body. As with all the ancients, Erasistratus' ideas were mixed with both correct and incorrect assumptions. For instance, he rejected the humoral theory and incorrectly believed that air was pulled into the body by the blood as the blood moved upward in the body. He rejected the belief that the body was filled throughout with three kinds of vessels (veins, arteries, and nerves), and in the extremities of the body these vessels became so fine that they were invisible (today, they are known as capillaries). He was correct in his general theories of how organs in the body functioned, but his concepts of the actual physiology were often wrong. The many mistaken concepts of anatomy and physiology of these ancient times were the result of the acceptance of the writings of others, theosophical reasoning, or simply inaccurate observation. One example was his correct belief that the veins transported blood, but that the arteries and nerve tissues transported "animal spirits."

Erasistratus was the first person to make a distinction between the main parts of the brain. He identified the cerebrum as the larger portion and the cerebellum as the smaller part. He was also the first to note that the human brain had more folds or convolutions and thus more surface area than did the brains of other animals. From this he correctly concluded that this larger surface area was partially responsible for the greater intelligence of humans when compared with animals. His theory that the brain was the seat of intelligence was at odds with Aristotle who considered intelligence to be in the heart.

Erasistratus also disavowed the occult and its forces. Rather, he explained the existence of matter in terms of "atoms" (as proposed by Democritus of Abdera). An example: To explain bleeding from arteries, he believed that a vacuum was formed in the artery as the blood flowed out and that the artery was filled with blood again by the connecting veins. In other words, a vacuum cannot exist—it must be filled with something (atoms of blood), and in this case more fluid from the veins. In addition, Erasistratus' concept of disease was opposite of the long-held beliefs of Hippocrates. Erasistratus' concept of disease was mechanistic in the sense that the blood, fluids, spirits, food, and so forth provided a means of blocking any inflammation of the vessels in the human body.

See also Aristotle; Democritus; Galen; Hippocrates

ERATOSTHENES' MATHEMATICAL CONCEPTS: Astronomy: *Eratosthenes of Cyrene* (c.276 BCE–194 CE), Greece.

Eratosthenes' theory of prime numbers: *From a list of ordered numbers (1, 2, 3, 4, 5, 6, 7, 8, 9,...) strike out every second number after 2, every third number after 3, every fourth number after 4, and so on. The remaining numbers in the original list will be prime numbers.*

Eratosthenes of Cyrene, a poet, historian, and mathematician, developed the system of filtering, which became known as the *sieve of Eratosthenes*. Using his "sieve" procedure, the prime number is a positive integer that has no divisors except the integer itself and the first number selected. For example, select 2, for which the next two numbers are 3 and 4. Strike out 4, which is an integral multiple of the original number 2. A prime number is a positive integer having no divisors except itself and the integer (where the integer is any number, except zero, used for counting).

Eratosthenes' concept of measuring the circumference of Earth: *At the summer solstice (June 21) when the sun is at its zenith in the city of Syene (Aswan), it will be 1/50th of a full circle when measured by the angle of the sun at the city of Alexandria at the same time on the same day.*

Eratosthenes of Cyrene knew that on June 21 the sun cast no shadow in the bottom of a water well in Syene. Therefore, on this date and at this point, the sun was at its zenith. At the same time, he measured the angle of the shadow from a stick placed upright in the ground at the city of Alexandria, which he knew was five thousand stadia from Syene (*stadia* is the plural for the Greek *stadium*, the unit of measurement based on the length of the course in a stadium. It is equal to about 607 feet). Eratosthenes knew how many stadia a camel can walk in one day, so to estimate the distance between the cities he multiplied the distance a camel walks in one day by the number of days it took a camel caravan to make the journey. We now know that Syene (now called Aswan) is about 800 kilometers or 500 miles southeast of Alexandria in Egypt. On June 21 at Alexandria the angle of the stick's shadow was 7°12', which corresponds to about 1/50th of a 360° circle. Multiplying 5,000 stadia by 50 equals 250,000 stadia as Earth's circumference. Eratosthenes' calculation was very close to today's accepted equatorial circumference of 24,902 miles.

Using similar measurements Eratosthenes was able to calculate the tilt of Earth to its axis (the ecliptic, which is the inclination of Earth's equator to its orbital plane) as 23°51'20", which is also close to the modern figure of 23.4 degrees. Modern calculations still use Eratosthenes' geometric and algebraic methodologies to arrive at the current figures.

Using his method it was also estimated that the distance between the earth and the sun was 804,000,000 stadia and the distance to the moon was 780,000 stadia.

ERNST'S THEORY OF THE MAGNETIC MOMENT OF ATOMIC NUCLEI: Chemistry: *Richard Robert Ernst* (1933–), Switzerland. Richard Ernst received the 1991 Nobel Prize in chemistry.

> *Atomic nuclei have a* ***magnetic moment*** *that will align with strong magnetic fields, thus submitting the nuclei to specific pulsating frequencies of radio waves that can alter the nuclei's magnetic moment.*

In the 1940s the Austrian–American physicist Isidor Rabi, Swiss physicist Felix Bloch (1905–1983), and the American physicist Edward Purcell developed the technology of nuclear magnetic resonance (NMR) to probe and study characteristics of the nuclei of simple molecules. Nuclei have a natural polarity that align themselves with strong magnetic fields. By exposing them to selected frequencies of radio waves, nuclei realign themselves in a new energy state. When the radio waves are removed, they return to their original energy state, giving off specific radiation that can be used to identify the nuclei.

Richard Ernst subjected larger protein (organic molecules) to pulsating high-energy radio waves, which provided a means to produce images of living tissue. The process was originally called nuclear resonance because it "excited" the nuclei of atoms. However, the name was changed because people still mistakenly connect the nuclei of atoms resonated by NMR with the nuclear energy released by the atomic bomb. Once the process was improved, with better imaging techniques that could view cross sections of the human body, a similar process became known as magnetic resonance imaging (MRI, which is still based on the magnetic moment of atomic nuclei). Magnetic resonance imaging provides a better series of images than X-rays. With no danger of radiation exposure, MRI is safer than X-rays because the radio radiation used is of a much longer wavelength and lower frequency than X-rays. The improved images have greatly assisted diagnostic procedures for the medical profession because of its ability to detect various abnormalities in the body more accurately than X-rays.

See also Mansfield

ESAKI'S THEORY OF TUNNEL DIODES: Physics: *Leo Esaki* (1925–), also known as Esaki Reona or Esaki Leona, Japan. Leo Esaki shared the 1973 Nobel Prize for Physics with Ivar Giaever and Brian David Josephson.

> *By doping semiconductors with selected impurities, the quantum wave-like nature of electrons could "tunnel" through barriers resulting in diode-like properties. In effect as the voltage increases, the flow of electrons in a circuit will decrease, and vice versa.*

Some background on Esaki's theory of tunnel diodes may help explain the theory and the tunneling device and how it works. When Thomas Edison was experimenting with his light bulbs, he observed that carbon from the filaments would darken the inside of the glass bulbs. He experimented with several solutions. One was to place a piece of metal alongside the filament (close, but not touching each other) with the expectation that the metal strip would collect the carbon rather than its being deposited on the inside of the glass bulb. This piece of metal was called a "plate," and when it received a positive charge (the filament was negatively charged by the battery), the current continued to flow even though there was a small space between the plate and filament. However, he noticed that if the plate received a negative charge instead of a positive charge, there was no flow of current. This phenomenon was later called the

"Edison effect" and was the only new scientific principle he discovered. Edison patented this phenomenon but never exploited his discovery. Later, others used this effect in devising what is known as the cathode ray tube where a hot filament gives off electrons. When a positive plate was introduced into the tube, the electrons would flow toward the plate similar to a completed electrical circuit even though there was no connection between the filament and plate. If the plate was negatively charged, there would be no flow of electrons to it by the filament, thus it would act as an "on-off" switch to control the flow of current (*see* Figure C6 under Crookes). In the 1880s John Ambrose Fleming "rediscovered" the Edison effect and called the device a "valve." Because it had two parts—the filament and plate—it was also called a diode because it had two elements and soon became familiar as the old-fashioned radio vacuum tube. In 1906 Lee De Forest (1873–1961), the American inventor, added a third filament made of fine wires in the form of a "grid" placed between the two-diode elements. This third filament grid could be supplied with its own flow of electrons and thus act to control the flow of electrons between the two filaments. Thus, the vacuum tube with three filaments became known as a "triode." In a diode, the electrons either flow or do not flow, similar to an on-off switch, whereas in a triode the flow of electrons can be "modulated" enabling the triode vacuum tube to act as an amplifier of current electricity, and as a receiving device to pick up and amplify radio and TV signals. Herein lies the gist of the next phase of the story of how the Edison effect led to the development of the tunneling diode switch at the quantum level by Leo Esaki.

In 1958 Esaki "doped" the element germanium. Doping is a process of adding specific impurities to elements that alter their semiconducting characteristics at the quantum level. Thus, as the voltage increases, the current (amount of electricity) decreases, thus allowing the flow of electrons to "tunnel" through what is known as the narrow p-n (positive-negative) junction barrier created by the electrons' valence state. Classical mechanics states that wave-like matter cannot pass through such barriers, but *by doping the semiconductors with selected impurities the quantum wave-like nature of electrons could "tunnel" through this barrier resulting in diode-like properties. In effect as the voltage increases, the flow of electrons in the circuit will decrease, and vice versa.* In other words the semiconductor now acts as a diode, but with several advantages over the vacuum radio tube. The tunnel diode is much, much smaller than the vacuum tube (today this type of diode is known as a semiconducting chip). They consume much less power, produce much less heat, and are much more reliable and last much longer. Other improvements in quantum-level microchips, transistors, and integrated circuits, and so forth have made the concept of tunneling diodes indispensable for today's many types of electronics with fast speeds, small sizes, and reduced internal "noise" created by the circuits as they use much less electric power.

See also Bardeen; Edison; Shockley

EUCLID'S PARADIGM FOR ALL BODIES OF KNOWLEDGE: Mathematics: *Euclid* (c.330–260 BCE), Greece.

> *It is postulated that all theorems must be stated as deductions arrived at as self-evident propositions or axioms for which a person can use only propositions already proved by other axioms.*

First, some definitions:

- A *paradigm* in geometry is the general plan for the development of the logical statement. In science, it is referred to as a "ruling theory" or a "dominant hypothesis."
- A *postulate* claims something is true or is the basis for an argument, such as in geometry. Euclid set out five postulates: 1) A straight line can be drawn between two points. 2) A straight line can be drawn in either direction to infinity. 3) A circle can be drawn with any given center and radius. 4) all right angles are equal. 5) A unique line parallel to another line can be constructed through any point not on the line (parallel lines never meet).
- A *theorem* in mathematics is a proven proposition.
- *Deduction* is a method of gaining knowledge. A deduction is inferred in the statement "if-then" (from the general to the specific).
- A *proposition* is a statement with logical constraints and fixed values (e.g., if proposition A is true, *then* proposition B must also be true).
- An *axiom* is a self-evident principle that is accepted. Several equivalent synonyms for *axiom* are *primitive proposition*, *presupposition*, *assumption*, *beginning postulate*, and *a priori*. An example of one of Euclid's axioms is 1) things that are equal to the same thing are equal to each other; 2) if equals are added to equals, the wholes are equal; 3) if equals are subtracted from equals, the remainders are equal; 4) things that coincide with one another are equal to one another; 5) the whole is greater than any one of its parts. These five axioms can be summarized as, "The whole is equal to the sum of its parts."

Euclid's paradigm for knowledge led to his great achievement in the field of plane geometry. He brought together the many statements related to geometry into a logical, systematic form of mathematics. Euclid's *elements*, written in about 300 BCE, included thirteen books of what was then known in the field of geometry to which he applied his paradigm. It is still valid today.

EUDOXUS' THEORY OF PLANETARY MOTION: Astronomy: *Eudoxus of Cnidus* (c. 400–350 BCE).

To account for the irregular motion of the planets, earth must be at rest and surrounded by twenty-seven celestial spheres.

Eudoxus of Cnidus was one of the first ancient astronomers to attempt to account mathematically for the irregular motions of the planets and still maintain Earth as the center of the universe. His system required not only a motionless Earth, but also twenty-seven crystal-like celestial spheres. The sun and moon each had three spheres, and each of the known planets required four spheres to account for their motions. The outermost twenty-seventh sphere contained all the fixed stars; beyond that were the heavens (*see* Figure P5 under Ptolemy.) Eudoxus was able to describe mathematically the rising of the fixed stars and constellations over the period of one year.

See also Aristotle; Euclid; Ptolemy

EULER'S CONTRIBUTIONS IN MATHEMATICS: Mathematics: *Leonhard Euler* (1707–1783), Switzerland.

Euler's three-body problem: *The motions of an object moving three ways simultaneously can be predicted using Newton's three laws of motion.*

One example of how a body can move in three different directions at the same time is Earth's rotating on its axis, revolving about the sun, and proceeding as part of the solar system toward a distant galaxy. Leonhard Euler, interested in determining the motions of the moon, used analytical techniques that could be applied to the problem to derive a form of mechanics. He also devised a system to analyze how the three Newtonian laws of motion and gravity affected objects that exhibited three-way movements. Euler's equation was based on Newton's dynamics called the *mass point*, for a body that contains mass and is rotating about a point. Euler's equations of motion state that a set of three different equations will express the relationships between the 1) force of moments, 2) angular velocities, and 3) angular accelerations of a rigid but rotating body (e.g., Earth). From this he developed two theories for the motion of the moon that proved to be an asset for sea navigation. These motions of the moon were used before dependable clocks became available to determine longitude (*see also* Newton).

Euler's theory of notations: *It is possible to use notations such as sines, tangents, and ratios when the radius of a circle equals 1.*

Euler introduced notations in algebra and calculus that Lagrange, Gauss, Leibniz, Einstein, and others followed. He developed an infinite series of numbers that included notations such as e^x sin x, and cos x and the relation of $e^{ix} = \cos x + i \sin x$. Euler also wrote the first text on analytical geometry that explains such concepts as prime number theory, differential and integral calculus, and differential equations. The contributions made by Leonhard Euler are numerous, including several important mathematical equations, formulae, methods, constants, criteria, correlations, and numbers that were important to the development of mathematical theories by other mathematicians.

See also Einstein; Eratosthenes; Leibniz

EVERETT'S MULTIPLE-UNIVERSE THEORY OF REALITY: Physics: *Hugh Everett III* (1930–1982), United States.

> *The wave function of quantum mechanics describes alternate outcomes of events in the same universe.*

According to the Copenhagen interpretation of quantum mechanics, as proposed by Niels Bohr, the quantum mechanical wave function states that only a statistical probability is possible for any explicit event to occur. This traditional interpretation applies only to submicroscopic particles, not to the Newtonian macro world. Even Einstein had a problem with the quantum principle of uncertainty of determining a particle's position and momentum at the same time because at this level the principle of cause-and-effect may not apply. Hugh Everett proposed a different interpretation. He suggested that every possible outcome that may occur as an event could actually do so in the same universe, or possibly in multiple universes. Everett's interpretation, also referred to as the "many world interpretation" or the "relative-state model," discounts the Copenhagen interpretation of quantum wave function. His concept relates to as many large and small events and as many outcomes as one could possibly arrive at when measuring the universe. Some scientists discount his theory, but others are attempting to develop a new quantum theory that eliminates the special role of an observer from the process. It seems the observer may account for Heisenberg's uncertainty principle of indeterminacy. Therefore, if the observer is eliminated so might the uncertainty. Another possible means of justifying Everett's concept of a many-worlds universe is to use probability theory. The main objection to his theory is that it either requires many different outcomes from the same cause in one universe or it requires many parallel universes that do not communicate with each other.

See also Bohr; Feynman; Hawking; Heisenberg; Schrödinger; Schwarzschild

EWING'S HYPOTHESIS FOR UNDERSEA MOUNTAIN RIDGES: Geology: *William Maurice Ewing* (1906–1974), United States.

> *The thin crust of the ocean floor enables the sea floor to spread, producing the upward movement of basalt rock and the formation of massive, long, worldwide underwater mountain ridges.*

William Ewing, a geophysicist and oceanographer, employed seismic reflection technology to determine that the crust of the ocean floor is only 3 to 5 miles deep, as compared to the depth of 25 to 60 miles for the land crust. He was also aware that the Mid-Atlantic Ridge had been detected in 1865–1866 when the intercontinental communication cable was laid across the ocean floor. His theory extended the ocean ridge system to over 40,000 miles of underwater mountains worldwide. Ewing and the American geologist Bruce Charles Heezen (1924–1977) discovered the Great Global Rift, a split in one of the major submerged ranges that created a gap deeper and wider than the Grand Canyon. His theory led to the concept for the movement of continents and the six major tectonic plates, of which five support the continents while the sixth plate forms the Pacific Ocean, causing earthquakes and volcanoes to occur along the boundaries where these plates meet. Today the concepts of plate tectonics, continental drift, and midocean ridges and rifts are well-established phenomena which indicate that Earth is a dynamic planet.

EYRING'S QUANTUM THEORY OF CHEMICAL REACTION RATES: Chemistry: *Henry Eyring* (1901–1981), United States.

> *It is possible to determine the rate at which molecules break up and form new molecules by calculating the surfaces of the atoms and molecules involved as related to the temperature of the chemical reaction by using statistical quantum mechanics.*

It had been known for many decades that the rate at which a chemical reaction takes place is dependent on temperature (as well as some other factors). This relationship can be expressed by the following formula: $k = Ae^{-E/RT}$, where k is the frequency rate of the chemical reaction, Ae is the energy involved in the reaction, and T is the temperature of the reaction.

Henry Eyring's extensive work as a creative chemist was most evident in the field of chemical kinetics as related to quantum mechanics. Based on the work of other scientists who indicated that a relationship between energy levels of molecules and their surfaces could be calculated using quantum mechanics, Eyring determined that the energy required to start a chemical reaction must be overcome to begin the reaction. The formula that explains this concept is very complex and mainly involves temperature, but other stress factors may be "plugged" into the formula that determines the rate at which a particular chemical reaction will occur.

Henry Eyring, the son of American Mormon missionaries who became Mexican citizens, was born and raised on a cattle ranch in Juarez, Mexico. During one of Mexico's many revolutions, the family was forced to cross the border to El Paso, Texas, and soon moved to Pima, Arizona, where the young Henry finished high school. He entered the College of Eastern Arizona and in 1919 received a fellowship to the University of Arizona, receiving a degree in mining engineering and chemistry. He was an affable young man with great curiosity and high intelligence who decided to work on a PhD in chemistry at the University of California, Berkeley. From 1931 to 1949 he was an instructor, and later in 1938 a professor of chemistry at Princeton University. His last appointment was as dean of the Graduate School at the University of Utah where he remained until retirement in 1966.

The Eyring theory has become known as the "Eyring model" because it can be applied in many situations involving multiple stresses. It can also be used to determine the degradation and failure date of systems, and how temperature is related to failed materials and mechanisms. One drawback of the Eyring model is that it does not directly address other stress factors. Thus, many simplified versions of the formula have been devised to address specific types of stress other than temperature.

F

FABRICIUS' THEORY OF EMBRYOLOGY: Biology: *Girolamo Fabrizio* (1537–1619), Italy.

> *The chalazae (spiral threads that suspend the yolk inside the egg) produce the chick embryo, while the sperm, yolk, and albumen of the egg merely furnish nourishment for the forming embryo.*

Although Fabricius' theory about the formation of the chick embryo was incorrect, he based his ideas on direct, but misconstrued observations. He did, somewhat accurately, describe the development of the chick embryo in the egg from the sixth day of fertilization. However, he did not realize the importance of male semen in the fertilization process or the earliest periods of embryonic development. He also studied the early developmental stages of eggs and placenta tissues of many vertebrates including mammals, reptiles, and sharks, in addition to birds and chicken embryos. His research led to the publication of two books: *De formato foetu* (On the formation of the fetus) in 1600, and *De formatione ovi et pulli* (On the development of the egg and chick) in 1612. These books contained well-drawn, descriptive illustrations based on his many observations and are considered to be the beginning of embryology as a field in the biological sciences.

As was common in the past, he had several names. *Fabricius Hieronymus* was his Latin name; *Fabrizo d'Acquapendente* is the name he was given for the town in which he was born, and his Italian name was *Geronimo Fabrizio*. Fabricius was not only a student of the great anatomist Gabriello Fallopius but he succeeded him as head professor of anatomy at the University of Padua, teaching there for fifty years. While at Padua, he built the first known anatomy theater that was used for demonstrations and the teaching of anatomy. It still exists today and continues to be used for surgical demonstrations. He was also the mentor to William Harvey, the renowned English physician and anatomist.

Some of Fabricius' research involved studies of the larynx, esophagus, respiration, muscle reactions, the stomach and intestines, the eye, and ear, all of which were done with animals rather than humans. At this time in history, dissections on human cadavers were extremely limited, often unlawful, and rare. He was the first to demonstrate the existence of "valves" within veins, although he had no idea of the function of these folds of tissue within veins. His student William Harvey followed up on this discovery and became noted for his study and description of the body's circulatory system.

Although Fabricius retired from his position at Padua after fifty years as professor of anatomy, he continued his research until he died at age eighty-six.

FAHRENHEIT'S CONCEPT OF A THERMOMETER: Physics: *Daniel Gabriel Fahrenheit* (1686–1736), Germany.

The temperature required to reach the boiling point for a liquid varies as to the atmospheric pressure. Thus pressure affects the temperature reading.

Daniel Fahrenheit, glassblower and maker of scientific instruments, knew of Galileo's thermoscope, which used the change in the volume of air (density) to indicate changes in temperature (see Figure G3 under Galileo). The *thermoscope* was inaccurate because it relied on the effects of atmospheric pressure on the water encased in the instrument. The first closed water thermometer was designed in the mid-1600s by either Ferdinand II, the grand duke of Tuscany, or the astronomer Olaus Romer. Romer improved the design by using wine rather than water that provided some alcohol to prevent freezing when ambient temperatures were below the freezing point of water. These designs responded to temperature changes, but not atmospheric pressure, as did Galileo's air thermometer. Another problem was that the water and alcohol mixture still created internal pressure changes and froze and boiled at temperatures just beyond normal ranges, thus reducing its precision and usefulness.

In 1714 or 1715, Fahrenheit improved the design by enclosing mercury in a glass tube, similar to today's mercury thermometers. He also devised an improved scale by selecting one without fractional units. His design placed the mercury in a vacuum within a sealed glass tube, which eliminated the effects of atmospheric pressure, as well as normal freezing and boiling problems. Fahrenheit then combined ice and salt to determine 0° on his scale, which had each degree divided into four divisions. He then placed it in his mouth to determine human body temperature as 96°, eventually corrected to 98.6°. Later, his thermometer and scale were calibrated to establish 212° as the boiling point and 32° as freezing of water at sea level. The Fahrenheit scale is used only in English-speaking countries, particularly the United States. Scientists worldwide use the more appropriate metric Celsius and Kelvin scales.

See also Celsius; Kelvin

FAIRBANK'S QUARK THEORY: Astronomy: *William Fairbank* (1917–1989), United States.

Quarks originate from high-energy cosmic rays. Therefore, their electrical charge can be detected and measured.

In 1964, Murray Gell-Mann, a particle physicist, postulated the existence of **quarks** that are strange, basic, subnuclear particles composed of protons and neutrons. Each had an *antiquark* and needed fractional electrical charges of either −1/3 or +2/3 to produce other particles. Also, they were thought to be not artificially producible because they were beyond the energy range of particle accelerators.

In 1977, using a sensitive device similar to the Millikan oil drop experiment (*see* Millikan), Fairbank determined the electric charge of an electron. He placed a tiny ball of the element niobium (about 0.25 mm in diameter) between two charged metal plates that were kept at a temperature near absolute zero. As a cosmic ray passed through this device, a small electrical charge formed on the ball, which could be measured as a change in the electrical field between the plates. The strength of the charge was extremely small (−0.37) and may have been caused by sources other than cosmic rays. There is one theory that says quarks cannot be produced because they are not "free." The question of magnitude of the charge on a quark is still being investigated. Twelve different types of quarks have been discovered, and now physicists believe the quarks in protons and neutrons are "confined" or held together by the "strong force" of gluons.

See also Friedman; Gell-Mann; Glashow; Nambu

FAJANS' RULES FOR CHEMICAL BONDING: Chemistry: *Kasimir Fajans* (1887–1975), Poland and United States.

Rule #1: *When the number of electrons increases for an atom, its ions obtain a higher charge and thus are difficult to form. Therefore, they are more likely to form covalent bonds rather than ionic bonds.*

Rule #2: *Large cations are more likely to form ionic bonds with smaller anions rather than covalent bonds.*

These two rules apply to the similarity of elements that are close neighbors on the Periodic Table of Chemical Elements. The rules explain the difficulty and the ease with which atoms gain or lose electrons, as the atoms become ions. It is almost impossible to form large highly charged ions. Therefore, covalent (sharing) bonds with another ion are more likely as the number of electrons to be removed or donated increases. Conversely, ionic bonds are more likely to form as the number of electrons to be removed or donated from large atoms becomes greater. This results in the rarity of highly charged ions. The second rule is more self-evident in that it states that ionic valences are more likely to form between large cations (positively charged ions or radicals that are attracted to the positive "anode" in electrolysis); and small anions (negatively charged ions or radicals that are attracted to the positive "cathode" in electrolysis). Two examples evident on the Periodic Table are the Earth metals located in the upper left-hand area of the table, namely 1) Groups I and II [Lithium (Li), Beryllium (Be), and Magnesium (Mg)] and 2) Groups III and IV [Boron (B), Aluminum (Al), and Silicon (Si)].

Fajans' law that was conceived in 1913 states that elements that emit alpha particles (positive helium nuclei) will decrease in their atomic numbers (and weights) and thus become positive ions as well as isotopes, whereas elements that emit beta rays (high-speed negative electrons) will gain electrons and thus become negative ions. He also formulated the law explaining how radioactivity moves. This law explained valence in chemical bonding and was later independently discovered by the British chemist

Kasimir Fajans was born in Warsaw, Poland, and educated at several institutions located in Heidelberg, Zurich, and Manchester, England. He emigrated to the United States in 1936 to serve as professor of chemistry at the University of Michigan. Fajans' experience with the Nobel Prize Committee is an example of the fickleness in which that institution occasionally operates. In 1924 Fajans was a candidate for the Nobel Prize in Chemistry and Physics and was expected to be a "winner" by all concerned. Even a Swedish magazine asked Fajans for a photograph to be used with the announcement of his award. On the day before the winners were announced, the magazine had already published an article that Fajans had won the prize. The next day the Committee changed its mind and announced there would be no winner in this category for 1924, ostensibly because the Committee wanted to chastise the magazine for its indiscretion in printing the article. In future years Fajans was again a candidate several times but never did win a Nobel Prize even though Frederick Soddy did receive the Nobel Prize for Chemistry and Physics in 1921 for his research related to Fajans' discoveries related to radioisotopes. In future years, Fajans received many awards from other institutions and academies.

Frederick Soddy. It became known as the Soddy–Fajans Method. In cooperation with Otto Göhring (1879–1968), Fajans discovered the radionuclide of the new element protactinium in 1913. Fajans and Göhring also jointly discovered the formula that defined the conditions for the precipitation of radioactive materials that could be used to separate small amounts of radioactive substances.

FALLOPIUS' THEORIES OF ANATOMY: Biology (Anatomy): *Gabriel Fallopius* (1523–1562), Italy. In Italian, his name was Gabriele Fallopio of Modena.

The female reproductive anatomy includes the ovaries that connect into "trumpet" (tuba) shaped tubes leading to the uterus.

The detailed anatomical examinations of the female reproductive organs that were done by Fallopius led to better understanding of how this system functions. The "tubes" through which the eggs pass became known as the "fallopian" tubes. Although Fallopius did not understand exactly the functions of these organs, he did realize that they were related to the reproductive process. He was also the first person to describe in detail the clitoris, and he coined the term "vagina" which he knew received the sperm from males during copulation. He was the first person to record experimental tests of different type sheaths made of linen and parchments designed to fit over the penis that could prevent syphilis during sexual intercourse. One might say he was the first to scientifically conduct research of the efficacy of condoms. His main concern was not contraception but rather the prevention of disease. Several hundred years would pass before it became known and understood how ova (eggs) form in the ovaries and pass through the fallopian tubes to the uterus to become an embryo and develop into a fetus. There are many specific parts of the female reproductive anatomy that have acquired his name, such as, fallopian aqueduct, fallopian canal, fallopian arch, fallopian ligament, fallopian neuritis, fallopian pregnancy, aqueducts fallopii, and tuba fallopiana.

Gabriel Fallopius became a greater anatomist than his famous teacher, Andreas Vesalius, the famous Belgian anatomist. Fallopius' writings explain his work with the skull and, in particular, the ear and auditory system. Fallopius introduced the terms "cochlea" and "labyrinth." He also described the larynx, respiration, and how muscles perform within the body.

FARADAY'S LAWS AND PRINCIPLES: Physics: *Michael Faraday* (1791–1867), England.

Faraday's laws of electrolysis: *1) equal amounts of electricity will produce equal amounts of chemical decomposition* and *2) when using an electric current, the quantities of different substances deposited on an electrode are proportional to their equivalent weights.*

Michael Faraday, who was Sir Humphry Davy's laboratory assistant, continued Davy's work on the electrolysis of chemical substances by passing electricity through chemical solutions. Davy demonstrated that sodium and potassium metals were deposited on the two electrodes in a solution of sodium chloride and potassium chloride (both salts) through which an electric current was passed (see Figure A7 under Arrhenius). Faraday went one step further and measured the amount of electric current being used and its effect on the deposition of either the sodium or potassium on one or both electrodes. His hypothesis was that the chemical action of a current is constant for a proportional amount of electricity. The equivalent weight of a chemical is the *gram formula weight*—the sum of the atomic weights of the elements as expressed in the formula, related to a gain or loss of electricity (electrons). The amount of electricity required to cause a chemical change of one equivalent weight is the unit named a *faraday*. The faraday constant is equivalent to 9.6485309×10^4 coulombs of electricity. Faraday's laws of electrolysis have been used over the past one hundred years to produce all kinds of chemicals. For example, electrolysis can be used to produce hydrogen and oxygen gases by breaking down water molecules in a weak electrolytic solution. No relationship exists between Faraday's laws of electrolysis and the cosmetic process of electrolysis for hair removal.

Faraday's principle of induction: *An electric current can produce a magnetic field; conversely, a magnetic field can produce an electric current.*

Oersted and Ampère had previously demonstrated that when electric current flows through a wire placed over a compass, the magnetic needle of a compass is deflected. Faraday rejected the then-current belief that electricity was a fluid, and with great insight, he saw electricity as one of several "uniting forces of nature," which he included with magnetism, heat, light, and chemical reactions.

Faraday recognized a connection between the actions of electrical lines of force and the magnetic lines of force. He devised an iron ring with a few turns of wire wrapped around opposite sides of the ring. The wires did not touch each other because he used twine to keep them separated. First, he connected a battery to the two ends of the wire on one side and a **galvanometer** to the two ends of the wire on the other side. When the electrical connection was made on the side with the battery, the needle on the galvanometer on the other side of the iron ring moved. Next, he tried the same experiment without the electric battery by passing a bar magnet through a ring that had a coil of wire wrapped around it. Again, the needle of the galvanometer moved when attached to the ends of this coil. His interpretation was that lines of "tension," as he called the lines of force of the magnetic field, created an

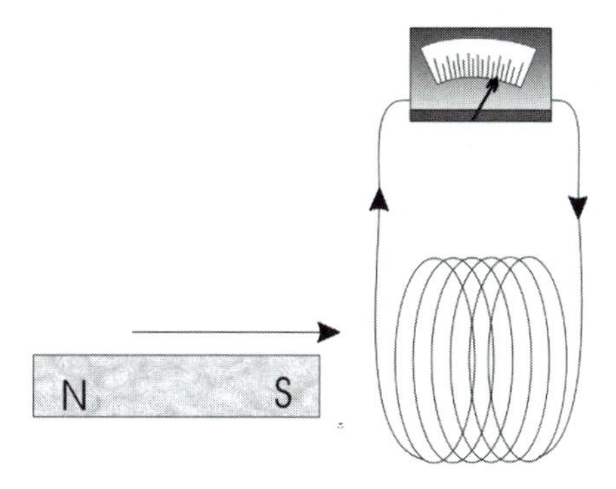

Figure F1. When the magnet is moved into and out of a coiled wire, an electric current is "induced" in the coil by the influence of the moving magnetic field. This is known as the *dynamo effect.*

electric current as the magnet moved through the ring (coil) of wire. Thus, an electric current was "induced" in the wire by the moving magnet (see Figure F1).

Faraday was not an accomplished mathematician. Others, particularly James Clerk Maxwell, developed the mathematics required to make Faraday's concept of induction into a viable field theory. By all and any scientific landmarks, recognizing electromagnetic induction was one of the most important human insights. The concept of induction resulted in the development of the dynamo, or electric generator, which produces electricity by mechanical means, and thus led to the modern age of electricity. Induction (brushless) electric motors have many modern applications, including the small induction motors that run the hard drives of personal computers (*see also* Edison; Henry; Tesla).

Faraday's principle of dielectrics: *The conductivity of different substances has different specific inductive capacities for the dissipation of electrical power.*

The *inductive capacity* refers to how much *permeability* and *permittivity* a substance exhibits; the *dissipation* of electrical power is the rate of heat loss within the system. This principle states that some substances are very poor conductors of electricity and that induction of electricity relates to the "dielectric" nature (the degree of insulating properties) of the substance. The dielectric strength (permittivity) of a substance is related to how much electricity can be passed through it without breaking down the material. Being able to calculate the dielectric nature of a substance is very important for many industrial uses, including the manufacturing of semiconductor computer chips. As an example, materials with high dielectric constants make excellent **capacitors,** which are important in the electronics industry because they can be made very small and still do the job. Knowing the dielectric properties of substances becomes important when looking for material suitable to make insulators, capacitors, and microelectronic components.

Faraday rotation effect: *The plane of polarization of polarized light will be rotated when passed through a magnetic field.*

Faraday's work with electricity and magnetism led him to explore relationships between light and magnetism. He demonstrated that polarized light can be altered when influenced by magnetic fields of force. This *Faraday effect* was developed for instruments used to study the molecular structures of many compounds and later was useful in explaining magnetic fields in the other galaxies of the universe. Faraday's other contributions include the discovery of benzene (C_6H_6) and two new chlorides

Michael Faraday was an excellent thinker and experimenter with the ability to relate his abstract ideas into understandable presentations for his audiences during his famous lectures. His most famous were his popular seasonal Christmas-time lectures for children that he began in 1826; the tradition has been continued ever since and attracts a full audience of young people.

As a young scientist, Michael Faraday was basically a theoretical and experimental chemist. When he was twenty-nine years of age, he synthesized several new chlorine compounds C_2Cl_6 and C_2Cl_4, and several years later he discovered benzene C_6H_6. However, the shape of the benzene molecule eluded him (the "ring" shape of the benzene molecule was later discovered by Fredrich Kekule; *see* Figure K1 under Kekule). Faraday's concepts related to electrolysis were based on his ability to theorize the process of electrolysis, but he also used experimental research to measure chemical and physical changes on the electrodes (cathode and anode). An important concept of electricity is related to how much is required to liberate one mole of individually charged ions. This led to what is known as the *Faraday constant* which is expressed as: $F = N_A{}^e$, where *F* is the Faraday constant, N_A is the Avogadro constant, and $^{e-}$ is the electrical charge on a single electron.

of carbon, as well as the system for liquefying several common gases, including chlorine.

See also Ampère; Maxwell; Oersted

FERMAT'S PRINCIPLES AND THEORIES: Mathematics: *Pierre de Fermat* (1601–1665), France.

Fermat's combination theory: *Combinations of units are based on the concept of probability.*

Pierre de Fermat and Blaise Pascal are credited with developing theories of probability. *Probability* is the ratio of how many times an event will occur as related to the total number of trials conducted.

A famous example of Fermat's theory of combinations follows. In a game where there are just two players, player Allen (A) and player Bill (B), player A wants at least 2 A's or more in a combination of four letters to "win" a point, while player B wants at least 3 or more *B*'s in a four-letter combination to gain points (Because 3 is greater than 2, Bill felt this higher number would win more combinations). There are 16 possible combinations for these two letters: *AAAA, AAAB, AABA, AABB, ABAA, ABAB, ABBA, ABBB, BAAA, BAAB, BABA, BABB, BBAA, BBAB, BBBA, and BBBB*. Every time an A appears at least two times or more in a combination, Allen will score a point. At the same time, player Bill requires *B* to appear at least three times or more within a combination to win a point. Note that there are eleven "wins" based on two or more A appearances within the sixteen combinations for player A; while there are only five cases containing at least three or more *B* appearances within the sixteen combinations for player B. The odds for Allen (A) winning the game over Bill (B) are eleven to five. Also, it is most likely (statistically probable) that A would win the game after only four random selections of combinations.

Fermat's last theorem: *For the algebraic analog of Pythagoras' theorem for a right triangle, there is no whole number solution for the equation* $a^n + b^n = c^n$ *(e.g.,* $3^2 + 4^2 = 5^2$*; or* $9 + 16 = 25$*) for a power greater than 2.*

Integers are positive or negative whole numbers that have no fractional or decimal components and can be counted, added, subtracted, and so forth. Pythagoras' theorem states that the square of the length of the **hypotenuse** in a right triangle is equal to the sum of the squares of the lengths of the other two sides of the triangle (see Pythagoras).

Although Fermat had an interest in the theory of numbers, as well as making several contributions to this field, Fermat's record keeping was poor. The equation for the "last

Pierre de Fermat did not publish much of his work during his productive years. Although a few fellow mathematicians knew about his ideas, his most important work did not become known until after his death. Proof of his "last theorem," also known as the "great theorem," obtained a reputation of insolvability that has intrigued amateur as well as professional mathematicians for ages. In the early 1900s over one thousand proofs were presented for his theorem; all were proven incorrect. After computers became available for number "crunching," an English mathematician Andrew Wiles (1953–) presented a proof that looked promising. Unfortunately, his proof had some problems that were finally worked out with assistance from a mathematician from Cambridge, England. Finally, in 1995 Wiles derived a version of the proof for Fermat's last theorem that seems to be correct. Wiles does not believe that Fermat had a secret "proof" that, for unknown reasons, he chose not to have published. Rather, at that time in history, neither the knowledge nor tools to solve the problem were available.

theorem" was written as: $a^n + b^n = c^n$ (or as $x^n + y^n = z^n$), and if the *n* is an integer greater than 2, there is no whole number solution. Fermat professed to have solved this problem, but his solution has not yet been found. For almost four hundred years, scholars have attempted to solve this mathematical conundrum. It seems that when *n* is 2, it is possible to express the value for *a*, *b*, or *c*, but once whole numbers greater than 2 are used for *n*, it cannot be calculated. Some claim to have arrived at a proof for $n = 3$, $n = 4$, $n = 5$, and $n = 7$, or even $n = 14$, but only when using prime and complex numbers.

Fermat's least time principle for light: *Electromagnetic waves (light) will always follow the path that requires the least time when traveling between two points. In addition, light will travel slower through a dense medium than one less dense.*

Fermat related light to mechanics in the sense that light followed mechanical principles, such as expressed in geometry and the physics of his day. For instance, the *principle of least action*, originally postulated by Aristotle as the *economy of nature*, was also used by Fermat to describe the behavior of light under different circumstances. He based his theory on analytical geometry, which showed that the path of light reflected from a flat surface always took the shortest distance, but for an elliptically curved surface, it took the longest path. Fermat's theory was later restated as the *wave theory of light* during the period when the principle of least action was applied to wave mechanics and quantum mechanics.

See also De Broglie; Descartes; Hamilton; Schrödinger

FERMI'S NUCLEAR THEORIES: Physics: *Enrico Fermi* (1901–1954), United States. Enrico Fermi was awarded the 1938 Nobel Prize for physics.

Fermi's theory for slow neutrons: *Since slow neutrons have no charge and less mass than alpha particles, they are capable of overcoming the positive charge on atomic nuclei, thus allowing them to enter (react with) the atomic nuclei to increase their atomic weight but not their charge, thus producing isotopes.*

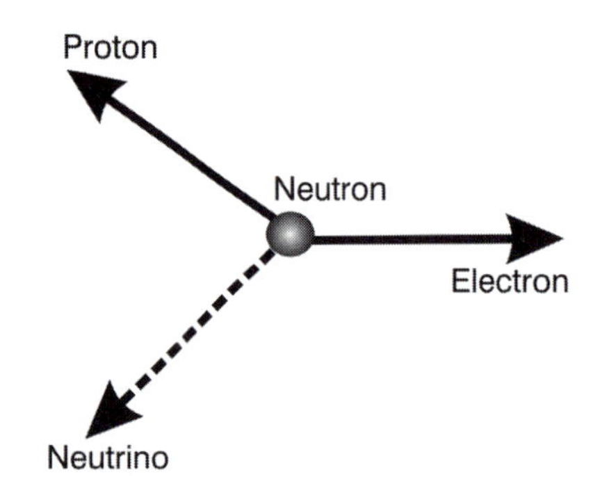

Figure F2. Enrico Fermi predicted that when a neutron disintegrated, the products formed were a proton, an electron, and a very small unknown particle with less than 1 percent the mass of an electron, or possibly no mass at all. This new particle was named the neutrino ("the little one") that became known as the "ghost" particles. In 1956, Frederick Reines and Clyde Cowan confirmed the existence of the neutrino.

Neutrons are found in the nuclei of atoms and have approximately the same atomic weight as protons but no electrical charges. Both protons (with +1 charge) and neutrons (with 0 charge) belong to the group of particles known as "baryons" that consist of quarks (*see* Gell-Mann). Neutral neutrons can be stripped from certain types of nuclei and used as particles to combine with other target nuclei, thus increasing the atomic weight and instability of the target nuclei. Alpha particles are the nuclei of helium atoms composed of two positive protons (++He). Thus, an alpha particle is approximately twice the weight of a single neutron. An isotope of an element whose atomic nucleus is composed of a specific number of protons is also defined by the number of neutrons in the nuceli. Thus, isotopes of an element have the same atomic number (protons) but different numbers of neutrons, thus different atomic weights.

Enrico Fermi considered using the neutron as the "bullet" to bombard atomic nuclei to produce isotopes since it has no charge and therefore little resistance to target elements. At that time he could not produce adequate numbers of neutrons for this action. On a hunch, he placed a thin sheet of paraffin (petroleum wax) between the neutron source and the target. To his astonishment, the production of neutrons increased about 100-fold because they were being slowed by the hydrocarbon molecules in the paraffin, and thus were more easily detected. Because more neutrons were produced and, more important, were slowed down, they would not "bounce" off the target nuclei, as did high-speed neutrons. Thus, it provided a better opportunity for the neutrons to interact with the nuclei of the target and produce isotopes of that element by adding neutrons to the target element's nuclei, thus increasing the element's atomic weight as well as creating an isotope of that element.

Fermi used his new technique to produce numerous radioisotopes of several elements. His theory of slow neutrons was of extreme importance to the new field of nuclear science, which soon led to the nuclear bomb, production of electricity by nuclear power plants, and the production of radioisotopes used in industry and medicine.

Fermi's theory of beta decay: *When a neutron decays, it is converted into a positive proton plus a negative electron and an antineutrino.*

The equation for this reaction is: ($n \rightarrow p + e^- + v$) where *n* is the neutron; *p* is the proton, *e* an electron, and *v* the **antineutrino** (see Figure F2).

This reaction led Fermi to speculate on a new force he named the "weak force" or "weak interaction," which is responsible for the beta decay process of the neutron. Beta decay occurs when the nucleus emits or absorbs an electron or positron, either increasing or decreasing the element's atomic number but not its atomic weight. The beta decay phenomenon resulted in additional research and discoveries related to the structure of the nuclei and radioactivity.

Fermi's theory for a self-sustaining chain reaction: *A self-sustaining chain reaction can be produced by stacking uranium oxide (uraninite ore) and graphite blocks (carbon as a moderator) into an "atomic pile" to produce slow neutrons that will interact with the small amount of fissionable uranium-235 (U-235) in the refined ore.*

After the discovery of the neutron in 1932, a number of scientists began to work with nuclei, neutrons, other elementary particles, and nuclear fission reactions. Familiar with this

On December 6, 1941 (one day before the bombing of Pearl Harbor by the Japanese), this ultra-secret Manhattan Project was tasked with demonstrating that a self-sustaining chain reaction was possible and could be used to convert a small mass of uranium into tremendous amounts of energy, as theorized by Einstein's famous equation, $E = mc^2$. For this purpose, an atomic pile was constructed beneath the stands at the University of Chicago's football stadium. It was designed to slow the neutrons to the extent they would produce adequate fission of the nuclei, which at some unknown point would enable the reaction to sustain itself. The pile was extra large to determine the amount of U-235 needed to reach a **critical mass**, resulting in fission of the U-235. The scientists had one problem. If this critical mass was reached, what would prevent the pile from becoming an exploding bomb? To solve this problem, Fermi and his colleagues devised a means of inserting cadmium rods into the pile to absorb the neutrons before the whole thing became unstable. As the rods were slowly removed, the number of slow neutrons needed to react with the U-235 could be controlled. There were also so-called delayed neutrons produced that kept the pile from going out of control by providing a brief period of safety before the rods needed to be reinserted. This happened on December 2, 1942. Then the neutron-absorbing rods were reinserted, and history was made.

science, Fermi was charged with the supervision of the Manhattan Project that was composed of a group of physicists attached to various universities. They became volunteer consultants who advised the U.S. government on the Manhattan Project and on the development of the atomic bomb and other classified military projects.

Today we know it takes only about 15 pounds of the rare form of U-235 to reach a critical mass, and less than 10 pounds of fissionable plutonium Pu-239 is required to form a self-sustaining chain reaction, and thus an explosive device. Enrico Fermi went on to assist other scientists with the construction of the atomic (nuclear fission) bomb and the H-bomb (thermonuclear fusion bomb).

See also Hahn; Meitner; Pauli; Oppenheimer; Steinberger; Szilard; Teller

FESSENDEN'S CONCEPT OF THE THERMIONIC DIODE: Physics: *Reginald Aubrey Fessenden* (1866–1932), Canada.

The heated negative cathode and the positive cool anode that make up a diode can accommodate a third electrode between them to form a triode. The third electrode can be used to modulate the amplitude of a wireless carrier signal.

The concept of thermionic emission within diodes was referred to as "tubes" in the United States and "valves" in England. The word "diode" is derived from the word di-electrode meaning two electrodes. The di-electric vacuum tube rectifier was first invented by the English electrical engineer and physicist Sir John Ambrose Fleming (1849–1945) in 1904. In essence, when the negatively charged metal cathode electrode inside a vacuum tube is heated and the positively charged anode is kept cool, the electrons that are "boiled" off the cathode form a cloud of electrons around the heated cathode that are attracted to the anode. This "boiling off" of electrons is called the "thermionic emission" because the electrons from the metal electrode are produced by the heat.

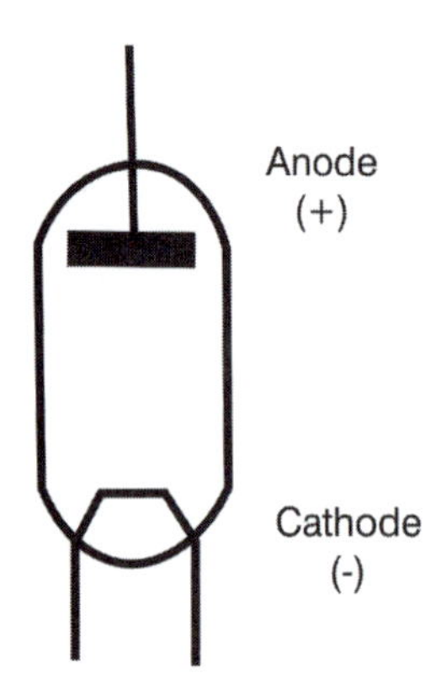

Figure F3. An artist's depiction of a simple diode.

Reginald Fessenden inserted a third electrode that could vary the flow of electrons from cathode to anode and thus vary the amplitude of a steady high-frequency radio signal. Thus the frequencies of sound waves could be modulated to correspond with the frequencies transmitted over the radio waves. This third electrode in the triode made it possible to modulate the flow of electrons from the cathode to the anode so the signal could act as a "carrier wave" for audio (sound) transmissions.

From early childhood Fessenden was interested in the possibility of a wireless telegraph and eventually developed an oscillating triode system that was the basis for his first successful self-sustaining wireless system that was a two-way (sending and receiving) system. On Christmas Eve in 1906 he transmitted the first historic audio radio broadcast from Brant Rock, Massachusetts. Reginald Fessenden was somewhat of a child prodigy and as a youngster became interested in how to send Morse code without the use of wires. He also witnessed a demonstration of Alexander Graham Bell's invention, the telephone. After training as an electrician, Fessenden worked with Thomas Edison in an attempt to find better insulations for electrical wires. He was a professor of electrical engineering at several universities, worked for the U.S. Weather Bureau, and tried to exploit his own inventions without much success. Fessenden formed the National Electric Signaling Company (NESCO) with two businessmen that provided him an opportunity to continue his research and the development of devices for wireless transmission of sound. A lifelong impediment to a more widespread recognition of his accomplishments was his personality that did not lend itself to compatibility. He had a fear, some say, of being "robbed" of his ideas and inventions. Eventually, his backers at NESCO fired him. Fessenden sued them and was awarded over $400,000 that resulted in the company's bankruptcy. Fessenden had many ideas and was a prolific inventor with over five hundred patents in his name.

See also Edison; Esaki; Tesla

FEYNMAN'S THEORY OF QUANTUM ELECTRODYNAMICS (QED): Physics: *Richard Phillips Feynman* (1918–1988), United States. Richard Feynman, Julian Schwinger, and Sin-Itiro Tomonaga shared the 1965 Nobel Prize for Physics.

> *Quantum electrodynamics joins three seemingly unrelated physical phenomena: Einstein's concepts of relativity and gravity, Planck's quantum theory, and Maxwell's electromagnetism.*

This theory is also known as the quantum theory of light, which synthesizes the wave and particle nature of light with the interaction of radiation and electrically charged particles.

Richard Feynman's theory grew out of the problem presented with the interactions between electrons and photons of light in which the strength of the charge on the electron mathematically came out to zero, while in actuality the basic unit of electricity is the electron. This did not coincide with known properties of electrons, so a new approach was required. Feynman's concept was based on the probability that if something does happen, it can also happen in many different ways. Thus, the probability of locating an electron in any particular space is like saying it would be in all the places it probably could be. To assist in understanding his version of the quantum theory, he designed what are known as "Feynman diagrams," which are simple sketches, similar to vector diagrams, that can be used to trace and calculate the paths of subatomic particles generated in nuclear bombardments that produce new particles and radiation.

At the age of twenty-four, Richard Feynman was the youngest scientist to work on the atomic bomb project during World War II. One of his main contributions took the form of four questions he proposed to the

> In addition to being a creative and excellent scientist, Richard Feynman was adept in explaining advanced physics concepts to the nonscience public. He presented many popular lectures which were published in 1963 as *Feynman Lectures on Physics*. Later in 1985 he presented a series of TV lectures followed by a book titled *Surely You're Joking Mr. Feynman*. This book made him a celebrity. He believed that science is not an impersonal enterprise, but rather follows rules, procedures, processes, and experimental tests—it is the results of these tests that are the science—not the individual involved. Even so, science is a rather recent human activity designed to explore the nature of the entire universe. He also was interested in and worked on other problems of physics besides the strong nuclear interaction. He explored superconductivity that led to a new model for the structure of liquid helium.

senior scientists on the project for which answers were needed to make the goals of the project achievable: 1) How much fissionable material is needed to achieve a critical mass? 2) What type of material would best make a reflector, or "lens," to focus neutrons on the uranium? 3) How pure must the uranium be? 4) What would be the expected extent of the damage the explosion would cause (by heat, shock waves, and radiation)? His questions focused the direction of the project and saved time and money in designing the final A-bombs.

See also Bethe; Fermi; Oppenheimer; Teller; Ulam

FIBONACCI'S NUMBERING SYSTEM: Mathematics: *Leonardo Fibonacci* (c.1170–1250), Italy. (Fibonacci was a nickname. His real name was Leonardo Pisano, and he also went by the name Bigollo.)

Fibonacci was one of Leonardo Pisano's nicknames, albeit his most famous one. Pisano called himself Fibonacci, short for *fillio Bonacci* ("Bonaccio's son"). His father's name was Guglielmo Bonaccio, also a nickname, meaning a "good, stupid fellow." Leonardo was also known as *Bigollo* which comes from the Italian word *bighellone* meaning "loafer" or "wanderer."

The "Fibonacci sequence" of numbers was based on the breeding of rabbits. If each pair of rabbits gives birth to another pair, and in one month this pair gives birth to another pair, and this pair gives birth to another pair in one month, and so on, and so on, and they are all alive after a reasonable time, the number sequence would look as follows:

F_n represents the Fibonacci numbers
$F_0 = 0$
$F_1 = F_2 = 1$
$F_n = F_{n-1} + F_{n-2}$
$F_{(-n)} = (-1)^{n-1} \bullet F_n$

The Fibonacci numbers are easier to understand when expressed as a series of numbers in which each successive number is the sum of the preceding two numbers, such as: $1 + 1 = 2$; $1 + 2 = 3$; $2 + 3 = 5$; $3 + 5 = 8$; $5 + 8 = 13$; $8 + 13 = 21$; $13 + 21 = 34$; etc. This way of calculating Fibonacci numbers for the sequential birth of rabbits can be depicted in a visual graph that shows the exponential nature of the sequence. This problem was first published in 1202 along with several other books that introduced solutions to problems in mathematics. Fibonnacci is credited with introducing the Hindu/Arabic decimal numbering system to Europe (1 to 9 plus 0). Up to this time calculations in European countries were accomplished with Roman numerals, which by any standard are difficult to use even in solving simple problems in arithmetic. He most likely learned the Arabic/Hindu numbering system as he traveled widely over the Byzantine Empire with his father who was a customs officer and merchant. Fibonacci spent the rest of his life studying mathematics and writing several important books, including *Liber abbaci* (1202, 1228); *Practica geometriae* (1220/1221); *Flos* (1225); and *Liber quadratorum* (1225). Because this was before the invention of the printing press, his books had to be copied by hand, which makes it unusual that some of these original volumes still exist.

An interesting mathematical property of the Fibonacci sequence of numbers is the ratio between the numbers, that is, 1/1; 2/1; 3/2; 5/3, and so on (by dividing the first number by the second. Two examples: 5 divided by 3 = 1.666; or 25 ÷ 13 = 1.615), which as the sequence increases the ratio become close to the number 1.618 that is known as the *golden ratio*.

The golden ratio is defined as an irrational number that is approximately 1.618. It is the ratio of a diagonal of a pentagon to its side and appears in numerous metrical properties of 12- and 20-sided polygons, just as the square root of 2 appears in the metrical properties of a 6-sided square or cube. Euclid included the first calculation of the golden ratio in his book *Elements* written in about 300 BCE. The golden ratio essentially defines the balance between symmetry and asymmetry in various shapes in nature and is used in art design, as well as in psychology, metaphysics, and history to represent those forms that are aesthetically pleasing, particularly in Western culture.

There are a number of examples in nature where this ratio is exhibited, for example, petals on flowers, seeds on sunflowers, and the ratio of your height to the distance from your belly button to the ground. It is also found in areas of science including symmetry as a constant in physical laws (e.g., cosmology). Admirers of Fibonacci and his mathematics founded a society in 1962 that publishes *The Fibonacci Quarterly* that seeks solutions to mathematical problems using Fibonacci's methods.

See also Euclid; Fermat; Riemann

FICK'S LAWS OF DIFFUSION: Physics and Physiology: *Adolf Eugen Fick* (1829–1901), Germany.

Fick's first law of diffusion: *The flux (J) of a fluid with the concentration (C) across a membrane is proportional to the concentration differential across the plane of the membrane. (Note: flux is the rate of flow across a given area perpendicular to the flow.)*

Diffusion is really the intermingling of a number of particles of a substance (or units of electromagnetic radiation). Diffusion may be thought of as a mechanism by which individual types of particles in a mixture are moved within the mixture by means of random molecular movement known as the *Brownian motion* (*see* Einstein). On the other hand, permeability is the ability of a substance to pass through a body based on its diffusion coefficient (D) as well as its solubility coefficient.

If there is no volume change on either side of a barrier plane, the rates of diffusion are equal and opposite. For high-density substances across a fixed barrier, the rate of diffusion is low and thus only one equation is needed. The equation for Fick's first law of diffusion can be expressed in various forms. One of the easiest to understand follows:

> $J_x = -D\, dC/dx$: *where* J_x *is the flux of the two types of diffusing fluids, and* $-D$ *is a proportional constant or diffusion coefficient, and* dC/dx *equals the changes in the concentration in relation to the distance between the fluids.*

Fick's second law of diffusion: *The rate of change of the concentrations and volumes of a membrane in the diffusion field is expressed as* t *for the time involved in the exchange.*

The temperature or kinetic energy factor greatly influences the rate for the entire diffusion process as an increase in temperatures results in a "speeding up" of the molecular motion within the different types of diffusing substances.

Fick introduced the law of diffusion as a way to understand how gas is diffused across a fluid membrane that he used in 1870 to measure cardiac output. Since then, the equations for Fick's law have been adapted to measure the transport process of foods, polymers, pharmaceuticals, and as a means of controlling the "doping" of semiconductors to increase their efficiency.

Fick is also credited with making and using the first contact eye lens in 1887. He made an *afocal scleral* contact shell-shaped lens made of heavy brown glass that he first tested on rabbits. He then wore a pair of lenses himself before finally using his new invention on a group of volunteers.

FISCHER'S PROJECTION FORMULAS: Chemistry: *Emil Hermann Fischer* (1852–1919), Germany. Emil Fischer was awarded the 1902 Nobel Prize for Chemistry.

Projection formulas can be used to describe the spatial relationships of atoms in large organic molecules that have the same structural formula.

Emil Fischer demonstrated he could separate and identify sugars, such as glucose, mannose, and fructose, having the same empirical formula. In other words, once he determined the molecular structural formula for one type of organic substance, he could then project this information to synthesize other similar large organic molecules. In addition to his work with carbohydrates (sugars), he contributed to the understanding of the structures of purines, peptides, proteins, and caffeine alkaloids. Fischer used this theory to synthesize polypeptides that contained 18 amino acids. He also devised the "lock and key" explanation for high-molecular-weight compounds such as enzymes. Because Fischer did comprehensive groundbreaking work in the fields of purines, sugars, and peptides, which are the basis for biochemistry, he became known as the father of biochemistry. In 1882, based on his knowledge of structural formula for various organic compounds, he synthesized a number of important substances including several types of sugars.

Emil Fischer was somewhat of a child prodigy who excelled in school to the extent that he did not need to take the school's exit examination. Afterwards, because he was too young to enter the university, he worked for his uncle in the timber industry. His father was a "jolly" fellow who ran the family grocery store, spinning mill, and brewery, and it seems Emil took after his father's personality. His family thought Emil was not smart enough to be a businessman so they assumed he should just be a student. Emil set up his own chemistry laboratory where he worked during the day. In the evenings he played the piano and drank excessively. He, in essence, was self-taught in organic chemistry and performed important research in purines, sugars, dyes, and indoles (crystallized perfume compounds). Some of his chemicals caused terrible odors that were imbedded in his clothes that made him repulsive to people. He continued to smoke and drink to the point where he had to take time off each year to recuperate. He later became a professor in Berlin, which interfered with his research. After twelve years he returned to his research that led to the second Nobel Prize awarded in chemistry in 1902.

FITZGERALD'S CONCEPT OF ELECTROMAGNETIC CONTRACTION: Physics: *George Francis Fitzgerald* (1851–1901), Ireland.

When the light from a body is moving relative to an observer's position, the light contracts slightly in the direction of the observer's motion.

George Fitzgerald and Hendrik Lorentz independently concluded that a fast-moving body appears to contract according to its velocity as measured by an instrument (or observer). This effect, known as the

Lorentz–Fitzgerald contraction, described the effect the "**ether**" (in space) had on the electromagnetic (light) forces binding atoms together. Fitzgerald's research was concerned with the study and understanding of electromagnetism based on James Clerk Maxwell's equations, as well as the existence of radio waves as predicted by Heinrich Hertz. The Lorentz–Fitzgerald contraction theory proposed to explain the observations made by Albert Michelson and Edward Morley that the speed of light did not depend on the motion of the detector. Therefore, this idea could not be used to determine the speed of movement of Earth through space. The Lorentz–Fitzgerald contraction was an opposite and alternate explanation to Einstein's theory of relativity. Einstein used this altered version of the "contraction of light" concept in developing his special theory of relativity. His theory accepted the concept of space as a vacuum; as such, matter, even "ether" could not exist in space. The concept differs somewhat from the classical Doppler effect and the redshift.

See also Doppler; Einstein; Fizeau; Maxwell; Schmidt

FIZEAU'S THEORY OF THE NATURE OF LIGHT AS A WAVE: Physics: *Armand Hippolyte Louis Fizeau* (1819–1896), France.

If the speed of light is known, it can be demonstrated that light travels faster in air than in water or substances denser than air.

Armand Fizeau was the first to measure the speed of light using something other than subjective observations or astronomy. After constructing a device consisting of a rotating disk into which two "teeth" or gaps were cut, he set up a mirror on one hill to return the light sent from another hill about 5 miles distant upon which the "toothed disk" instrument was placed. He sent a light through the gaps in the disk, which acted like a rapid on-off switch that dissected the light into small "bits," similar to a series of light dots. As Fizeau increased the speed of the rotating disk, the reflected light from the mirror on the other hill was blocked off by the solid portion of the disk, but some light would shine through the toothed gap. The faster the disk rotated, the dimmer the light became, until it was blocked entirely by the solid parts of the slotted disk. Conversely, as he slowed the disk, the light would again brighten. By measuring the speed of rotation and the brightness of the light coming through the disk and knowing the distance between the two hills, he calculated the speed of light in air (*see* Figure M5 under Michelson). Fizeau and Jean Foucault are credited with proving that light behaves as waves. With an improved instrument, they measured the speed of light in water and compared it with the speed of light in air to confirm that light travels more slowly in denser mediums (*see* Figure F5 under Foucault). Fizeau also used his instrument to determine that the Doppler effect is the change in the wavelength of light relative to speed. This is known as the *Doppler–Fizeau shift*.

See also Doppler; Einstein; Fitzgerald; Foucault; Maxwell; Michelson; Schmidt

FLEISCHMANN'S THEORY FOR COLD FUSION: Chemistry: *Martin Fleischmann* (1927–), England.

Nuclear fusion can be achieved at room temperatures by the process of electrolysis using palladium as an electrode in an electrolyte of heavy water.

In 1989 while at the University of Utah Martin Fleischmann and the American electrochemist Stanley Pons (1943–) announced they had sustained a controlled fusion reaction in a laboratory setting at room temperatures that produced 100% more energy than was used by the electrolytic process. If this proved to be possible, it was estimated that the discovery would be worth at least $300 trillion and provide an unlimited supply of energy to the world. Other laboratories around the world tried to duplicate this experiment. All failed. Fleischmann claimed other scientists were not using the correct materials or procedures and that he was not going to reveal the exact nature of his experiment. The majority of scientists do not believe cold fusion is possible—at least at this time and at the current state of technology. Nuclear fusion requires an extremely high temperature and pressure such as is achieved in the sun and thermonuclear H-bombs to "fuse" the light hydrogen nuclei to form heavier helium nuclei, producing a great deal of energy in the process. The debate continues, but only a few scientists believe Fleischmann's cold nuclear fusion can take place as he described the process. Pons and Fleischmann parted ways in 1995. Reportedly, Pons is no longer in the research field, and Fleischmann now works in the private sector.

See also Bethe; Teller; Ulam

FLEMING'S BACTERICIDE HYPOTHESIS: Biology: *Sir Alexander Fleming* (1881–1955), England. Sir Alexander Fleming, Baron Florey, and Ernst Boris Chain shared the 1945 Nobel Prize in Physiology or Medicine.

> *If the* Penicillium notatum *mold growing in a laboratory dish can destroy staphylococcus bacteria, then it can be tested to destroy other selected harmful bacteria.*

Sir Alexander Fleming, a bacteriologist in the Royal Army Medical Corps, was familiar with the use of chemicals in the treatment of wounds. He devised *lysozyme*, an enzyme found in human tears and saliva, which proved to be a more effective bactericide than the chemicals available to him at the time. Even so, lysozyme was limited in its effectiveness. A few years later while examining a dish containing a culture of staphylococcus bacteria, he noticed several clear areas where the bacteria did not grow due to the presence of a mold identified as *Penicillium notatum*, which seemed to kill the harmful bacteria. Based on his experience with bacteria, Fleming recognized the potential for a new antibiotic but, at first, neglected to follow up on this discovery. It had been known for many years that some molds kill "germs," but no one had acted on it until Fleming finally recognized the importance of this phenomenon. Fleming's mold was isolated, developed, and tested for effectiveness by other scientists in Great Britain. By 1943 Fleming's discovery led to the production of limited amounts of penicillin, which saved the lives of many wounded Allied servicemen and women during World War II. Since then, a large number of similar "molds" have been identified and produced to provide a wider selection of antibiotics useful in the treatment of a variety of diseases.

See also Florey

FLEMING'S RULES FOR DETERMINING DIRECTION OF VECTORS: Physics: *John Ambrose Fleming* (1849–1945), England.

Fleming's right-hand and left-hand rules: *The right-hand rule (also known as the generator rule) is related to the Cartesian coordinate system for the three-dimension x, y, and z*

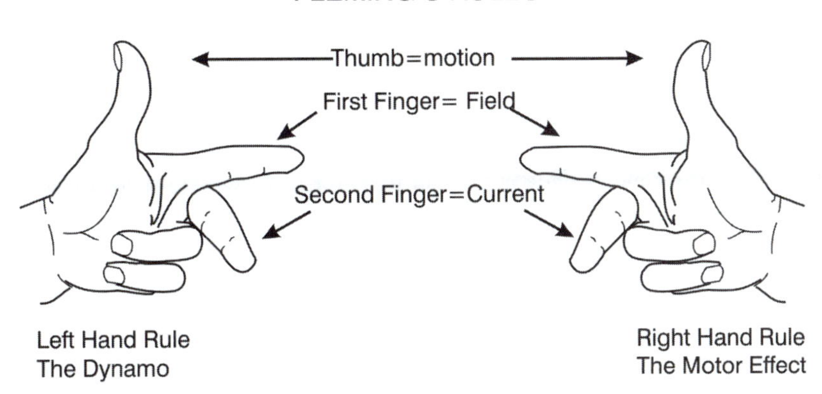

Figure F4. The right-hand and left-hand rules are designed as an easy way to remember the direction of the current, the magnetic field generated, and the motion of either the electric motor or generator.

axes for electric dynamos; while the left-hand rule (also known as the motor rule) is used to determine similar x, y, and z dimensions for electrical motors (*see* Figure F4).

Essentially, Fleming's right-hand and left-hand rules were conceived as easy-to-remember representations of electromagnetism and a means for determining 1) the direction of the flow of the electric current, 2) the direction of the field around the conductor carrying the electric current, and 3) the direction of the force involved in the operation of electric motors and electric generators. As shown in Figure F4, if the forefinger, second finger, and thumb of the *right hand* are extended at right angles to each other, the forefinger indicates the direction of the field, the second finger the direction of the current, and the thumb the direction of the force and in the dynamo (electric generator). Conversely, the *left hand* illustrates the same positions of the fingers and represents the same characteristics, except for electric motors. In other words, a person's hands are convenient tools to help remember the vectors associated with the directions of flow of current, the current's field, and force for the motor and dynamo. The diagrams help visualize the directions of magnetic forces and fields produced by conductors carrying electric currents.

Although the motor and dynamo are similar electrical devices, they perform opposite functions. The dynamo uses mechanical force to turn an armature to cut through the magnetic fields to produce a flow of an electric current, whereas a motor uses electric current to turn an armature in the magnetic field to produce a mechanical force.

In early life, John Ambrose Fleming showed evidence of being a child prodigy when, at the age of thirteen, he gave his first lecture on electricity and magnetism. He was educated primarily at the University College in London, the premier college of London University where he received his BS degree in 1870. He taught school for a few years to support his continuing education and later studied under the famous British physicist James Clerk Maxwell at Cambridge University. In 1885 he created and headed a new department of electrical engineering at University College in London. Fleming was a popular lecturer and consultant, who also conducted research for wireless signal transmissions across the Atlantic Ocean. He designed the transmitter that made the first successful wireless transatlantic transmission in 1901.

Fleming's theory for rectifying alternating current (AC) to direct (DC) current: *By adapting and using the Edison effect (see Edison) to construct a two-electrode vacuum-tube rectifier to act as an oscillation valve (also called a thermionic valve, vacuum diode, kenotron, and thermionic tube), it is possible to convert an alternating current into a direct current.*

The Fleming valves were the beginning of the electronic age and were used in early radio and radar instruments. The American inventor Lee De Forest added a third electrode to the Edison/Fleming diode that acted as a "grid" that could be used to control the flow of electrons across the gap from the hot negative electrode to the cool positive electrode. This formed what was known as a triode that could be used to amplify electronic signals as well as modulate sound signals that led to long-distance wireless communication systems of today, including radio, television, radar, computers and, the Internet.

FLEROV'S THEORY OF SPONTANEOUS FISSION: Physics: *Georgii Nikolaevich Flerov* (1913–1990), Russia.

Uranium nuclei will fission spontaneously, and the process requires no bombardment of the nuclei by neutrons to accomplish the splitting of the nuclei.

In 1941 Flerov observed spontaneous fission where the nuclei of uranium nuclei break into nuclei of smaller, lighter elements. This fission occurs naturally without the bombardment of the uranium nuclei by additional neutrons. Flerov was familiar with scientific journals that indicated that scientists from other countries were working with this natural fission, as well as attempting to cause artificial fission by bombarding nuclei of heavy elements with slow neutrons to cause them not only to fission, but also to cause a chain reaction. He urged the Soviet government to initiate a research program to investigate this phenomenon.

Georgii (or Georgy) N. Flerov was the Soviet Union's leading nuclear physicist and soon became the leader of a team of scientists at the Joint Institute for Nuclear Research (JINR) at Dubna in Russia. In 1942 he wrote to Joseph Stalin, the infamous Communist leader of the Soviet Union, about the progress being made in developing a nuclear bomb in the United States and Germany. His efforts eventually led to the Union of Soviet Socialist Republics' (USSR) production of their own atomic bomb.

Flerov's team at Dubna claimed in 1967 to have isolated the isotopes of the elements with the mass number (atomic weights) of 260 and 261 by bombarding the isotope americium-243 with the isotope neon-22 ions. They had created the element with an atomic number of 105 (number of protons in the nucleus), for which they suggested the name *nielsbohrium* in honor of the Danish physicists Niels Bohr. The reaction follows: $_{95}$Am-243 + $_{10}$Ne-22 $\rightarrow$ $_{105}$ nielsbohrium-260 and -261.

The American group under the direction of Albert Ghiorso at the Lawrence Berkeley National Laboratory in Berkeley, California, could not confirm the Dubna results. However, by using different techniques they synthesized an isotope of element 105. They bombarded californium-249 nuclei with nitrogen-15 ions and created an isotope with an atomic weight of 260 and a half-life of 1.6 seconds, which was long enough to positively identify it as a new element with atomic number 105. It was then named *hahnium* in honor of the physicist Otto Hahn. Over the next few years two other disputes about the name for element 105 arose. In 1994 a committee of the International Union of Pure and Applied Chemistry (IUPAC) named 105 *joliotium* (Jl-105) after the French physicist Frederic Joliot-Curie. Still later in 1997 an international agreement between Russia and the United States led to the acceptance of the name *dubnium* (Db or Unp) for the new artificially produced element with 105 protons in its nuclei. This name was decided upon because of the original work done with this element by the Dubna group in Russia. Since then, the Russian and American nuclear laboratories (as well as one in Germany) went on to produce isotopes of additional heavy elements up to and beyond element 118.

See also Curies; Hahn; Seaborg

FLOREY'S THEORY OF MUCUS SECRETIONS: Biology: *Howard Walter Florey (Baron Florey of Adelaide)* (1898–1968), Australia. Baron Florey shared the 1945 Nobel Prize in physiology or medicine with Sir Alexander Fleming and Ernst Chain.

Mucous secretions that contain lysozyme can destroy the cell walls of bacteria.

Following up on Sir Alexander Fleming's work with the enzyme lysozyme, Baron Florey determined how this enzyme found in tears and saliva destroyed the cell walls of bacteria. This research led to Ernst Chain's (1906–1979), the German-born British biochemist, and Baron Florey's idea for developing *penicillium*, an antibiotic discovered earlier by Fleming. Although Fleming discovered *penicillium* and recognized that it killed staphylococcus bacteria, he did not pursue its commercial development. Chain's and Florey's efforts brought to fruition the success of this and other similar effective antibiotics used to treat numerous types of bacterial infections.

Baron Florey continued his work on other antibiotics as well as experimental procedures involving pathology and the lymphatic and vascular systems. Although he lived most of his life in England, in spirit he remained an Australian.

See also Fleming (Alexander)

FLORY'S THEORY OF NONLINEAR POLYMERS: Chemistry: *Paul John Flory* (1910–1985), United States.

Polymer molecules have a definite size and structure consisting of multiple macromolecules with different chain lengths that are branching as well as linear.

Macromolecules are an aggregate of two or more molecules that are not combined by regular chemical bonds, but rather by intermolecular forces acting between the molecules themselves (*see* the Van der Waals force). A distinction between regular smaller molecules and macromolecules is that smaller molecules are easily dissolved in a liquid whereas macromolecules need some type of assistance to dissolve. Rarely do macromolecules occur individually, but they are more likely to assemble into *macromolecular complexes*. However, proteins are considered *subunits* of such macromolecular complexes. The IUPAC (the International Union of Pure and Applied Chemists) prefers the term "macromolecules" for grouping of individual molecules. The term "polymer" is something that consists of many macromolecules with either a linear or multiple branching structures. There are many polymers whose molecules consist of many thousands of atoms formed into long linear chains. But not all polymers are linear. Most polymers are branched chains where both ends of the branched chains are attached to a backbone chain of separate molecules, and it is possible for backbone chains to become attached to each other to form very large polymer molecules. Some of these are cross-linked polymers that are groups of molecules large enough to pick up in your hand, such as synthetic rubber. Liquid crystals and DNA are examples of polymer-type macromolecules. For example, chromosomes of DNA can be tens of millions of base pairs in length.

Paul John Flory used statistical methods to determine the length and branching of chains of molecules from which he developed his theory of cross-linking between molecular chains to form linear chains as well as branching liquid macromolecules that became known as polymers. Most plastics, artificial rubber, and other materials are composed of polymers. For example, bowling balls are considered a single polymer macromolecule, whereas computer cases, telephones, tires, and many other everyday items, including synthetic motor oil, artificial blood, and new types of liquid crystals used in screens for some electronic instruments, are all forms of branching polymer macromolecules.

FOUCAULT'S THEORIES OF LIGHT AND EARTH'S ROTATION: Physics: *Jean Bernard Leon Foucault* (1819–1868), France.

Foucault's wave theory of light: *If light is a wave, it will travel faster in air than in water; if light is a particle, it will travel slower in air than in water.*

To test his theory, Jean Foucault required a more accurate means of determining the speed of light than resulted from Armand Fizeau's rotating "toothed disk" device experiment conducted some years earlier. He devised a system using a rotating mirror, which provided a measurement very close to that later achieved by Albert Michelson. Foucault's mirrors were located on hills 22 miles apart. This greater distance made his measurements more accurate. The turning mirror reflected the light back at a slightly different angle to another mirror. This slight angle of reflection could be measured and compared with the rate of rotation of the mirror to give the approximate speed of light. The formula is $v = c \div n$, where v is the speed of light through a particular medium, c is the constant for the speed of light in a vacuum, and n is the index of refraction of the medium through which the light is being measured. He repeated a similar procedure with light projected through water and found that the reflection angle, and thus the speed of light through water, is less than when it travels through air (*see* Figure F5). For instance, the speed of light is approximately 186,000 miles per second, or exactly 299,792,458 kilometers per second in a vacuum. To find the velocity of light through water, divide 186,000 by the index of refraction of water, that is 1.33, which results in the speed of light through water at about 140,000 miles per second. This was the proof he needed to arrive at his theory for the wave nature of light, which upset many scientists of his day because of their belief in Newton's theory that light was composed of minute particles (photons). Later scientists accepted the duality of light as having both a wave and particle nature (*see also* Fizeau; Huygens; Michelson).

Foucault's theory for the rotation of Earth: *Because a pendulum swings in an unchanging plane, its apparent progressive movement out of the plane must be caused by the rotation of the earth beneath the pendulum.*

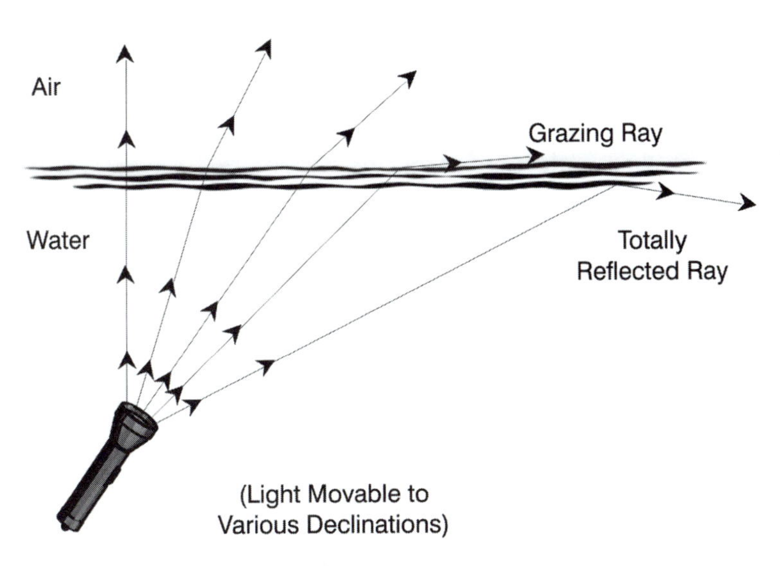

Figure F5. Light rays are "refracted" as they pass through substances of different densities at an angle to the normal (the line perpendicular to the surface). They will change directions when going through the boundary between the two different substances according to their angle of entrance to the boundary. The index of refraction is the ratio of light's speed (in a vacuum) to its speed in a given material.

Jean Foucault was familiar with the work of Galileo, who applied the periodic motion phenomenon of a pendulum to measure time. Galileo used his heart pulse to time the frequency of the pendulum "bob," which he determined is inversely proportional to the length of the string suspending the bob. This means that the shorter the string, the greater the frequency of the swinging bob, and conversely, the longer the string, the lower the frequency. Needing an accurate timing device to measure the speed of light, Foucault considered a pendulum. He knew the pendulum executed a form approximating simple harmonic motion where the force that "drove" it was outside the system (i.e., gravity). He noticed the pendulum appeared to stay in the same plane (compass direction) even when the platform holding the pendulum apparatus was rotated. Recognizing the significance, he determined that because the pendulum always moves in the same plane, it must be Earth beneath the swinging pendulum that turns. He performed a demonstration during which he suspended an iron ball weighing 28 kilograms from a 67-meter wire attached to the top of the inside of Le Pantheon in Paris. Sand was spread out on the floor under the pendulum and a needle was attached to the bottom of the ball weight of the pendulum. When set in motion, the needle slowly inscribed a pattern in the sand, proving that Earth was moving under the pendulum.

This was the first Earth-based demonstration that proved that Earth actually rotates on its axis, but not at the same rate at all latitudes. For instance, at the equator, an east-west swinging pendulum will show no rotation motion of earth because Earth and pendulum are moving in the same east-west direction while Earth rotates once every

Figure F6. Foucault's Pendulum.

twenty-four hours. Also, at the North Pole, Earth makes a complete rotation every twenty-four hours under a pendulum swinging in the same plane. Thus, an east-west swinging pendulum will stay in the same plane while the Earth turns under it. The effect differs with the latitude from the equator to the poles. The formula for determining the period of rotation at latitudes other than at the Earth's poles or at the equator is $P = 23^h\ 56^m \div \sin$ (latitude). The pendulum used to measure Earth's rotation is called *Foucault's pendulum.*

Pendulums are also used to measure acceleration due to gravity and velocities (the *ballistic pendulum*), as well as for accurate clocks. Seventy years after Galileo's study of the pendulum, Christiaan Huygens incorporated it into an accurate timekeeping instrument. Called the grandfather's clock, it used a weight system to continue the "force" to overcome friction and thus maintain the pendulum's constant swing. There is also a *compound pendulum* where a rigid body swings around a central point. One model, designed by the British physicist Henry Kater (1777–1835), measured acceleration due to "free fall," which can then be used to calculate the force of gravity.

See also Galileo; Fizeau; Hooke; Huygens

FOURIER'S THEORIES OF HEAT CONDUCTION AND HARMONIC WAVE MOTION: Physics: *Jean-Baptiste-Joseph, Baron Fourier* (1768–1830), France.

Fourier's theorem of heat conduction: *The rate at which heat is conducted through a body, as related to that body's cross section area, is proportional to the negative of the temperature gradient existing in that body.*

Baron Fourier worked out the mathematical expression for conduction of heat through different types of solid materials. The theory explains why excellent heat conductors are also good electrical conductors. Conduction relates to the motion of free electrons in solid matter when a temperature difference exists from one end of the matter to the other. The proportionality constant of his equation is referred to as the *thermal conductivity* of the solid. For instance, glass, wood, paper, and asbestos all have thermal conductivities of a much lower value than metals. They are referred to as *insulators* of heat and electricity. Conversely, most metals have a high value of thermal conductivity and are excellent conductors of heat and electricity. Conduction is one of three forms of heat transfer described in physics. The other two are convection and radiation. Modern technology and industry use the mathematics of heat transfer and the temperature differential for various materials. Fourier's theory was the beginning of dimensional analysis, which requires any expression of unit quantities to be balanced in equations just as are numbers. One of his main contributions was the establishment of linear partial differential equations that are used as a powerful tool in mathematics and to determine boundary values in physics problems such as heat conduction in different substances. Today, computers calculate the equations for these properties of different substances.

Fourier's harmonic analysis: *Any periodic motion (wave pattern) can be separated mathematically into the individual sine waves of which the pattern is composed.*

Robert Hooke's law of elasticity states that the change in size of an elastic material is directly proportional to the stress (force applied per unit area) applied to the material. Hooke applied this concept to thermal expansion and wave motions of metal spiral springs. Baron Fourier believed that complicated wave motions were not all that complex and could be solved mathematically. His work, referred to as *wave analysis*, is applied to music. Pythagoras was the first to determine that certain musical notes blend together to produce pleasant sounds. He also ascertained these notes represent a ratio of small whole numbers. Pythagoras used strings of different lengths to define "nice" sounds (e.g., 1:2; 2:3; and 3:4 string-to-length ratios). Musical sounds consist of a number of separate sine waves that, when combined, display some order of interrelationship. These ordered sets of sine waves, which may or may not reinforce each other, produce musical notes. If the sine waves are randomly selected and combined or interfere with each other it results in dissonance we call *noise*. Fourier's wave analysis techniques established a firm physical and mathematical basis for modern music and the development of musical instruments.

See also Hooke; Pythagoras

FOWLER'S THEORY OF STELLAR NUCLEOSYNTHESIS: Physics: *William (Willy) Alfred Fowler* (1911–1995), United States. Willliam Fowler shared the 1983 Nobel Prize in Physics with Subrahmanyan Chandrasekhar.

Thermonuclear fusion reactions in stars produce enough kinetic energy to overcome the electrostatic repulsion of hydrogen nuclei to form helium nuclei. Additional fusion reactions produce even more kinetic energy to overcome the repulsion between other nuclei, thus producing the heavier elements.

A number of scientists, including Georges Lemaître, George Gamow, and Sir Fred Hoyle have proposed their versions of the big bang theory. William Fowler expanded on this theory, which propounds that the explosion of a tiny "seed," "point source," or "singularity" of energy, which created tremendous radiation and energy, forming in less than a few seconds mostly hydrogen atoms, resulted in the formation of the entire universe soon after this singular beginning. None of the scientists have claimed to know what caused the "explosion" or the origin of the "seed," or the singularity of energy/matter. The big bang theory further states the colossal amount of energy provided the force for two hydrogen nuclei (^{+}H + ^{+}H) to combine to form a helium nucleus (^{++}He) creating **thermonuclear** energy in the stars. This reaction occurred at temperatures of about 10 to 20 million kelvin (K). The reaction did not produce enough energy to overcome the mutual electrostatic repulsion of the double-charged helium nuclei (alpha particles); thus this temperature was insufficient to form nuclei of the heavier elements. Following the lead of several other physicists, Fowler developed the theory that this temperature in the core of the sun became much greater, up to 100 to 200 million K, and produced enough kinetic energy to overcome the mutual repulsion of the positive alpha particles, which then combined to form the nuclei of heavier elements. First, a fusion reaction occurred, which combined three alpha nuclei (3 ^{++}He) to form a carbon nucleus with 6 positive protons ($^{++++++}C$). This thermonuclear reaction increased the temperature to over 500 million K, which provided the energy to fuse an alpha nucleus (^{++}He) with a carbon nucleus ($^{++++++}C$) to form an oxygen nucleus ($^{++++++++}O$) with eight protons. These stellar thermal nucleosynthesis (fusion) processes continued, reaching over one to three or four billion degrees, thus forming the other heavier elements that make up Earth and the universe. Fowler's work aided the understanding of the composition of stars, our solar system, and the nature of the universe, that is, its age and future. He and Fred Hoyle cooperated and published an important paper "Nucleosynthesis in Massive Stars and Supernovae" in 1965 that explained how heavier elements were formed as the temperature increased inside stars.

See also Gamow; Hoyle; Lemaître

FOX'S THEORY OF PROTEINOID MICROSPHERES: Biochemistry: *Sidney Walter Fox* (1912–1998), United States.

Using heat, thermal proteins will self-organize into protocell-like microspheres.

Since the beginning of time humans have wondered about their biological origins, life, death, and the reason we are on Earth. The Greeks referred to this particular curiosity as *abiogenesis*, a Greek word meaning "a genesis of nonbiological origins." Later in history it was known as the spontaneous generation of life. In the Middle Ages it was thought that rats and mice were "generated" from grain or old hay that was stored over long periods of time, and that maggots/flies came from spoiled meat and food. Even Aristotle believed that small insects grew out of the dew from the flowers and leafs of plants. These ancient beliefs were not corrected until 1668 when Francesco Redi conducted an experiment that proved that maggots did not come from spoiled meats. He placed a gauze barrier over the mouth of jars of meat that prevented flies from touching the meat (*see* Redi for details of this experiment). Numerous other famous scientists and researchers devised other theories for the generation of life. In the 1930s the Soviet biologist Alesksandr Ivanovich Oparin and others produced molecules that could "grow" and fuse together and form into daughter cells. John Burdon Sanderson Haldane, the British geneticist and evolutionary biologist, believed that life formed in a "soup" containing organic substances. Many other scientists theorized about the origin of living things, including Leeuwenhoek, Spallanzani, Darwin, and Pasteur. This concept became known as "biopoiesis," which is the formation of biological substances, or as "autopoiesis" that is defined by American biologist Lynn Margulis as the self-organizing of living systems that have properties that maintain their own boundaries. The concepts involve the "spontaneous" formation of simple inorganic chemicals into more complex organic molecules on Earth. The concept of "panspermia" is the belief that life came to Earth from some other planet or asteroid in the universe. The astronomers Sir Fred Hoyle and the Sri Lankan astrobiologist Chandra Wickramasinghe (1939–) believed in panspermia. In 1953 Stanley L. Miller and Harold C. Urey while at the University of Chicago placed a "primeval soup" composed of water vapor, methane, ammonia, and hydrogen, but no oxygen, in a closed container. They then shot electrical sparks though this "soup" to represent lightning which they hypothesized might provide the energy required to form organic molecules and thus might represent the environment on the early Earth. They did demonstrate that complex molecules can be formed from simple amino acids, but alas, they did not create life.

> Sidney Fox was one of several biochemists to have conducted experiments on the origins of life. In 1984, 1985, and 1990, Fox was invited to discuss his work related to the creation of life before Pope John II and other papal scientists in Rome in a forum sponsored by the Italian Academia dei Lincei, IBM, and the National Foundation for Cancer Research. It appeared that the Pope wanted to be as well informed as possible before issuing the Roman Catholic Church's famous statement in support of the scientific theory of organic evolution. Fox assumed there were three reasons for the Pope's interest in his research on the origins of life and subsequent evolution of plant and animal species. Pope John II's interests appear to be related to where he and everyone else came from, how future evolutionary research might be related to Genesis; and last, the Pope did not want to repeat the mistakes that the early Church made regarding Giordano Bruno (burning at the stake), Copernicus (his book on the heliocentric universe remained on the Index for one hundred and fifty years), and Galileo (condemned by the Inquisition), as well as other less-famous philosophers and scientists who were either excommunicated or suffered similar fates. Fox's lectures were well received by the Vatican.

Sidney Fox's 1958 experiment did produce a type of polymer (long chain of molecules) of amino acids that he called "proteinoids" by the application of heat. In turn these proteinoids tended to form "microspheres" that resembled bacteria-like "protocells" which were about a micrometer in size. These tiny protocells developed a membrane on their surface that seemed to produce "buds" that, under certain conditions, divided. Since then others have attempted to "create" life by developing an understanding of the complexity of inorganic substances as they react and interact to form basic organic type molecules. If the question is whether simple chemicals and substances can form the molecules of life, the answer seems to be yes. However, to date, the "spark" that is life has not been scientifically identified and found.

See also Hoyle; Miller; Urey

FRACASTORO'S THEORY OF DISEASE: Biology and Medicine: *Girolamo Fracastoro* (1478–1553), Italy.

Diseases are transmitted by "seedlike" entities that are transferred from person to person.

Girolamo Fracastoro recognized that invisible "seed-like" particles transmitted infection. In 1546 Fracastoro wrote *On Contagion and Contagious Diseases* in which he anticipated the germ theory. In this book he mentioned three types of infection: 1) by direct contact of an uninfected person with a person infected by the disease; 2) by indirect contact with the infected person by some path, such as clothing, infected air, food, or other medium that could carry the infection; and 3) contagion from a distance, where there is no evident contact between the infected person and a noninfected person, such as a fever. In all three cases, the disease is transmitted by "seeds of contagion" that are capable of reproducing themselves and thus spreading diseases. He also stated that each disease has it own nature of contagion, that is, types of "seeds," that have their own rate of multiplying as tiny bodies that carry the disease. Unfortunately, his observations and theory made little impact on the actual status of medicine of his day, although it was influential for several centuries until the early nineteenth century when Agostino Bassi (1773–1856), the Italian biologist and bacteriologist, proposed his germ theory in 1825.

Before his contagion theory Fracastoro was known for his book *Syphilis, or the French Disease* published in 1530. This book introduced the term "syphilis" to Europeans. The Italians called it the French disease. The French called it the Italian disease. And all Europeans believed syphilis was imported from the New World by explorers returning to their homes in Europe. It was more likely in the 1500s that infected sailors and settlers from Europe infected Native Americans than the other way around.

See also Leeuwenhoek; Pasteur

FRANCK'S THEORY OF DISCRETE ABSORPTION OF ELECTRONS: Physics: *James Franck* (1882–1964), United States. James Franck and Gustav Hertz shared the 1925 Nobel Prize for Physics.

Only electrons at specific velocities can be absorbed by a medium (mercury) in precise (quanta) amounts.

In 1914 James Franck and the German experimental physicist Gustav Hertz (1887–1975) collaborated to demonstrate experimentally that energy is transferred in selected quantized amounts as it reacts with atoms and other particles. They used electrons at different velocities to bombard mercury atoms and discovered that electrons could be absorbed by the mercury atoms only and exactly at 4.9 electron volts of energy. If the electrons had less energy, they were lost on collision with the nuclei of the mercury atoms. If the energy was greater than 4.9 eV, they were not absorbed. It was only at the discrete 4.9 eV of energy that electrons were permitted to enter the orbits of the mercury atoms. This was the first experimental evidence for the *quantum* (Latin for "how much") theory of energy and was later confirmed for the "quantum leap" of electrons for other atoms. A simplified explanation of the quantum leap states that it is a tiny, discrete amount of energy emitted by an electron when it jumps from an inner orbit to an outer orbit (energy level). The closer an electron's orbit is to the nucleus, the greater is its energy. Therefore, as it jumps from an orbit of greater energy to an orbit of lesser energy (further from the nucleus), it must give up a "quantized" bit of energy. The energy emitted is a photon (light particle). Conversely, when an atom absorbs a specific level of energy, an electron in an outer orbit can take a quantum jump down to an inner orbit (*see* Figure D6 under Dehmelt). Franck's and Hertz's experimental proof of the quantum theory was an important step in understanding the physics of matter and energy.

See also Dehmelt; Heisenberg; Hertz; Planck

FRANKLAND'S THEORY OF VALENCE: Chemistry: *Sir Edward Frankland* (1825–1899), England.

The capacity of the atoms of elements to combine with the atoms of other elements to form molecular compounds is determined by the number of chemical bonds on the given atoms.

Sir Edward Frankland is considered the father of the concept of *valence* (Latin for "power") that is the number of chemical bonds (connections), atoms, or groups of atoms that can be exchanged or shared with other atoms. In his work with organic compounds, Frankland discovered that atoms of different elements would chemically bond within fixed ratios with other groups of atoms. From this observation in 1852, he developed an explanation for the maximum valence for each element. The theory of valence explained the relationships of atomic weights with the ratios of atoms combining with each other (*see* Figure S3 under Sidgwick). Although valence is the "combining power" of atoms to join with each other to form molecules, the electrovalence of an ion (atoms that lost or gained electrons) is the numerical value of the electrical charge on the ion. The concepts of valence and electrovalence are important for the understanding and advancement of chemistry. Later in 1867 Frankland and his colleague, the British chemist B.F. Duppa (dates unknown), identified and determined that the –COOH (Oxatyl) group is a found in all organic acids. This simplified the determination for the structure of many organic compounds.

See also Abegg; Arrhenius; Berzelius; Bohr; Dalton; Langmuir; Lewis; Sidgwick

FRANKLIN'S CONCEPT OF DNA STRUCTURE: Physics: *Rosalind Franklin* (1920–1958), England.

The complex organic DNA molecule is a helix structure with phosphate chemical groups situated on the outer boundaries of the helix spirals.

Rosalind Franklin, an expert crystallographer, made X-ray photographs of a form of DNA that clearly indicated the helix nature of its molecular structure.

As the story goes, in 1952 James Watson viewed her X-ray photographs and recognized the importance of the obvious helix structure of the DNA substance to the DNA research he and Francis Crick were conducting. Crick obtained copies of Franklin's X-ray photographs from her boss Maurice Wilkins (1916–2004) (some say without her

ARTIST'S DEPICTION OF FRANKLIN'S PHOTOGRAPH

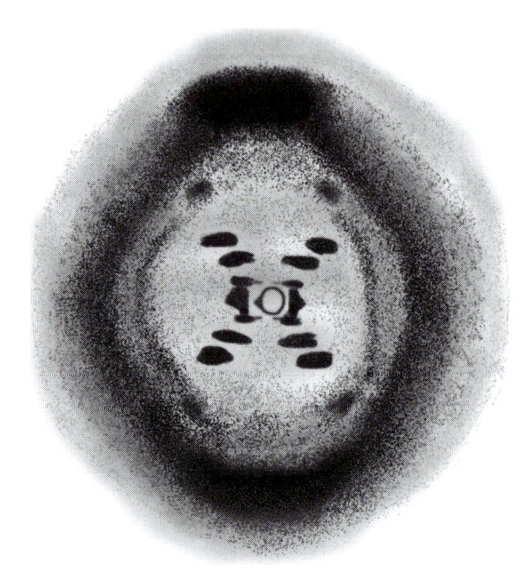

Figure F7. Rosalind Franklin's X-ray crystal photograph of the helix structure of the DNA molecule.

permission). These photographs clearly provided the information required for Watson and Crick to develop an acceptable helix structure of the double helix of DNA (*see* Figure C5 under Crick). Rosalind Franklin later wrote a paper in which she mentioned the structure for the DNA molecule that consisted of a double-chain helix. Although she did not recognize the complete structure nor did she recognize the inclusion of base pairs of nucleotides, she did indicate that the phosphate groups on the outside of the strands were responsible for holding the units together. In the publication that reported the structure of the DNA molecule, Franklin was not given credit for her important work or her X-ray photographs. The 1962 Nobel Prize for Physiology or Medicine was awarded to Watson, Crick, and Wilkins, but because Rosalind Franklin had died of cancer in 1958 she was ineligible for the award.

See also Crick; Watson

FRANKLIN'S THEORIES OF ELECTRICITY: Physics: *Benjamin Franklin* (1706–1790), United States.

Franklin's single fluid theory of electricity: *Electricity is a single fluid with both attracting and repelling forces.*

Benjamin Franklin knew of other scientific experiments that demonstrated that when substances (then called "electrics," and now are known as insulators) such as amber and glass, were rubbed with wool or silk, static electricity was produced. Conversely, when "nonelectrics," (now known as conductors) such as metals, were rubbed, no static electricity was produced. Based on this evidence, scientists concluded that because rubbing different substances could produce a repelling or attracting force, electricity must be composed of two different types of fluid. The French chemist Charles

Benjamin Franklin might be considered an eighteenth-century Renaissance man because of his eclectic interests and endeavors. Not only was he interested in many areas of science, he was also an accomplished diplomat, politician, publisher, and wit. Two of his famous publications, are the *Pennsylvania Gazette* from 1729 to 1733, which was followed by the more successful *Poor Richard's Almanac.* He also published works on heat, light, and oceanography. He made measurements of the temperature of the water in the Gulf Stream and determined how this moving body of water affected the northern part of the United States and Western Europe. His data was used to develop accurate maps of the Gulf Stream. Among his many inventions were a new type of home heating stove (now referred to as the "Franklin stove,") bifocal eyeglasses, and the rocking chair. He lived a long, productive life. He believed that life should be lived to its fullest potential.

Du Fay (1698–1739) proposed the concept of two kinds of electricity: positive and negative. He called the positive electricity *vitreous* and the negative *resinous*, which seemed to confirm the "two fluid" nature of electricity. In 1747, after experimenting with a **Leyden jar**, Benjamin Franklin advanced a single fluid concept of electricity, but he still considered it a "flowing" substance.

Franklin's concept of electric charges: *Because electricity is a single fluid substance, two types of forces must be present to cause attraction and repulsion.*

Benjamin Franklin, as well as Du Fay, is credited with the terms "positive" and "negative" to explain the attraction and repulsion characteristics of "fluid" electricity. Franklin's single fluid electricity was on the right track, even though he interpreted the terms incorrectly. He reasoned that "positive" would be the direction of the current flow, when in actuality the negative electrons determine the direction of the electric current toward the positive pole, which lacks electrons, and thus has a potential of gaining electrons. This has caused much confusion ever since.

Franklin's concept of lightning: *Lightning is a form of electricity that is more strongly attracted to points, particularly metal points at high altitudes.*

For many years, scientists' related lightning to static electricity because both produced a jagged spark of light and could cause shock. But there was no proof they were the same phenomenon. Benjamin Franklin's experiments with the Leyden jar illustrated that electricity was more strongly attracted to point sources than to flat surfaces. From this concept, he believed it possible to demonstrate that lightning was an electrical discharge by attracting it to a metal tip on the end of kite. The result was his famous kite experiment during a thunderstorm in the year 1752, when lightning was attracted to the kite and was conducted to a silk ribbon attached to a metal key. When he brought his knuckle close to the key, a spark jumped to his hand, producing a mild shock. This was a very dangerous experiment, and several scientists were electrocuted when trying to replicate Franklin's demonstration that proved lightning is electricity. However, this experiment forged Franklin's development of the lightning rod, which has prevented lightning damage to many homes and commercial buildings.

FRAUNHOFER'S THEORY OF WHITE LIGHT: Physics: *Josef von Fraunhofer* (1787–1826), Germany.

> *White light projected through a prism produces a continuous color spectrum that is crossed by dark lines.*

Josef Fraunhofer was an expert lens maker familiar with Isaac Newton's studies that proved white light is composed of colored lights when it is projected through a prism.

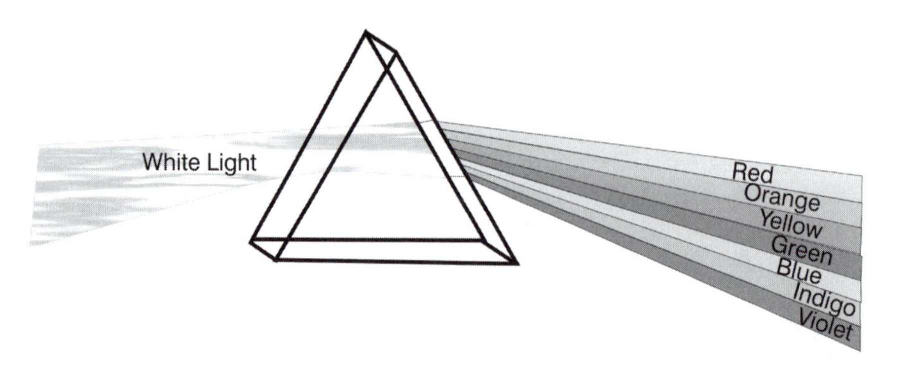

Figure F8. An artist's rendition of the electromagnetic spectrum produced by a beam of white light passing through a prism. The white light is composed of separated electromagnetic waves ranging from long infrared to short ultraviolet rays.

Fraunhofer was also familiar with the English chemist and physicist William Hyde Wollaston's (1766–1828) observation that dark lines within a spectrum were produced from a white light source. As Fraunhofer permitted light from a very narrow slit to pass through one of his excellent prisms, he observed a series of narrow bands of light of varying wavelengths. Some images of specific wavelengths (colors) were missing and produced the dark lines reported by Wollaston. Fraunhofer did not know what caused them nor did he realize their significance, but they were then, and still are, referred to as *Fraunhofer lines*. Fraunhofer went on to identify over seven hundred of these dark lines. Later, Robert Bunsen (inventor of the Bunsen burner) and Gustav Kirchhoff, using a spectroscope, determined that the dark lines were the absorption of specific wavelengths of light by a vapor between the source and the prism. This concept of radiation absorption was used to identify numerous new chemical elements by spectroscopy (spectro analysis). The same concept can be used to study the nature of light from the sun and stars to determine their chemical composition.

See also Bunsen; Kirchhoff; Newton

FRESNEL'S THEORY FOR MULTIPLE PRISMS: Physics: *Augustin Jean Fresnel* (1788–1827), France.

A single beam of light produces multiple interferences when split into multiple beams of multiple prisms.

Augustin Fresnel considered light to be similar in nature to sound waves. Based on this concept, he worked out the mathematics for light as transverse waves that explained reflection, refraction, and diffraction, which are related to the longitudinal waves for sound. His work with the interference of the beam of light by a prism caused him to consider what would happen if two prisms were used to split the beam of light into two parts. Fresnel conceived of a series of prisms formed as concentric circles on a circular glass lens. This resulted in the Fresnel lens, which, from the front, looks like concentric circles similar to a bull's-eye on a target, but from a cross section, these circles appear as a series of "sawtooth" tiny circular prisms.

The Fresnel lens, first developed to concentrate the light from lighthouses, is now used in a multitude of devices, from overhead projectors, to large-format cameras, to other devices with screens too large to make use of a heavy glass convex lens to concentrate light.

August Fresnel also used transverse waves to explain the phenomenon of polarization of light. The electric field **vector** oscillates in directions perpendicular to the direction of light. Light is polarized when the electric field oscillates just up and down or right and left. If the electric vector oscillates in all directions, the light is said to be unpolarized. (Today, sunglasses use lenses that transmit the one-directional polarized light waves, while blocking out most of the light with the opposite polarization.) This seemed to settle the controversy of the nature of light as a wave, at least until the light particle (photon) theory was developed.

See also Einstein; Fizeau

FRESNEL LENS

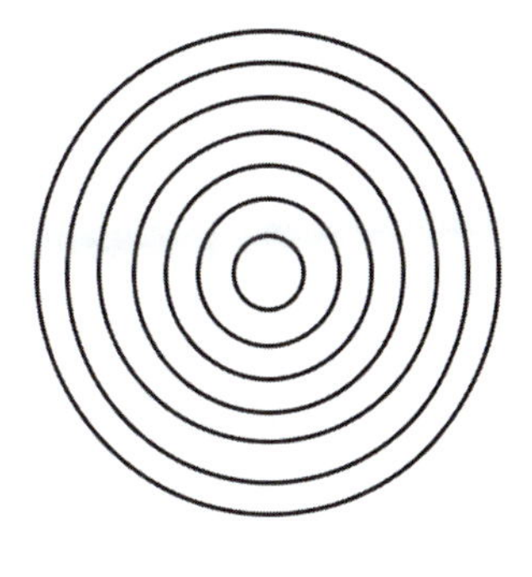

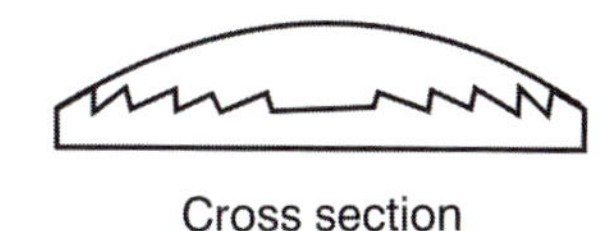

Figure F9. From the front, the Fresnel lens looks similar to a bull's eye, but from the side it appears as a series of "saw-tooth" circular prisms. It is used to concentrate light.

FRIEDMAN'S THEORY OF THE QUARK STRUCTURE OF NUCLEONS: Physics: *Jerome Isaac Friedman* (1930–), United States. Jerome Friedman shared the 1990 Nobel Prize in Physics with Henry Kendall and Richard Taylor.

> *The angular distributions of the scattering of electrons from point sources (**partons**, which are related to gluons and quarks) match the characteristics of Murray Gel-Mann's hypothetical "quarks."*

Experiments that Friedman conducted at Massachusetts Institute of Technology (MIT) with other colleagues demonstrated that protons that were struck with sufficient energy by deflected electrons were scattered at wide angles. These, as well as other factors resulting from the experiment, verified the nature of the hypothetical quarks. Friedman and other scientists, using the laboratory at the Stanford Linear Accelerator Center (SLAC) in California, as well as the equipment at MIT between the years 1967 and 1975, detected the scattering of electrons from the protons and neutrons. This provided the first evidence of the quark internal structure of nucleons (the particles within the nucleus, i.e., protons and neutrons).

Quarks are a particular type of tiny "bits" of energy or particulate matter that compose the **protons** and **neutrons** of the nucleus of atoms. Note: Baryons (heavy) are three-quark groups that compose protons and neutrons, that when considered together, are called nucleons. Protons and neutrons make up the mass of the nuclei of atoms. They have about the same mass. Each has $1/2$ spin, and they can be transformed into each other by giving up or receiving beta particles. Each proton and neutron consists of three quarks. Protons and neutrons are considered stable baryons (*see* Figure F10).

Six types of quarks have been identified and are usually referred to in terms of "pairs," as follows: 1) up and down quarks, 2) top and bottom quarks, and 3) charm

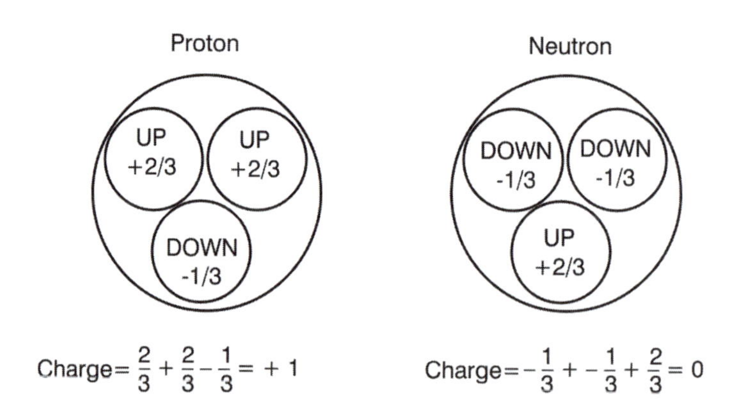

Figure F10. Quarks are bits of matter that compose the proton (with a +1 charge in an atom's nucleus) and neutrons (with a 0 charge, also found in an atom's nucleus). Together they make up most of the mass of the nuclei of all atoms.

and strange quarks. (Who says physicists don't have a sense of humor?) Protons consist of two "up" quarks and one "down" quark, while neutrons consist of two "down" quarks and one "up" quark. Unlike the proton with a +1 charge or a neutron with a 0 charge, quarks come with fractional electric charges as integers ranging from +1 to −1. There are several other "heavy" particles besides proton and neutrons that are composed of quarks. One example is the meson. Quarks are forever bound up inside the heavier particles found in the atoms' nuclei and are not found in a "free" state outside of the heavy subnuclear particles. There is some evidence that "mystery" particles called "gluons" are also found in the atoms nuclei. Their role is to hold together all the quarks found in the nuclei—thus the name "gluon." Otherwise all the positively charged protons would repel each other and the nuclei would come apart or just not exist.

See also Gell-Mann; Feynman

FRIEDMANN'S THEORY OF AN EXPANDING UNIVERSE: Astronomy: *Aleksandr Alexandrovich Friedmann* (1888–1925), Russia.

> *In opposition to Einstein's theory of a finite volume of an unchanging universe with a "saddle" shape, Friedmann's theory postulated a changing, and expanding universe with possible curvatures that may be zero, negative, or positive.*

Aleksandr Friedmann was educated at the University of Saint Petersburg in Russia where he received a degree in applied and pure mathematics in 1913. He became a manager in an aviation machine shop and later taught mechanics during World War I. After the war he began his career as an astronomer at the Pavlovsk Observatory in Saint Petersburg. He first became known for his work in the physics of the atmosphere and meteorology. He is best known for a paper he wrote in 1922 on his theory for the expanding, growing, changing universe.

His theory grew out of Einstein's theory of relativity to explain cosmology. Einstein's theory stated that the density is constant and thus must have a curvature. In other words, the universe did not change over time and had a limited volume. Friedmann came up with the idea of a universe of unlimited volume and a model of one that would change over time. Contrary to Einstein's theory, Friedmann considered the universe to be isotropic which means that all points in the universe move in all directions at the same rate. Therefore, the average density and size of the universe will change over time. The significance of Friedman's theory is that it later led to the big bang theory as an explanation for the creation and evolution of the universe.

Friedmann is also credited with coming up with different models for the shapes of the cosmos. Einstein proposed a "saddle" shape for the universe, while Friedmann claimed that different cosmological models would result in the curvature of space being either zero, positive, or negative. These various types of models for the universe are known as "Friedmann universes." Today, the theory of an expanding universe is accepted by most cosmologists based on evidence of rapidly receding galaxies, the "red-shift," and the rapid expansion of the universe since the big bang, rendering Einstein's "closed" universe theory less likely.

See also Dicke; Einstein; Gamow; Hale; Hubble; Lemaître

FRISCH'S THEORY OF A CHAIN REACTION: Physics: *Otto Robert Frisch* (1904–1979), England.

> *A sustainable nuclear chain reaction can be obtained by using just a few pounds of fissionable isotopes of uranium-235.*

In the late 1930s, Otto Frisch was involved in research with other scientists who discovered that uranium would decay into lighter elements when bombarded with slow neutrons. This work was confirmed and called *nuclear fission*, which was seen as a process capable of producing large amounts of energy. Frisch and Rudolf Peierls determined the rare U-235 was more likely to fission than other isotopes of uranium, such as heavier isotope U-238. Their additional calculations determined it would take only a few pounds of U-235 to reach a critical mass that would produce a sustainable chain reaction resulting in a massive explosion. This made the production of the atomic bomb a practical reality. Otto Frisch moved to the United States in the early 1940s to work on the atomic bomb project at Los Alamos, New Mexico.

See also Bohr; Hahn; Meitner; Peierls; Teller; Ulam

G

GABOR'S THEORY OF REPRODUCING THREE-DIMENSIONAL IMAGES: Physics: *Dennis Gabor* (1900–1979), England. Dennis Gabor (born Gábor Dénes in Budapest) was awarded the Nobel Prize in Physics in 1971.

> *A coherent light source will produce a two-dimensional holograph image that appears as a three-dimensional object on a photographic plate.*

Dennis Gabor was educated in Hungary and Germany followed by a career as a research engineer with Siemens & Halske AG, an electrical engineering company headquartered in Munich. One of his first inventions was the high-pressure mercury vapor lamp that is used for street lighting all over the world. With the rise of Adolph Hitler in 1933, Gabor left Germany. After a short period in Hungary, he moved to England where he found employment with the BTH (British Thomson-Houston) Research Laboratory in Rugby in 1934. In 1949 he joined the Imperial College of Science and Technology in London, eventually becoming a professor of applied electron physics until his retirement in 1967. Gabor always explained his work as "research serendipity" because his discoveries were based on previous research of other scientists, and much of the new information he reported grew out of their ideas.

He worked on improving the electron microscope that has a higher power of resolution than the light microscope but requires "light" with shorter wavelengths to be useful in observing things, such as the structure of crystals. To improve the image, Gabor used a method of positioning the light wave's cycle and intensity to build a fuller image of the objects being viewed by the electron microscope. He worked with micrographs produced by the electron microscope using coherent light, which is light that consists of wavelengths (frequencies) where the frequencies were exactly in phase and exhibited the same intensities. Therefore, because the light was of the same wavelength, it was a single pure color. He named the process *hologram* which is derived from the Greek words *holos* meaning "whole" and *gamma* meaning "message." The hologram expressed the idea that the resulting image contained "all" the information about the object being viewed. He used the mercury vapor lamp as the source of the coherent light even though it does not exhibit a high degree of coherent light (light of a single frequency or color).

Gabor learned of the work of the French physicist Gabriel Lippmann (1845–1921) who experimented with methods of recording the colors of nature so they could be reproduced more realistically than was possible with black-and-white film. In other words, Lippmann experimented with coherent light waves, as did Gabor, but merely to improve the technique of color photography. Gabor's work enabled the viewing of flat, two-dimensional photographs as three-dimensional images that acted somewhat like a stereoscopic image that stored all the information on the film, giving it a three-dimensional orientation. The problem of finding a pure single frequency light source was solved in 1960 when the laser was developed. Using a ruby crystal the laser produces a pure, intense, single frequency, coherent beam of red light that made the technique of holography a more common reality. In 1962 two engineering professors at the University of Michigan, Emmett Leith (1927–2005) and Juris Upatnieks (1936–), were the first to produce holograms that used laser light. Lasers have found many applications beyond the electron microscope, including side-reading radar, in the production of three-dimensional images that can be sent over wireless systems, laser light displays for entertainment purposes, and the all-important usage in various medical practices that have either replaced or surpassed standard surgical procedures.

GALEN'S THEORIES OF ANATOMY AND PHYSIOLOGY: Biology: *Galen* (c.130–200 BCE), Greece.

Galen's theory of the circulatory system: *The arteries and veins carry blood, not air, and the veins and arteries carry blood.*

Until Galen's time, Erasistratus' (c.300–260 BCE) theory that the essential body elements were "atoms" that were vitalized by air (*pneuma*) that circulated throughout the body by the arteries was accepted. Galen, one of the early experimenters who paid attention to his own observations, studied the structure and functions of organs and attempted to disprove this "air" theory. Experimenting with various small mammals, he discovered that blood, not air, flowed through the arteries. But Galen considered the liver to be the main organ of the circulatory system, and his theory stated blood was distributed to the outer parts of the body from the liver by the veins, and from the heart by the arteries. Galen also believed blood "seeped" through the intraventricular septum (central wall) of the heart through minute pores and that the heart had three chambers, each with its own function: the anterior or lateral ventricles (sensory information), the middle or third ventricle (cognition and integration), and the posterior chamber or fourth ventricle (memory and motor motion). He did not understand the role of the lungs in the circulatory system and believed the venous system (not the arteries) responsible for the distribution of food from the stomach to all parts of the body. Galen is considered by some historians to be the first to use the pulse of the heart as a diagnostic aid.

Galen's theory for the nervous system: *The brain controls the nervous system.*

Through the dissection of animals (never humans, except wounded gladiators), Galen demonstrated the distinction between sensory nerves (soft) and motor nerves (hard) and correctly placed the medulla as part of the brain rather than as part of the arteries. He correctly identified the nerves responsible for breathing and speech and demonstrated that specific nerves in the spinal cord control various muscles.

Galen's concept of the kidneys: *The kidneys, not the bladder, produce urine.*

Up to this time it was believed the bladder produced urine. By tying off the ureter, Galen proved the bladder did not produce urine but was merely a holding area for it. He also diagnosed several illnesses, including liver disease, by observing the urine of patients.

Galen's philosophy: *Nature does nothing in vain. God endowed every organ with a special purpose to perform special functions.*

Galen's philosophical outlook on nature was responsible for his success as a physician and scientist. He believed that the form of an organ was designed by a supreme being to perform a specific function, now known as "form follows function." Galen's medical knowledge and writings were accepted for over fifteen hundred years. Although his medical knowledge was advanced for this time in history, later physicians accepted his teaching without question and did little further investigating of the human body. Galen was the first to understand and use the pulse beat of the heart as a diagnostic aid. He proposed many theories concerning blood formation and flow, the nervous system, digestion, excretion, and so forth. His written works included over five hundred articles on his medical concepts that were translated by Arab scholars in the ninth century and later used during the Renaissance period in Europe that became the basis of medical theories and practices until the sixteenth century. This is why many historians believe that this respect for Galen's authority (the so-called tyranny of Galen) impeded medical progress for several centuries.

Although eclectic in his acceptance of the doctrines of earlier philosophers, Galen's main beliefs were based on the humoral pathology of Hippocrates (c.460–377 BCE) and Aristotle. For example, he based his theory of circulation on a three-part system of the liver, heart, and brain, each with its own spirits: natural, vital, and animal. His concept of preventive medicine was based on hygiene as well as "critical days" that were days when treatment would be more successful. He believed prevention was better than treatment and thought that diseases could be prevented if the "critical" days were observed. He was an excellent diagnostician for his era and was able to discern the source of many complaints. In addition to prescribing many different types of drugs, he used cold to treat hot diseases, and hot to treat cold diseases, and often used bleeding, purges, and enemas.

See also Hippocrates; Townes

GALILEO'S THEORIES: Physics: *Galileo Galilei* (1564–1642), Italy.

Galileo's theory of falling bodies: *Discounting air resistance, two bodies of different sizes and weights will fall at the same rate. Both will increase in speed of descent and land at the same time.*

From the time of Aristotle, it was believed a force could not act on a body from a distance. In other words, for an object to continue to move something physical needed to continue to push it; otherwise its movement would cease. In addition, Aristotle and others believed a body of greater weight would fall faster than a body of lesser weight, but they had never experimented with bodies heavy enough to overcome air resistance. Very light objects, such as a feather, would descend more slowly than would a rock, which seemed proof enough. It is most likely a myth that Galileo dropped objects of different weights from the Leaning Tower of Pisa. It may have been the Flemish mathematician and engineer Simon Stevin (1548–1620), not Galileo, who first dropped two rocks of different weights simultaneously from the Tower of Pisa to determine if they would land at the same time.

What we do know is that Galileo contrived his method of using an inclined plane made from a long wooden board to make accurate measurements and arrive at a reasonable explanation for the phenomenon of free-falling objects. We also know that he could not have dropped the balls from the Leaning Tower of Pisa because his inclined plane experiment was conducted while he was living in the town of Padua, not Pisa. Galileo assembled an inclined plane that allowed two balls of different weights to roll slowly down the incline, which enabled him to measure their rates of descent by using the pulse of his heartbeat. The only other timing devices available at that time were sundials, time candles, and dripping water clocks. None was accurate enough for Galileo's purposes. He also ensured that the balls were of sufficient, but of different weights so the resistance of air or the surface of the wooden planks of the inclined plane would minimally affect them. His measurements confirmed that not only did the balls roll down to the bottom of the plank in equal time, but also their rates of descent increased as they passed equally spaced marks on the planks. When he experimented with planks raised higher and lower to form different degrees of inclination, he discovered an interesting factor: No matter at what angle the planks were positioned, the balls covered a single unit of distance on the plank for the first unit of time based on his heartbeat and a water clock that used a slow dripping stream of water. But for the second unit of time, the balls rolled three times faster than the first unit's distance. He discovered that the ratio of distances covered by the balls increases by odd numbers. This means that for the total time of descent of four seconds, the balls covered a distance sixteen times greater than is covered in one second. This relationship of the ratio between time and distance is further explained as acceleration acting uniformly on a falling body, where the descending distance covered is directly proportional to the square of the time (*see* Figure G1).

From these data, Galileo formulated the law that states $s = \frac{1}{2} at^2$, where *s* is the distance the ball travels, *a* is the acceleration, and *t* is the time lapsing of the ball's descent. Galileo's experimental results illustrated the uniform accelerating force of gravity, which Sir Isaac Newton later developed as part of his concept of inertia and

INCLINED PLANES

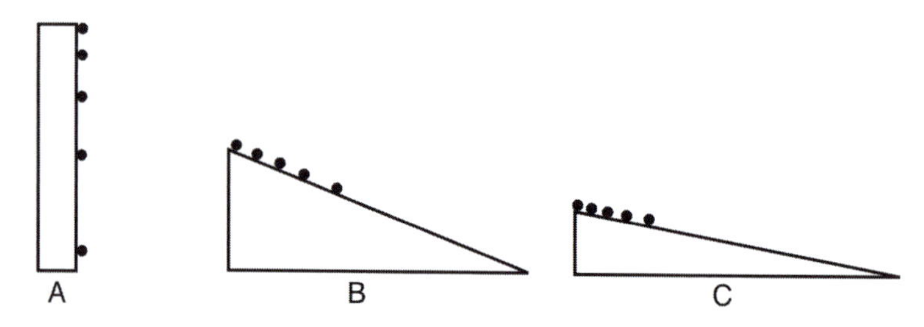

Figure G1. The balls represent the ratio of the distance of their descent to the square of the time of their descent.

the three laws of motion. The other consequence of this experiment was that Galileo was now able to correct the Aristotelian idea that the push of "angles" was required to maintain planetary motion. Once friction was removed from consideration, the constant pull of the sun's gravity sustains the planets' orbiting the sun.

An interesting bit of modern history: During the 1991 Apollo 15 moon landing, astronaut David Scott dropped a feather and hammer at the same time, and they hit the surface of the moon at the same time. Apparently he wanted to prove that Galileo was correct. The same experiment with a feather and rock can be performed on Earth within a large vacuum chamber where most of the air has been removed.

Galileo's concept of the pendulum: *The square of the period (oscillation) of a pendulum varies directly with the length of it suspending string.*

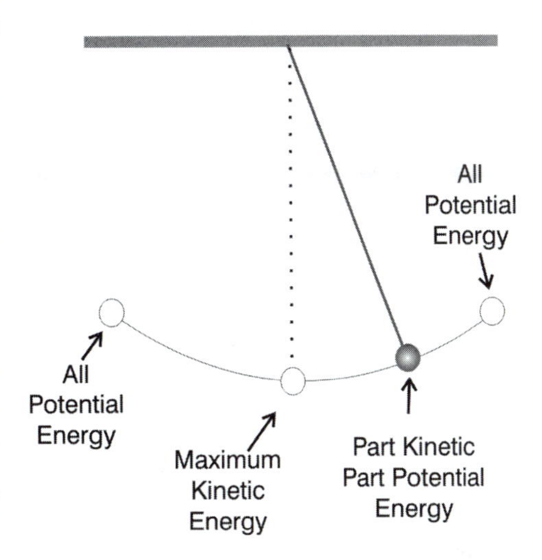

Figure G2. A typical pendulum as used by Galileo to establish his theory of pendulums.

While studying medicine at the University of Pisa, a youthful Galileo was attending church services in the town's cathedral when he noticed that a large chandelier swayed in the breeze. Sometimes the chandelier swayed in longer arcs and sometimes in smaller arcs; the time period of the swing seemingly was the same regardless of the sweep of the chandelier. The pulse of his own heartbeat that he used to count the time that lapsed for each swing provided him with an idea for an experiment. Upon returning home, he designed a pendulum with a bob on a short string and another bob of a different weight on a longer string (*see* Figure G2).

Again timing them with his pulse, he confirmed his theory. He summarized his ideas as follows: 1) Air resistance (friction) prevents the pendulum from returning to its exact starting position. However, if there is no air resistance, the bob will always return to its original position. Thus, sooner or later, all pendulums come to rest. Pendulums with lighter bobs come to rest sooner than those with heavy bobs. 2) The period of swing or sweep of the pendulum is not related to the weight of the bob. 3) The time period for each sweep of a pendulum is not dependent on the length of its sweep (this observation was later proved incorrect). 4) The square of the period for a pendulum is directly proportional to the length of the pendulum.

Figure G3. An artist's depiction of Galileo's air thermometer.

Once set in motion, a pendulum oscillates with a constant frequency that is inversely proportional to the length of its string. Although Galileo recognized the importance of this phenomenon, he was unable to develop his pendulum into a practical timepiece (he continued to use a water clock and his pulse). However, it became an important concept in the design of accurate clocks. Some years later, Christian Huygens fabricated a workable pendulum clock similar to a grandfather's clock that used weights to maintain the movement of the pendulum (*see also* Huygens).

Galileo's concept for the measurement of temperature: *There is a direct relationship between the temperatures of air and water and their volumes.*

From the beginning of time, people understood the concepts of hot and cold, but until Galileo, there was no objective way to measure the exact temperature for either. Galileo devised an air thermometer, or thermoscope, a crude, and not very accurate, instrument for measuring temperature. He used a long, thin, stalk-like tube of glass, open at one end and with a closed bulb at the other end. Placing his hands on the bulb until it was warm; he then inverted the open tube into a pan of water. As the bulb cooled, some water was drawn up in the narrow tube toward the bulb. Also, as the temperature of the surrounding air changed, so did the level of water in the tube. This furnished Galileo the means to measure the level of water in the tube and to make some calculations (*see* Figure G3).

His instrument, however, was quite inaccurate due to the effects of atmospheric pressure on the water in the pan, which was open to the air. Even so, this was the first thermometer, which Galileo later redesigned. He enclosed the water in a sealed tube containing "floats" constructed of small, hollow glass balls adjusted for different water densities. As the temperature changed, so would the density of the water, causing one or more balls to rise or fall, thus indicating the air temperature. Today these fascinating instruments are sometimes referred to as *thermometro lentos* (*see* Figure G4).

THERMOMETRO LENTO

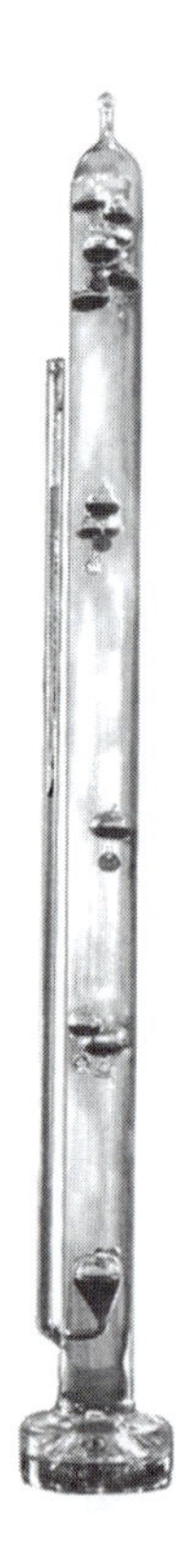

Figure G4. A modern version of Galileo's Thermometro Lento.

Galileo's astronomy theories: *1) Dark "spots" on the surface of the sun appear to move around the sun; therefore the sun must rotate, and so must Earth and other planets revolve around the sun. 2) Jupiter has several of its own moons similar to Earth's moon. 3) Saturn has bulges on its side as well as its own moons. 4) The Milky Way is composed of a multitude of stars clustered together.*

The telescopes that Galileo constructed enabled him to view objects never before seen by humans, and thus he conceived many theories about the planets and stars. The credit for the development of the first telescope is usually attributed to either of two Dutch spectacle makers, Hans Lippershey (1570–1619) or Zacharias Janssen (1580–1638). Janssen is also credited with inventing the microscope in 1608. Galileo learned of this "secret" device and then constructed his own telescope. An excellent lens maker, he improved the curvature of his lenses to reduce optical aberration. Galileo built three telescopes, the last of which was improved to approximately 30-power, or about the power of a good pair of modern binoculars.

One of Galileo's first telescopic viewings was of the surface of the sun, where he observed the movement of darker areas or "spots" and concluded the sun must be rotating. Based on knowledge of moving bodies, he surmised the planets and Earth are not only spinning on their axes, but are also revolving about the sun in circular paths. This was the first confirmation of the Copernican heliocentric concept of the solar system. (See Figure G5.)

Galileo disagreed with, or ignored, Kepler's laws that state that planets move in ellipses. Because the concept of gravity was unknown in his time, Galileo believed the paths of planets were based on inertial circular movement. This erroneous concept prevented him from completely developing his law of uniform acceleration into the Newtonian-type laws of motion (*see also* Copernicus; Kepler; Newton).

Using the telescope he constructed, Galileo observed two tiny objects that appeared to move around the planet Jupiter, and he tracked and recorded the changes in their position. Later he discovered two other moons of Jupiter, for a total of four larger moons. (A total of sixteen satellites of Jupiter have been subsequently discovered.) His records of the eclipses of Jupiter's satellites aided sailors in determining longitudes at sea.

GALILEO'S HELIOCENTRIC SOLAR SYSTEM

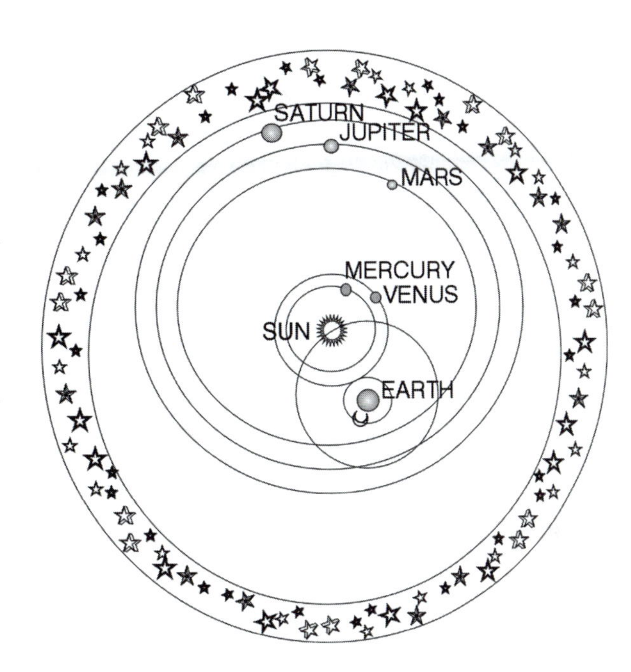

Figure G5. Galileo's heliocentric solar system was the first confirmation of Copernicus' original concept of a sun-centered solar system.

After viewing Saturn at different times, Galileo noticed "bulges" on each side of the planet that periodically became larger, then smaller. His telescope was not powerful enough to resolve these "bulges" into the many rings around the planet that change their apparent shape as the orientation of the planet changes when viewed from Earth. Galileo also identified several of Saturn's moons.

Always fascinated by the multitude of stars that could be seen with his telescope, Galileo observed that, when aiming it at the Milky Way, it became obvious that this huge area of the sky was composed of many millions of stars. He recorded there were more stars, some very faint, in this area of the sky than in all the other areas combined. Up until this time, the Milky Way was considered to be just a large cloud in the sky.

Although Galileo did not completely understand gravity or inertia, he had a firm concept of the mechanics of force, and his theories concerning falling bodies were a forerunner to Newton's three laws of motion. His "thermoscope" was a precursor to more accurate instruments for temperature and pressure measurements, including the modern mercury thermometers and barometers.

His work with fluid equilibrium, as in a working siphon, led to a new concept of pumps. At one time it was believed that by reducing the air pressure above water, the water would be "sucked up" into the pump (similar to a drinking straw). Galileo understood that normal air pressure outside the pump "pushed" the water up into the pump

Most likely Galileo will be remembered for his theory that challenged Ptolemy's Earth-centered universe concept, which was then accepted by the Church of Rome. Because of his belief in the theory of a Copernican sun-centered system, Galileo was secretly denounced to the Inquisition for blasphemous utterances. He was later forced to recant and was sentenced to house arrest until his death at age seventy-eight. Near the end of the twentieth century, the Roman Catholic Church removed the charges and exonerated Galileo.

Galileo was one of the first persons to make a distinction between religious beliefs related to the physical world and the results of experimental evidence related to the nature of the universe. Even today, there continue to be misunderstandings between religious beliefs based on faith and those related to the nature of the universe that are based on experiments and factual evidence.

to the area of reduced air pressure (just as normal air pressure "pushes" the liquid up a straw used for drinking).

He also developed the mathematics that explained the flotation of solids in liquids and studied the magnetism of lodestones that would influence later scientists such as William Gilbert. Galileo was the first to demonstrate that a magnet broken into smaller pieces retained its magnetic properties because each piece, no matter how small, is still a magnet with its own north and south poles.

Galileo, along with Janssen, is credited with developing the first practical microscope after adapting his concept of a telescope to produce a crude but workable microscope. In addition, he tried to measure the speed of light by flashing a lantern positioned on one hill to an assistant who flashed his lantern back from another hill. Although he could not detect the speed of light, nevertheless he was convinced that light travels with great and measurable speed.

At six-month periods, as Earth revolved around the sun, Galileo attempted to measure the parallax of stars to determine their distance from Earth. Although his instruments were not accurate enough to accomplish the task, his concept was correct because he used parallax to measure the distance of the moon to Earth.

See also Copernicus; Fahrenheit; Gilbert; Kepler; Newton; Ptolemy

GALLO'S HIV-AIDS THEORY: Biology: *Robert Charles Gallo* (1937–), United States.

The HTLV-3 retrovirus suppresses the immune system lymphocytes, thus causing acquired immune deficiency syndrome (AIDS).

Robert Gallo was familiar with the process of how, when under attack, the body's immune system produces interleukin-2, which stimulates special lymphocytes identified as T-cells to fight viral infections in leukemia patents. In the early 1980s he surmised a virus was responsible for a similar suppression of the immune system that led to opportunistic (AIDS) infections. Gallo then hypothesized the retrovirus he identified as HTLV-3 was reacting with the immune system in a similar manner as it did for the blood cancer disease leukemia. At the same time, Luc Montagnier, the French virologist and researcher, made a similar deduction and in 1983 sent Gallo a sample of his virus, which he called lymphadenopathy associated virus (LAV). Gallo's assistant discovered a particular T-cell type that could be invaded but not killed by these viruses, which then could be used to develop a test for the virus in AIDS patients. Gallo proceeded to secure a patent for his new AIDS test. This resulted in an international argument about the discovery of the AIDS virus because the American HTLV-3 and French LAV viruses were the same. It was also claimed that the French LAV virus was used in Gallo's laboratory to develop the test. It was settled in 1986 when Gallo and Montagnier agreed that both their names would appear on the patent document and 80% of all royalties would be given to an AIDS research foundation. The HTLV-3 and LAV virus were renamed the HIV virus by an international committee. Between 1981 and 1990 Gallo published over four hundred papers related to his research on HIV/AIDS.

See also Baltimore; Montagnier

Recently, several scientists have claimed that a vaccine for the HIV virus can be developed. In 1997 Dr. Robert Gallo, now the head of the Institute of Human Virology at the University of Maryland's Baltimore campus, questioned the possibility of developing such a vaccine. His concerns are related to the pathogenesis of the HIV virus, including the large number strains of the HIV virus, the lack of good animal models for testing the vaccine, and the fact that the HIV virus invades basic DNA cells of the immune system. In addition, for such a vaccine to be effective, it must be the only viable vaccine in the patients' bodies. Since then, several experimental vaccines have been tested but to date none seem to be 100% effective.

More recently it was reported in the September 22, 2007, issue of the *New York Times* of the failure of AIDS vaccine tests. "A much-heralded H.I.V. vaccine has failed to work in a large clinical trial, dealing another serious setback to efforts to stop the AIDS epidemic." Despite this failure, efforts continue to develop an effective AIDS vaccine, although researchers acknowledge that the quest will be difficult and in all likelihood not viable in the foreseeable future. There is continuing debate about how to eradicate HIV infections and AIDS from the human population. Although this disease is preventable and somewhat predictable, human nature is not.

GALTON'S THEORY OF EUGENICS: Biology: *Sir Francis Galton* (1822–1911), England.

> *The human race can be improved by selective and controlled breeding, as is done with domesticated plants and animals.*

As an anthropologist, Francis Galton's extensive travels enabled him to observe varied cultures and races and subsequently to conduct research in the areas of human heredity. Galton knew that since the beginning of farming, humans selected not only the best grains as seeds to plant, but also selected animals with the most desired characteristics to breed, therefore, improving the quality and quantity of agricultural products, just as can be done today with genetic engineering. He was also aware of the theory of organic evolution proposed by his cousin, Charles Darwin. Based on this background, Galton considered controlled breeding a means to improve the human species just as it had for many species of plants and animals. Galton is credited with coining the term "eugenics," meaning "good genes" or "good breeding." The first to study identical twins, he discovered that though identical twins have similar patterns of whorls and ridges in their fingerprints, there are just as many differences. In other words, everyone's fingerprints are unique. He also developed the first system for classifying and identifying fingerprints, which expanded the field of **forensic** science. Using his research on identical twins, he attempted to resolve the distinction between environmental versus inherited factors that influence intelligence. His research, the premise of which was to determine what is most important in the development of intelligence—nature or nurture—continues today. Evidence indicates that nature may be responsible for over 50%

The "science" of eugenics has a checkered reputation due to the moral and ethical implications of controlling the selection of who shall be born into the world. Many people associate and relate eugenics with the Nazis' program for developing a "super race" during the 1930s and early 1940s. Today, "ethnic cleansing" might also be an application of eugenics. Even so, there are instances where forms of eugenics are used today without that stigma. When a couple (the husband is infertile) who wants to have children select frozen sperm for fertilizing the wife's eggs, they may have the option of selecting the physical characteristics of the sperm donor. The same type of selection can be made when deciding on the implantation of a **zygote** into the womb of a female. Another example is **amniocentesis** that is used to determine the genetic health of a fetus, and thus this information can be used to make the decision to abort or continue the pregnancy. Another more negative example is the use of ultrasound to specifically determine the sex of the unborn child that may lead to the abortion of a child of the "wrong" sex. On a more subtle level, when men and women—consciously or unconsciously—select a mate, they are "discriminating" as to how that mate should look and behave to make a good parent.

of a person's intelligence, but environmental factors during gestation and after birth are also important to the development of intelligence. Galton was the first to use new quantitative methods for eugenics research, including statistical correlation coefficient and regression analysis (statistical methods for comparing similarities between variables). His 1888 statistical techniques were sound but somewhat inadequate as compared to the statistics used for data analysis today. Although Mendel's work with genetics was not yet "rediscovered," Galton's research indicates an understanding of its basic principles. His methods were used mainly for analyzing the results of experimental medical research. His invention of correlation coefficient is probably more important than his work on eugenics. Interestingly, Galton's research led to the development of fingerprinting based on the unique swirl patterns for each individual's fingertips that are used for identifying individuals.

See also Darwin; Mendel; Wallace

GALVANI'S THEORIES OF GALVANIZATION AND ANIMAL TISSUE ELECTRICITY: Physics: *Luigi Galvani* (1737–1798), Italy.

Galvani's theory of galvanization: *Small electrical currents can be used to "coat" metals that are easily oxidized (iron rust) with other metals that resist oxidation (zinc).*

Although Galvani was trained in medicine/anatomy and experimented with electrophysiology, he was honored for his discovery of the galvanization process. He called his discovery the *metallic arc* because it used an electric current to bind a coating of zinc to iron. Galvani's process is similar to electroplating of metallic and nonmetallic items. Today, galvanized iron or steel also can be produced by dipping the item made of iron into a hot bath of molten zinc or by "spraying" very small zinc particles onto hot iron or steel. The resulting iron products will be more rust resistant than nongalvanized metal.

Andre Ampère named the *galvanometer*, an instrument used to measure small electric currents, after Galvani who is also credited with the discovery of current electricity. Galvani pursued his work with electric currents to include investigating whether animal nerves carry electricity.

Galvani's theory of animal tissue electricity: *Electricity is present in animal tissue that can be discharged when in contact with two different metals.*

Luigi Galvani's electrophysiology experiments involved the touching of a dissected frog's leg with a spark from a machine that produced static electricity. However, he made a famous experimental error with regard to animal tissue electricity. During a thunderstorm, Galvani clamped onto an iron railing the brass hooks that he inserted into a dissected frog's spinal cord. The frog's muscles twitched, as they did also when two different types of metals touched the spinal cord. He incorrectly concluded that the electricity was generated by the frog's tissue, while rejecting the possibility that the electricity that caused the twitching came from another source. Later, Alessandro Volta demonstrated that the electricity was not derived from the tissue but rather from the brass and iron coming into contact with each other under moist conditions.

See also Ampère; Franklin (Benjamin); Ohm; Volta, Watson (William)

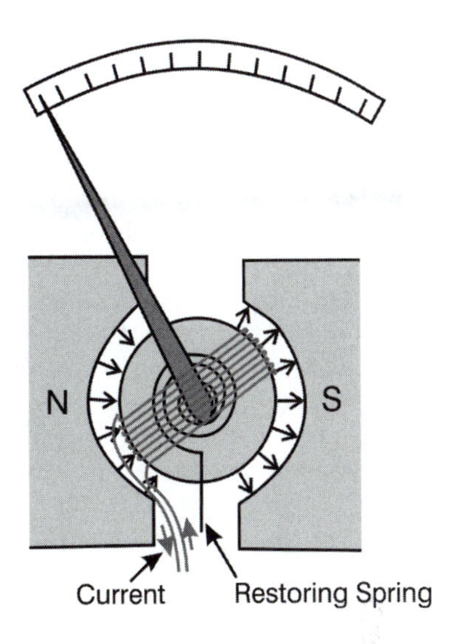

Figure G6. Andre Ampère named the *Galvanometer* after Luigi Galvani. It is used to measure small amounts of current electricity.

GAMOW'S THEORIES OF THE UNIVERSE AND DNA: Physics: *George Gamow* (1904–1968), United States.

Gamow's big bang theory: *The universe was created more than fourteen billion years ago from a single point source in space and time.*

Many civilizations over many generations theorized that the genesis of the universe was based on the egg or seed concept, where everything "grew" from a minute and rudimentary source. Also, persistent through the ages are religious concepts for the origin of Earth and/or the universe. The big bang theory proposes that an incredible singularity event occurred where a dense point or source of energy "exploded" and rapidly expanded in all directions in microfractions of a second, to form all the energy and matter in the known universe. There are several current theories for the creation, nature of, and demise of the universe—the static universe, the ever-expanding universe, the regeneration concepts, and the possibility of multi-universes of which ours is just one of many. Many scientists proposed theories similar to the big bang for an expanding universe that they based on Einstein's theory of relativity, despite the fact that Einstein proposed a static universe as evidenced by the "saddle-shape" for his unchanging universe. In 1948 Sir Fred Hoyle proposed that matter was continually generated by existing matter and is continually spread throughout the expanding universe. This is considered the model that George Gamow used to revise earlier theories and advance another based on mathematical concepts. Gamow believed there was evidence not only of an expanding universe but also that if the universe is continuing to expand, it must have had a beginning. In other words, if the universe is forever expanding in all directions, then it must have started at a central point from something extremely small that contained all the energy and matter required to form it, more recently referred to as a "**singularity**." As evidence to substantiate his big bang theory, Gamow answered Olbers' paradox, which raised the question of why the sky of the universe appears dark rather than full of light, as does the area illuminated by our sun. Gamow explored several possible answers to this paradox but felt the best is that the universe is ever

George Gamow was educated at the University of Leningrad where he received his PhD in 1928. He spent time in Copenhagen, Denmark, and Cambridge, England, before moving to the United States in 1934. He was a professor of physics at George Washington University and later at the University of Colorado. He had wide research interests, including a later interest in molecular biology. In addition to his work with the big bang theory, he determined that the heavier elements were all formed in the hot thermonuclear interior of the stars. He also demonstrated how our sun is warming, not cooling, which may account for the cyclic nature of Earth's cool and warm periods. The sun's temperature cycles may partially account for today's slight warming of Earth's oceans and atmosphere. George Gamow was well known and respected not only as a physicist/cosmologist and microbiologist but also as a popular science writer. He was the author of several excellent easy-to-read books on modern physics that are still in print.

expanding. Thus, stars cannot shine enough light to fill up all that space because their radiation is not in equilibrium with their surfaces. One of the main factors supporting an expanding universe and opposing a static universe is that a self-contained, nonexpanding universe is incapable of disposing of the energy produced by all the stars that would result in a very hot static universe, whereas an ever-expanding universe would ultimately reach a balance for star energy. Thus, **equilibrium** would be established and a stable temperature would exist, or there possibly could be a decrease in overall temperature. According to the second law of thermodynamics (entropy), the final temperature would be absolute zero, and an absolute equilibrium would exist. George Gamow calculated that the leftover uniform background radiation from the big bang explosion is equal to about 5 kelvin. In 1964 two AT&T scientists Arno Penzias and Robert Woodrow Wilson (1936–) discovered this "leftover" energy by detecting the **primordial** microwave type radiation. This discovery is the best current evidence for the big bang theory. Even so there are new theories that may lead to a reconsideration of an infinite universe; that an ever-expanding universe is an illusion, and that it is finite after all (*see also* Doppler; Hale; Hubble; Lemaître).

Gamow's theory for the beginning of life: *The amino acids, which form proteins, are constructed from the four nucleic acid bases of DNA.*

George Gamow based his theory for life on Francis Crick's and James Watson's proposed structure of the DNA double helix. The nucleic acid–based pairs connecting the two sides of the double helix are the nucleotides of adenine plus thymine (A + T) and guanine plus cytosine (G + C). Gamow realized this sequence of codes for these four nucleotides (A, T, G, and C) could produce only four amino acids, not the twenty or more existing in humans, and thus would be inadequate to produce the multitude of proteins necessary for life. Therefore, he concluded there needed to be at least three sequences of the base pairs present to produce codes necessary for the required number of amino acids. Using this code of the nucleotides, at least sixty-four amino acids could be produced ($4 \times 4 \times 4 = 64$). Gamow's mathematical code explained the sequencing required for amino acids to produce proteins.

Gamow's theory of the living cell: *Cells in plants and animals are structured to carry out functions analogous to those processes and procedures related to running a factory.*

George Gamow used the analogy of an industrial factory to explain the functioning of a living cell. The manager's office represents the nucleus of the cell, whereas the chromosomes are the file cabinets where information, production plans, and diagrams are stored. When a new cell (factory) is to be opened, the secretary and staff produce an exact copy of what is in the file cabinets. As the new factory grows, a new manager's office takes over, and the process is repeated. The "workers" and "machinery" of the factory are abundant and represent the enzymes, protoplasm, and other cell components. The chromosomes and their genes, which are stored in the file cabinets, are very limited but can be used to replicate the factory and start all over.

See also Crick; Einstein

GARROD'S THEORY OF CONGENITAL METABOLIC DISORDERS: Biology and Medicine: *Sir Archibald Edward Garrod* (1857–1936), England.

> *The rare metabolic disorder* alkaptonuria *is not a bacterial infection of the urinary system but rather a genetic defect related to the lack of an enzyme in the chemical breakdown of a crucial protein.*

Alkaptonuria is an uncommon disorder where urine turns dark brown when exposed to air. At first it was thought to be due to a bacterial infection, but Garrod demonstrated that it was a genetic disorder—not a disease caused by bacteria. This genetic condition is rare in the general public but is common in the offspring of first-cousin marriages. Garrod demonstrated that alkaptonuria followed the pattern explained by inheritance of recessive genes as described by Mendelian genetics.

Archibald Garrod, an English physician who discovered the nature of several congenital metabolic disorders, determined that the condition of alkaptonuria was due to the presence of large amounts of homogentisic acid (also known as alkapton) that is excreted in the urine due to a deficiency of several amino acids. In other words, Garrod, who was exploring the field of biochemistry, understood that this condition was due to the lack of an enzyme responsible for the breakdown of a protein that resulted in the buildup of the chemical that darkens the urine.

Garrod was fifty years ahead of his contemporary biochemists' understanding of the implications of this theory for the genetic nature of metabolic disorders. Although the conditions of alkaptonuria (dark urine) are visible, not all metabolic disorders are that obvious. Garrod wrote about and gave lectures about these conditions that he referred to as "inborn errors of metabolism." His belief that genetics was involved in the process was evident when he wrote that there are many variations in humans that are determined by genetics, and that no two individuals are alike either chemically, biologically, or structurally. Garrod identified several other congenital metabolic disorders including *cystinuria* (an inborn defect involving an excess secretion of several amino acids), *pentosuria* (a congenital urinary defect in the oxidation of glucuronic acid that is a condition principally in those of Jewish heritage), and porphyria (a form of inherited insanity that caused the English King George III's illness).

GASSENDI'S THEORIES: Physics: *Pierre Gassendi* (1592–1655), France.

Gassendi's atomic theory: *God created the atoms as immaterial souls that could exist and interact in a void. He then gave them to man.*

Believing that atoms could exist only in a void in which the tiny particles could interact with each other and religious spirits, Gassendi tried to make the atomism of Lucretius, Epicurus, and Democritus agreeable with Christianity, but he opposed

In 1624 the Paris Parliament passed an ordinance declaring that any person would be put to death if he taught or held any doctrine opposed to Aristotle. In spite of this law, Gassendi published his *Dissertations against Aristotle* which attacked many of the ideals and teachings of not only Aristotelianism, but also the many beliefs of scholasticism. Because he was a doctor of theology who was ordained in 1617 and later a professor of mathematics at the College Royal in Paris, he escaped punishment by the Parliament. His views on atomism were expressed in his *Observations on the Tenth Book of Diogenes Laertius* in which he insisted that atoms were created by God and that God created the void of space so atoms could exist and thus interact with each other. This concept was later expanded to a variety of theories related to the atomic and chemical nature of the universe (*see* Atomism Theories). Gassendi was interested in astronomy and recorded many viewings of eclipses, comets, and the planets. In his book *Mercury in the Face of the Sun* he recorded the first transit of Mercury that supported Johannes Kepler's theories on the motions of planets.

Aristotelianism regarding these matters. Gassendi's concept of a void was very much like the modern concept of the vacuum of space. He disagreed with Aristotle's belief that a void did not, and could not, exist. Gassendi was a believer in the Epicurean views of Lucretius' doctrine of atomism. Galileo, Robert Boyle, and later Isaac Newton, as well as other scientists, were influenced by Gassendi's "Epicureanism" philosophy. This corpuscular concept states that for atoms to exist, a vacuum must surround them. Thus, if the atoms were removed, only the vacuum would remain (*see also* Aristotle; Atomism Theories; Boyle).

Gassendi's theories for falling bodies, sound, and astronomy: Pierre Gassendi, an early philosopher, propagated his ideas by incorporating his moderate skepticism with some experimentation that influenced his philosophy.

Gassendi was the first to test Galileo's contention that a ball dropped from the mast of a ship would fall at the base of the mast, not at some distance aft of the mast. Ancient sailors who dropped rigging tools could attest to this fact. Oddly, no empirical experiment had been conducted prior to Gassendi's.

It is also reported that Gassendi was one of the first to measure the speed of sound. It is unclear how he made his measurements, but it is assumed he fired a cannon while someone on a far hill at a known distance from the cannon timed the smoke pouring out the barrel until the sound was heard. His figure of 1,473 feet per second was about 50% greater than the current figure of 1,088 to 1,126 feet per second in dry air at sea level. The speed of sound depends on the density of the substance through which the sound is traveling. For example, sound travels at the speed of 4,820 feet per second in water, 11,500 feet per second in brass, and 16,500 feet per second in steel.

Gassendi studied **comets** and eclipses and recorded the first observed transit of Mercury in 1631. He was the first to describe as well as name the northern lights *the aurora borealis*.

See also Descartes; Galileo

GAUSS' MATHEMATICS AND ELECTROMAGNETISM THEOREMS: Mathematics: *Karl Friedrich Gauss* (1777–1855), Germany.

Gauss' theory of least squares: *A circle can be divided into a heptadecagon by using Euclidean geometry.*

Karl Gauss, a child prodigy in mathematics, was considered a "human calculator" who could solve all kinds of complicated problems in his head. Gauss demonstrated this seventeen-sided polygon (heptadecagon) could be drawn using only a compass, ruler, and pen. All seventeen sides were of equal length when laid on arcs of a circle. Earlier Greek mathematicians could never accomplish this exercise, which was considered an advancement in geometry. Gauss also demonstrated there were a limited number of polygons (many-sided figures) that could be constructed using these tools. An example of a polygon that cannot be so constructed is the heptagon (a seven-sided polygon) (*see also* Archimedes).

The International System of Units (SI) for this flux density is called a *gauss* in his honor. In the SI, CGS units (using centimeters, grams, and seconds rather than MKS—meters, kilograms, and seconds), a unit area of 1 square centimeter with a flux density of 1 *maxwell* per square centimeter equals 1 *gauss*. The gauss is equal to 1 maxwell per square centimeter, or 10^{-4} *weber* per square meter or 10,000 gauss equals 1 weber. Gauss and Weber developed the magnetic-electric telegraph and a new instrument called the magnetometer. Magnetic field strength is rated in gauss units, an important concept for modern technology utilizing all types of magnets.

Gauss' theory of errors: *Successive observations and measurements made of the same event by the use of instruments are never identical, but their mean value can be calculated.*

This theory is related to probability. Gauss claimed that the distribution of errors for the **mean** differences of measurements by observations (particularly astronomical observations) is as accurate as the probability (odds) when throwing dice. This statistical technique has been, and still is, used by most scientists who make a series of measurements and calculate the means of these measurements. They can be reasonably certain that the difference between two means is a meaningful representation of their observational measurements. Gauss' work in statistical probability distributions is referred to as *Gaussian* statistical distribution. This concept is used for the statistical treatment of data for most research experiments.

Gauss' theory of aggregates: *Properties of individual units of populations can be accurately observed and studied in large groups (aggregates).*

This is another theory used by most scientists and involves the study of large populations of particles, such as atoms, molecules, chromosomes, and genes. An example is Brownian motion, which is the observed movement of tiny microscopic particles of a solid, such as pollen, that is caused by molecular motion in solution. The concept of aggregates explains the kinetic theory of gases as well as the gene theory for inheritance. Although this theory is based on and accepted as an assumption, it does work (*see also* Ideal Gas Laws).

Gauss' law of the strength of electric and magnetic flux: *The greater the closeness (density) of the lines of force of an electric field (or magnetic field), the stronger the field.*

Gauss and the German physicist Wilhelm Weber (1804–1891) collaborated on studying the nature of electric and magnetic fields. They calculated the number of lines of force and the closeness of those lines, which determine the "flux density" representing the strength of the electric field. Electrical flux is a measure of the number of lines in an electric or magnetic field that passes through a given area. Gauss' law states the relationship between electrical charge and an electrical field. It is easy to picture by considering that the field is stronger if these lines of force are crowed together, and the field will be weaker if the lines of force are further apart. In some ways, his statement for the relationship between an electric charge and an electric field is another way to explain Coulomb's law.

See also Coulomb; Faraday; Maxwell; Weber

Gay-Lussac, a French engineer, physicist, chemist, and accomplished experimenter, made several other contributions to science. He collaborated with several other Frenchmen on a number of projects, including one where he used balloon flights for scientific purposes. In 1804 he and Jean-Baptiste Biot ascended 4 miles (about 7 km) in a balloon, the highest altitude attained by humans as of that date. They made the first high-altitude measurements of atmospheric pressure and Earth's magnetism. Gay-Lussac discovered the poison gas cyanide (HCN), and in 1815 he made cyanogen (C_2N_2), a toxic univalent radical used for the production of insecticides. His experiments with compound radicals were a precursor to the development of organic chemistry. Gay-Lussac and the French chemist Louis Jacques Thenard (1777–1857) produced small amounts of the reactive metals sodium and potassium. When Gay-Lussac mixed metallic potassium with another element, it exploded, wrecking his laboratory and temporarily blinding him. He also discovered a new **halogen** similar to chlorine. He named it *iode* (iodine), which means "violet."

GAY-LUSSAC'S LAW OF COMBINING VOLUMES: Chemistry: *Joseph-Louis Gay-Lussac* (1778–1850), France.

The volumes of gases that react with each other, or are produced in chemical reactions, are always expressed in ratios of small, whole numbers.

An example: When one volume of nitrogen gas (N_2) is combined with three volumes of hydrogen gas ($3H_2$) the result will be exactly two volumes of the gaseous compound ammonia ($2NH_3$). Their respective volumes are the exact ratio 1:3:2. Gay-Lussac determined the existence of the law for combining volumes of gases, but he had no idea of why this law applied. An explanation had to wait until Amedeo Avogadro established the law explaining that equal volumes of all gases contain the same number of molecules (at the same temperatures and pressures), regardless of the physical and chemical properties of the gases.

In addition to the law of combining volumes, Gay-Lussac discovered that all gases expand equally when the temperature rises. This is a modification of Charles' law. Both of these gas laws, including Boyle's gas laws, are considered the *ideal gas laws* because they are really approximations. Gases exhibit only the relationships of P, T, and V (P = pressure, T = temperature, and V = volume) as expressed in the laws, at ordinary (moderate) temperatures. In 1808 Gay-Lussac published his law, usually called the law of combining gases, when referring to chemical reactions where the number of atoms is constant. This law confirmed the work of Dalton, who missed the importance of the relationship between temperature and volume of gases. In essence, the law states that for any gas, the temperature and pressure are directly related at a constant volume for that gas. The equation is $P/T = K$, where P is the pressure directly related to T, the temperature for K, a given constant (i.e., volume). Conversely, if the gas is heated, its volume increases as long as the pressure on the gas is constant; and if the pressure increases, so does the temperature of the gas for a given volume of a contained gas. Another way to state it is that the volume of gases expands equally when subjected to the same changes of temperature provided that the pressure remains the same.

See also Avogadro; Boyle; Charles; Ideal Gas Laws

GEIGER–NUTTER LAW (RULE) FOR DECAY OF RADIOACTIVE ISOTOPES: Physics: *Hans (Johannes) Wilhelm Geiger* (1882–1945), Germany.

There is a linear relationship between the logarithm of the strengths of alpha particles and the particles' rate of decay from their source nucleus.

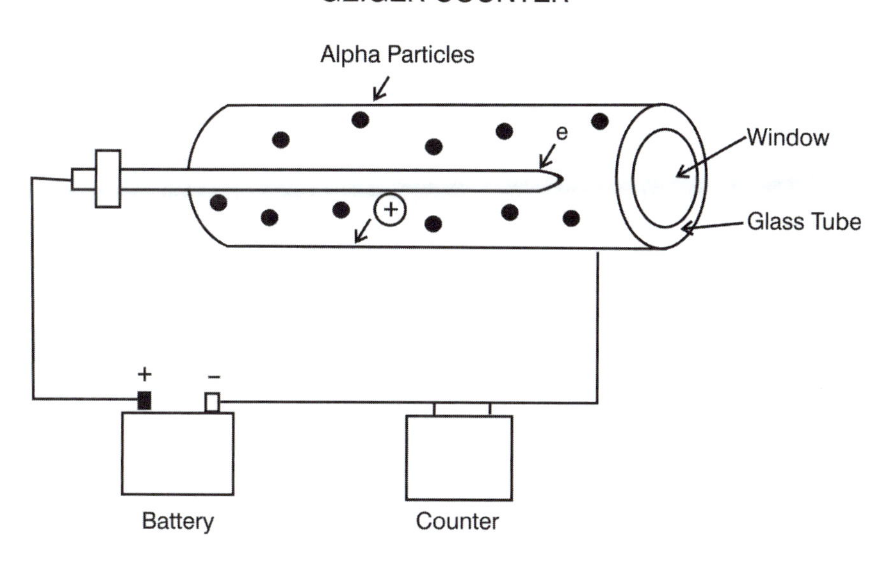

Figure G7. The *Geiger-Müller Tube* contains a high voltage wire that runs down the central axis of a glass tube. It is filled with a gas that becomes ionized, resulting in pulses of electric current that are detected by a sensitive meter and counter that registers continuous readings of the strength of radiation being detected.

A more simple way of stating this law/rule is to say that the short-lived radioactive isotopes emit alpha particles (helium nuclei) more energetically than the longer-lived radioactive isotopes.

Hans Geiger worked with Ernest Rutherford as they performed Rutherford's famous experiment on radioactivity and transmutation in the early 1900s. Geiger's task (*see* Rutherford) was to devise a way to detect and count the number of alpha particles resulting from radiation that caused ionization. To accomplish this task Geiger and Rutherford devised a counter in 1908 that was able to count the number of alpha particles as well as other forms of ionizing radiation. Over the next several years Geiger improved the accuracy and sensitivity of his counter and in 1928, with Walther Müller (1905–1979), a graduate student of his, produced the modern instrument known as the Geiger–Müller counter. This instrument has a glass tube with a wire that carries a high voltage running down the central axis of the tube. It is filled with a gas that becomes ionized when a form of ionizing radiation passes through the tube and ionizes the gas, resulting in

> Hans Geiger's birth name was Johannes Wilhelm Geiger. He attended the University of Munich and the University of Erlangen in Germany where he earned a PhD in 1906 for work with the nature of electrical discharges on various gases. During the years 1907 to 1912 he worked with the famous English physicist, Ernest Rutherford. Rutherford made his famous discovery on the structure of the atom based on work done by Geiger and Ernest Marsden (1889–1970), a physicist from New Zealand, who in 1909 actually set up the experiment that detected the scattering of alpha particles by a sheet of very thin gold leaf. Following his work with Rutherford, Geiger held several administrative positions and by 1925 became professor of physics at Kiel University in Germany.

a pulse of electric current that is detected by a sensitive meter connected to the glass tube.

See also Rutherford

Margaret Geller's childhood and early education was not typical for a girl raised following the period of World War II. At that time there were not as many women in the fields of mathematics and astronomy as there are today. Currently, there are a higher percentage of women attending college than men. And, the enrollments of women in prestigious universities, such as Massachusetts Institute of Technology, Harvard, Princeton, Cal Tech., and University of California at Berkeley, are now at an all-time high.

As a small child, Margaret was interested in mathematics and science. She attended the University of California at Berkeley and later was awarded a PhD at Princeton. Upon completion of her doctorate, she studied astronomy at Cambridge, England, and in 1980 moved to Harvard where she became a professor of astronomy. Margaret Geller is also on the staff of the Smithsonian Astrophysical Observatory where she continues to be involved with research in astrophysics and cosmological theory with the expectation that new models of the universe will be forthcoming.

GELLER'S THEORY OF A NONHOMOGENEOUS UNIVERSE: Astronomy: *Margaret Joan Geller* (1947–), United States.

> *A map of the redshifts of the light from galaxies indicates a nonuniform distribution of galaxies in specific sections of the observable universe.*

By using the Doppler effect, astronomers Margaret Geller and John Huchra (1948–) observed the distribution of over fifteen thousand galaxies while at the Harvard-Smithsonian Center for Astrophysics in Cambridge, Massachusetts. They recorded the longer light rays toward the red end of the spectrum, indicating that the galaxies are receding. Light from some of the galaxies started its journey to Earth about six hundred and fifty million years ago, and these galaxies are still receding from us, as well as from each other. According to the big bang theory of an ever-expanding universe proposed by Lemaître, and George Gamow, the universe should be rather uniform, or at least galaxies should be randomly distributed throughout all sections of the heavens. When Geller plotted her data for one section of the sky, she discovered very large groups or clusters of galaxies rather than a random or uniform distribution. Some clusters were many hundreds of millions of light-years across in size. She also noted there were a few galaxies between these clusters, but the clusters contained the majority of all visible galaxies. The implication of this information is unclear, as it relates to future cosmological theory. More recently, superclusters composed of clusters of galaxies have been discovered, which seems to support Geller's theory of a nonhomogeneous universe. Geller suggests a revision may be needed for the current big bang model.

See also Doppler; Gamow; Hubble; Lemaître

GELL-MANN'S THEORIES FOR SUBATOMIC PARTICLES: Physics: *Murray Gell-Mann* (1929–), United States. Murray Gell-Mann received the 1969 Nobel Prize for Physics.

Gell-Mann's quark theory: *The heavy particles of atoms (protons and neutrons) are composed of three fundamental entities called quarks.*

Quarks, as proposed by theoretical particle physicists, are considered the most fundamental building blocks of matter yet discovered. String theory proposes a more basic

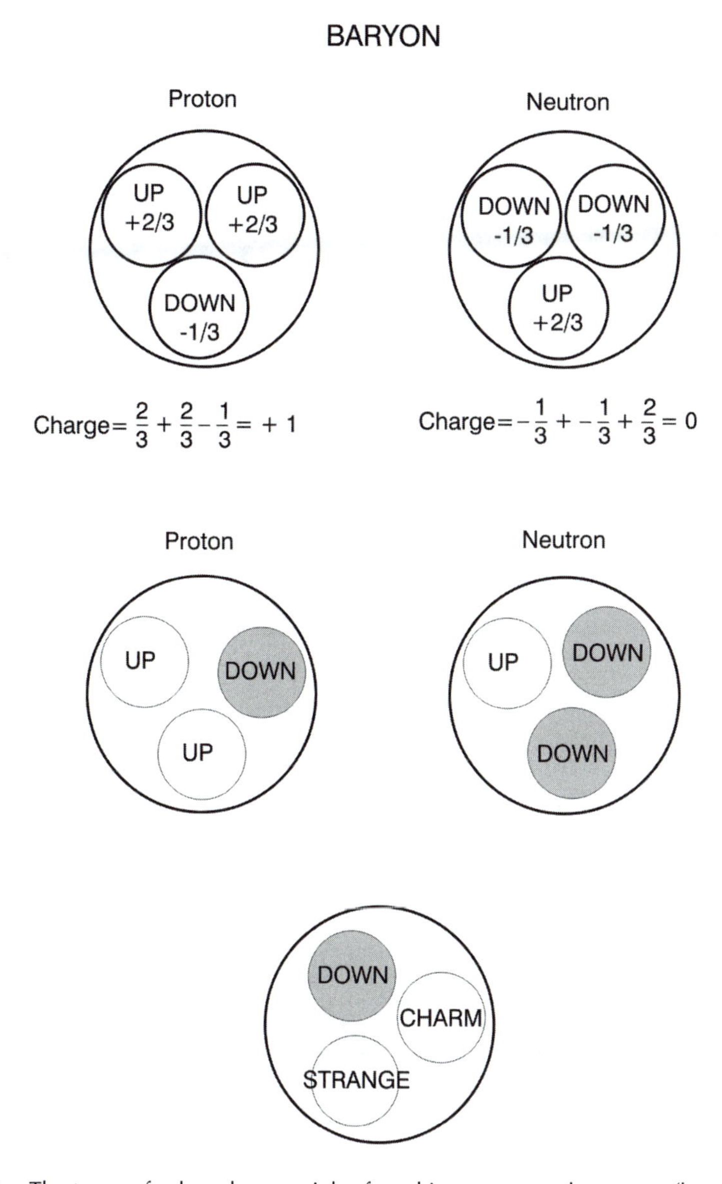

Figure G8. The types of subnuclear particles found in protons and neutrons (baryons).

particle, but it has yet to be discovered, and although there is mathematical justification for the extra dimensions required for the string theory, there is no empirical evidence that they exist. Quarks come in groups of threes (originally named red, blue, and green by Gell-Mann), have a fractional electric charge and have not yet been detected in a free, uncombined state in nature but are being investigated by using a supercollider (RHIC) at Brookhaven National Laboratory on Long Island, New York. By "smashing" nuclei of gold atoms together at 99.9% the speed of light, researchers hope to create some "free" quarks. Being submicroscopic, quarks have never been observed, but they are also considered to be bound up within the interior of the **subatomic particles**. For

FLAVOR AND CHARGE OF QUARKS

Quark Flavor	Symbol	Charge	Relative Mass
Up	u	$\frac{2}{3}$	1
Down	d	$-\frac{1}{3}$	2
Strange	s	$-\frac{1}{3}$	40
Charm	c	$\frac{2}{3}$	300
Bottom (beauty)	b	$-\frac{1}{3}$	940
Top (truth)	t	$\frac{2}{3}$	34,000

Figure G9. The descriptive names given to the flavors and charges of quarks.

instance, the two heavy subatomic particles in the nucleus of atoms belong to a class of particles known as **baryons** (meaning heavy). These two baryons are the positive proton that is composed of two up quarks and one down quark held together by *gluons*, and the neutral neutron which is composed of two down quarks and one up quark (*see* Figure G8).

Gluons and their quarks are responsible for the strong interactions that hold nuclei together. There are three types of quark pairs (for a total of six types of quarks): the *u* quark has a charge of $+^2/_3$, and the *d* quark and *s* quark have a charge of $-^1/_3$; thus symmetry is preserved (*see* Figure G9.)

Looking for some way to express the triad nature of these theoretical particles, Murray Gell-Mann borrowed the word "quark" from James Joyce's book *Finnegans Wake* (1939), "Three quarks for Muster Mark!" (This might be interpreted as three quarts of ale for Mister Mark for a job well done.) Gell-Mann liked the sound of this, and it seemed to fit the concept of his triplet of quarks: *u* for up, *d* for down, and *s* for strange, which was intended to explain the organization of the myriad subatomic particles. Since the time that Gell-Mann first advanced his three quarks theory, many other subatomic particles have been proposed. The **hadrons** are a series of heavier quarks, referred to as the *c*-quark (for *charm*), which is many times heavier than, but related to, the moderately heavy *s*-quark (for *strange*). Hadrons are a group of particles such as the baryons and mesons that lead to symmetry. Quarks never exist as individual particles but rather in groups called "hadrons" which, in combinations, are known as "quark confinement." Today the six quarks are named *up, down, charm, strange, top,* and *bottom.* The *u*-up and *d*-down quarks are thought to be just about massless but make up almost 100% of all the matter (protons and neutrons) in the universe whereas the others, produced in particle accelerators, are unstable and have a very short existence.

Gell-Mann's theory of strangeness: *All fundamental particles are characterized by the property of strangeness.*

The concept of strangeness evolved because of the odd or "strange" manner in which some elementary particles strongly interacted. Strangeness is conserved in the strong and electromagnetic interactions of hadrons and the *s*-quark, but not so for weak interactions. If for ordinary particles we assign $S = 0$, then we can allow $S \neq 0$ to represent "strange." Therefore, *S* equals "strangeness." This concept of strangeness led to the development of a new concept for the physical principle of symmetry, used to classify subatomic particles that interact strongly, such as the *c*- and *s*-quarks. The new concept of symmetry, also referred to as the *eight-fold-way*, resulted in the discovery of several new particles, including the omega minus.

See also Feynman; Freidman; Glashow; Nambu

GERHARDT'S TYPE THEORY FOR CLASSIFYING ORGANIC COMPOUNDS: Chemistry: *Charles Frédéric Gerhardt* (1816–1856), France.

Organic chemistry consists of four "types" or derivatives related to water, ammonia, hydrogen, or hydrogen chloride.

Gerhardt's type theory had a revolutionary effect on the field of organic chemistry. Previously Jöns Berzelius' proposed a *dualistic system theory* of chemistry that was unsatisfactory in explaining the nature of chemical reactions, whereas, Jean Baptiste Dumas' *substitution theory* was more acceptable, and thus was more-or-less combined by Gerhardt into a *theory of types* for the structure of organic compounds. Gerhardt did much to reform the system of chemical formulation by stressing the distinctions between atoms, molecules, and equivalents into a single type system.

Gerhardt considered that organic chemistry was formulated from four types of inorganic substances: (Type I) H_20 (water); (Type II) NH_3 (ammonia); (Type III) HCL (hydrochloric acid); and (Type IV) H_2 (hydrogen). His theory stated that all *organic* compounds were derived from these four types of *inorganic* substances. He further classified other substances according to these four types as 1) the sulfides, tellurides, oxides, acids, bases, salts, ethers, alcohols, and so forth, belong to Type I, the water types; 2) nitrides, phosphides, arsenides, and other related chemicals belong to the Type II, the ammonia types; 3) the chlorides, bromides,

In 1853 Charles Frédéric Gerhardt, a French chemist, neutralized salicylic acid by buffering it with sodium and acetyl chloride which created acetylsalicylic anhydride, an effective pain reliever now known as aspirin. Up to this time, and as far back as the fifth-century BCE, willow bark was used to relieve pain, fever, and chills. It was later found that when oxidized the extract of the willow bark became salicylic acid. Another extract from meadowsweet flower produced an extract that was just as effective but caused digestive problems. After Gerhardt buffered the extract, he lost interest and had no desire to commercialize the product. Since that time, salicylic acid, better known as aspirin, was marketed by Friedrich Bayer & Co. in Germany along with another pain killer, called heroin, which was much more effective but after some use proved to be addictive. Aspirin was first sold as a powder and was widely used. In 1915 Bayer & Co. produced aspirin in tablet form as it is used today. Several people and companies have claimed to have invented acetylsalicylic acid (aspirin), but Bayer was the company that marketed it most effectively, and Felix Hoffman, a research assistant at the Bayer Co., is generally recognized as its "official inventor." The name "aspirin" was patented but the patent has been abused for years, and aspirin is now considered a common, over-the-counter (OTC) drug.

iodides, and other halogens belong to Type III, the hydrochloric acid types; and 4) most metals and metallic hydrides belong to Type IV, the hydrogen types.

For some time organic compounds were referred to using this classification of system of "types." This system was particularly useful when the characteristics of certain organic substances were converted into different, more useful compounds. This was done during the reaction when a specific organic compound had one or more of its hydrogen atoms replaced with an atom different from the hydrogen atom (or a group of different atoms known as a "radical.") By replacing the hydrogen atom(s) a different organic compound was created. One of the great advantages of Gerhardt's scheme was that unknown, and undiscovered chemical compounds could be created, but predicted by using this theory of classification. This system led to a plethora of the many different organic compounds produced today.

GIAUQUE'S THEORY OF ADIABATIC DEMAGNETIZATION: Physics and Chemistry: *William Francis Giauque* (1895–1982), United States. William Giauque received the 1949 Nobel Prize for Chemistry.

By cooling an already cold substance in liquid helium within a magnetic field, and then removing the magnetic field (demagnetization), strong entropy occurs, thus greatly lowering the temperature to near absolute zero of the substance being cooled.

William Francis Giauque in 1927 and Peter Debye in 1926 independently arrived at the theory for **adiabatic** demagnetization as a means for obtaining temperatures that approach a small fraction of absolute zero.

The magnetic field is used to control the entropy of a sample substance consisting of paramagnetic salts that are referred to as the "refrigerant." The magnetic field aligns the **dipoles** of the molecules in the refrigerant while it is kept at a constant temperature with what is called a "heat sink" consisting of liquid helium (a "heat sink" is any substance that absorbs heat and/or shields something from heat). The heat sink removes most of the heat from the "refrigerant" and protects it from absorbing more heat. At this point, the magnetic field is turned off which causes a change in the dipole arrangement of the substance's molecules' positions slowing its molecular motion, thus the temperature of the refrigerant is cooled below that of the liquid helium heat sink with a temperature just 4K, which means it is just four degrees above absolute zero. (Remember, heat is a form of kinetic energy related to the degree of motion of the molecules composing a substance. Reducing the temperature, in essence reduces molecular motion and thus the heat of a substance.) This process increases entropy, whereas the paramagnetic salts' molecules are trapped at a lower energy state (reduced molecular motion) with temperatures as low as 0.0015K, which is just above absolute zero. Zero kelvin is the theoretical temperature point at which all molecular motion ceases.

Since the days of William Giauque's and Peter Debye's experiments, more elaborate low-temperature refrigeration techniques and equipment have been developed. These are referred to as nuclear demagnetized refrigeration (NDR). This method uses adiabatic nuclear demagnetization instead of electric demagnetization to control the molecules' nuclear spin (aligning nuclear dipoles). Temperatures as low as 0.000016 degrees kelvin have been reached as a result of using this system.

William Giauque, in cooperation with Ohio State University chemistry professor Herrick L. Johnston (1898–1965), discovered the presence of the oxygen isotopes O-17 and O-18 and that these two heavier isotopes were mixed with the lighter, more abundant O-16 in Earth's atmosphere. At one time, physicists set the mass of the oxygen-16 isotope at 16.000 as the base for determining the masses of all other elements. After the discovery of oxygen-17 and oxygen-18 in 1929, the new mass figure for oxygen was set at 16.0044. Although this change in the standard unit for atomic mass was small, it caused many problems because, at the time, scientists were using different scales for atomic weights. In 1961 physicists and chemists compromised and set the isotope carbon-12 as having a mass of 12.0000 as the standard. Under this system, oxygen now has an atomic mass of 15.9994. Today, the atomic mass of an element is considered the *average* of the mass numbers of all the isotopes of that particular "natural" element (meaning the isotopes of elements that have been on Earth for eons of time).

Giauque is also well known for his work in the field of chemical thermodynamics. He clarified the influence of atomic and molecular structures on entropy and how this related to the laws of thermodynamics.

See also Carnot; Kelvin; Maxwell

GIBBS' THEORY OF CHEMICAL THERMODYNAMICS: Mathematics and Physics: *Josiah Willard Gibbs* (1839–1903), United States.

> *Mathematics can be applied to determine the interrelationship between heat and chemical reactions, as well as the physical changes of state in the field of thermodynamics.*

Chemical thermodynamics is the interrelationship of heat with chemical reactions (exothermic and endothermic), and with the physical change of state (solid, liquid, or gas) within the parameters of thermodynamics. In essence, this means that mathematical methods can be used to explain the relationships of heat to chemical reactions.

Gibbs was one of only a few famous physical scientists from the United States in the 1700 and 1800s. Others were Benjamin Franklin, Joseph Henry, and Henry Augustus Rowland (1848–1901). However, Gibbs is considered to be the only truly theoretical physicist and chemist in United States at that time. He was also known as a linguist and mathematician. He received the first PhD in engineering awarded by Yale University. Today, the Gibbs Professorship in physical chemistry at Yale is named after him. His work, in essence, established the fields of chemical thermodynamics and statistical mechanics expressed in esoteric mathematical forms. His research papers were difficult to understand, even for other scientists, and in some ways, he was better known and respected by European scientists than those in the United States.

Through writing a series of important papers, Josiah Willard Gibbs established the field of chemical thermodynamics. The paper "On the Equilibrium of Heterogeneous Substances" written in 1876 expressed his famous "Gibbs phase rule" that explained the conditions required for a chemical reaction to take place. Later, he published several important papers including his concept of "Gibbsian ensembles" that explained how a large number of macroscopic entities that have the same heat properties are related statistically.

He developed a concept called "Gibbs free energy" in the field of thermodynamics that is basically a complicated mathematical expression of thermodynamics. He also considered that the energy involved in thermodynamic systems is available to do work. It is expressed in the following formula:

G = H-TS where in metric units:
G is expressed in the unit known as joules (Gibbs energy)
H is expressed in *joules* for *entropy*
T is the temperature given in *kelvin*
S is the entropy expressed in *joules per kelvin.*

This equation basically states that every chemical and physical system seeks to achieve a minimum of free energy. It is important for determining the thermodynamic functions of such systems to establish the equilibrium constants. This applies for any reversible chemical reaction, for example, $N_2 + 3H_2 \leftrightarrow 2NH_3$. (The $\leftrightarrow$ double arrow indicates a forward and reverse chemical reaction.) Another example that uses the *Gibbs free energy* formula is measuring the output voltage from an electrochemical cell.

See also Carnot; Giauque; Maxwell

GILBERT'S THEORY FOR DNA SEQUENCING: Biology: *Walter Gilbert* (1932–), United States. Walter Gilbert shared the 1980 Nobel Prize for Chemistry with Frederick Sanger and Paul Berg.

> *Chemicals can be used to modify DNA by sequencing base pairs of DNA in either single strands or double strands of DNA that then can be used to study the interactions of proteins with the DNA.*

Walter Gilbert was educated as a physicist at Harvard University and at Cambridge University in England where he received his degree in 1957. After studying the fields of chemistry and physics, he developed an interest in biochemistry, primarily after meeting with and learning of the research and experiments that James Watson (of Crick and Watson fame) was conducting. Consequently he switched to the field of molecular biology in 1960 and in 1968 became a Harvard professor of microbiology and later became department chairman. Gilbert worked in the United States with Allan M. Maxam, a graduate student, while Frederic Sanger worked independently in England to make use of new techniques, such as electrophoresis for the analysis of results achieved by their method of multiplying, dividing, and fragmenting a large section of DNA strands. They used chemicals to break large strands of DNA into smaller fragments along the bases of (A) adenine; (G) guanine, (C) cytosine, and (T) thymine. The main difference between Gilbert's method of sequencing DNA and Sanger's was that Sanger's method only worked with single strands of DNA, whereas Gilbert's methods were effective for either single or double strands of DNA (*see* Figure C5 under Crick for a diagram of the DNA molecule.) The techniques for sequencing DNA devised by Gilbert can be used to read up to thirty thousand base pairs.

Over the past fifty years a number of methods for sequencing DNA have been developed. In addition to "chemical sequencing," Sanger developed a "chain termination method." Another procedure is the "dye terminator sequencing." Just as important as the techniques for sequencing DNA is the improvement for the preparations of the samples and developing automated procedures for the sequencing operations. Currently, the number of sequences of short lengths of DNA is limited due to the power of resolution of the systems used. The magnitude of the problem is realized because even simple single-cell bacterium can have a genome of about a million base pairs while the human genome has more than three billion base pairs in their DNA molecules. Recently to overcome these problems, several techniques were developed to get a "reasonable" reading of the human DNA genome. One technique is to clone a sample and "grow" copies of the desired DNA at a rate of thousands of pairs at the same time. Another method is called "shotgun sequencing" that uses small samples of DNA and then assembles them into a connected sequence. No doubt, in the future, improved faster methods will be developed to determine the complete genome of any plant or animal, including a more detailed map of the human genome.

See also Crick; Sanger; Sharp; Watson (James)

GILBERT'S THEORY OF MAGNETISM: Physics: *William Gilbert* (1544–1603), England.

Gilbert's theory for electric and magnetic forces: *The amber effect (static electricity), which can attract small particles when certain materials are rubbed with certain types of cloth, such as silk, is not the same phenomenon as natural magnetism, which exists in lodestones (magnetic iron ore).*

The phenomenon of rubbing amber with cloth to cause the amber to attract bits of straw and other small particles was known since the days of the Greek philosophers. They related it to some magic or spirit, not to static electricity. William Gilbert experimented with amber to produce static electricity and lodestones to magnetize iron bars. He was the first to distinguish these two forces of attraction and the first to use the terms "electric attraction" and "magnetic attraction" to make this distinction.

Gilbert's theory for the rotation of Earth: *Because a magnetized needle will swing horizontally as it points to the poles of Earth and also dip down toward the vertical, Earth must be a giant spinning lodestone.*

William Gilbert's experiments formed his magnetic philosophy, eliminating much of the superstition and false information about magnetism existing at that time. He constructed a globe from a large lodestone to demonstrate how a compass needle behaves on the lodestone and then related this to Earth. Because of the action of a compass needle, he assumed that the "soul" of Earth was also a spherical lodestone with a north and south pole. In addition, he demonstrated that the compass needle would dip down at different angles as related to the different latitudes, and the needle would point straight down at the north pole of his lodestone globe. Thus, it would do the same for the North Pole of Earth. Sailors had already observed this "magnetic dip" phenomenon, but Gilbert was the first to relate Earth's magnetism to latitudes. Gilbert concluded that Earth acts like a large, spherical bar magnet, which is spinning on its axis once every twenty-four hours. However, he continued to believe Earth was the center of the universe. His theory was the first reasonable explanation for a rotating Earth, but Gilbert did not go as far as Copernicus, who claimed Earth moved through the heavens around the sun. Up to this time, scientists believed Earth was stationary, and the canopy of stars was in motion. Historically, the magnetic compass was a reliable instrument that aided in navigation. Gilbert's magnetic philosophy, which included the belief that Earth's magnetic influence affected everything in the solar system, led to the modern concept of gravity. One gilbert (Gb), a unit of electromotive force, named for him, is equal to the magnetomotive force of a closed loop of wire with one turn in which the flowing current is 1 ampere. In the CGS (centimeters, grams, seconds) system, 1 Gb is equal to 10/4π ampere turns.

See also Ampère; Coulomb; Faraday; Maxwell; Oersted

GLASER'S CONCEPT OF A BUBBLE CHAMBER FOR DETECTING SUBNUCLEAR PARTICLES: Physics: *Donald Arthur Glaser* (1926–), United States.

> *High-energy ionized particles that cannot be detected in a Wilson cloud chamber can be detected by leaving a trail in a depressurized fluid bubble chamber, and the trail representing characteristics of the particles can be captured by high-speed photography.*

Donald Glaser received his PhD degree in physics in 1950 from the California Institute of Technology. He taught and did research in various areas at the University of Michigan, and was promoted to the rank of full professor in 1957. In 1959 he moved to the Berkeley campus of the University of California where his interests spread to the

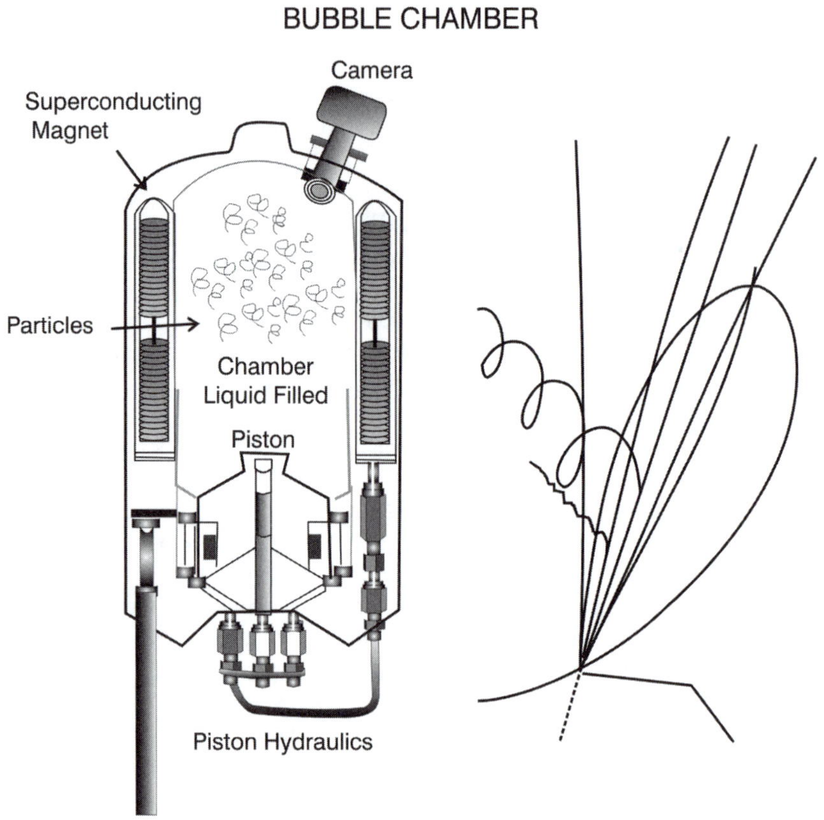

Figure G10. The super cooled liquid in the chamber will create small bubbles as charged particles interact with the bubbles to form tracks that are photographed.

area of biology as well as physics. His early work with the Wilson cloud chamber that was used to detect cosmic rays led him to realize that the cloud chamber was an inadequate instrument to detect subnuclear ionized high-energy particles. While at the University of Michigan, he became aware that high-energy particles passing through a superheated fluid would produce small bubbles along a specific trajectory. His design for the bubble chamber consisted of a large cylinder filled with a liquid that was heated almost to its boiling point. The cylinder is surrounded by a magnetic field with a high-speed camera positioned at the top and focused down into the chamber. A piston that can be moved up and down is located at the bottom of the cylinder. As the piston is rapidly lowered, the pressure in the cylinder's chamber is reduced and the liquid becomes supercooled. This supercooled liquid will create a series of tiny bubbles as the charged particle interacts with an atom of the liquid. At that moment, the camera captures the image of the bubble track that is then used to determine the decay modes, lifetime, spin, cross section, and other characteristics of the subnuclear particle (*see* Figure G10 for an artist's depiction of a bubble chamber and particle tracks).

Today's bubble chambers are much larger than the original one invented by Glaser in 1952. They use liquid hydrogen or liquid helium for the fluid. Another type of bubble chamber that requires heavy liquids to slow down the particles uses organic compounds. But the principle of the cloud chamber and bubble chamber are basically the same. The use of these unique research instruments has resulted in the identification of many new types of elementary particles.

Donald Glaser has received many honors for his contributions in the fields of physics and biology. More recently, Glaser has been interested in applying physics to problems related to molecular biology. His current position is professor of physics and neurobiology in the Graduate School at Berkeley.

See also Compton; Millikan; Wilson (Charles)

GLASHOW'S UNIFYING THEORY OF THE WEAK FORCES: Physics: *Sheldon Lee Glashow* (1932–), United States. Sheldon Glashow shared the 1979 Nobel Prize of for Physics with Steven Weinberg and Abdus Salam.

> *The unification of the electromagnetic interactions with the interaction of leptons (electrons and neutrinos) can be extended to include baryons (a heavy particle) and mesons (elementary particles with a baryon number of zero) by establishing a new, fourth "charm" quark to add to Murray Gell-Mann's three-quark theory.*

Sheldon Glashow was born and raised in Manhattan. He attended the Bronx High School of Science, along with another future famous scientist Steven Weinberg. He attended Cornell University and received his PhD degree from Harvard University in 1959. He was granted a National Science Foundation (NSF) fellowship to Russia but never received the required visa from the Soviet government. Rather he spent his fellowship at the Niels Bohr Institute in Copenhagen where he did his original work on the structure of the electroweak theory. Later back in the United States, Glashow and two other scientists predicted that "charm" would be discovered, as well as realizing that many of the theories related to subatomic particles, forces, and fields were more-or-less merging into a future single theory of all universal physical principles. Glashow's

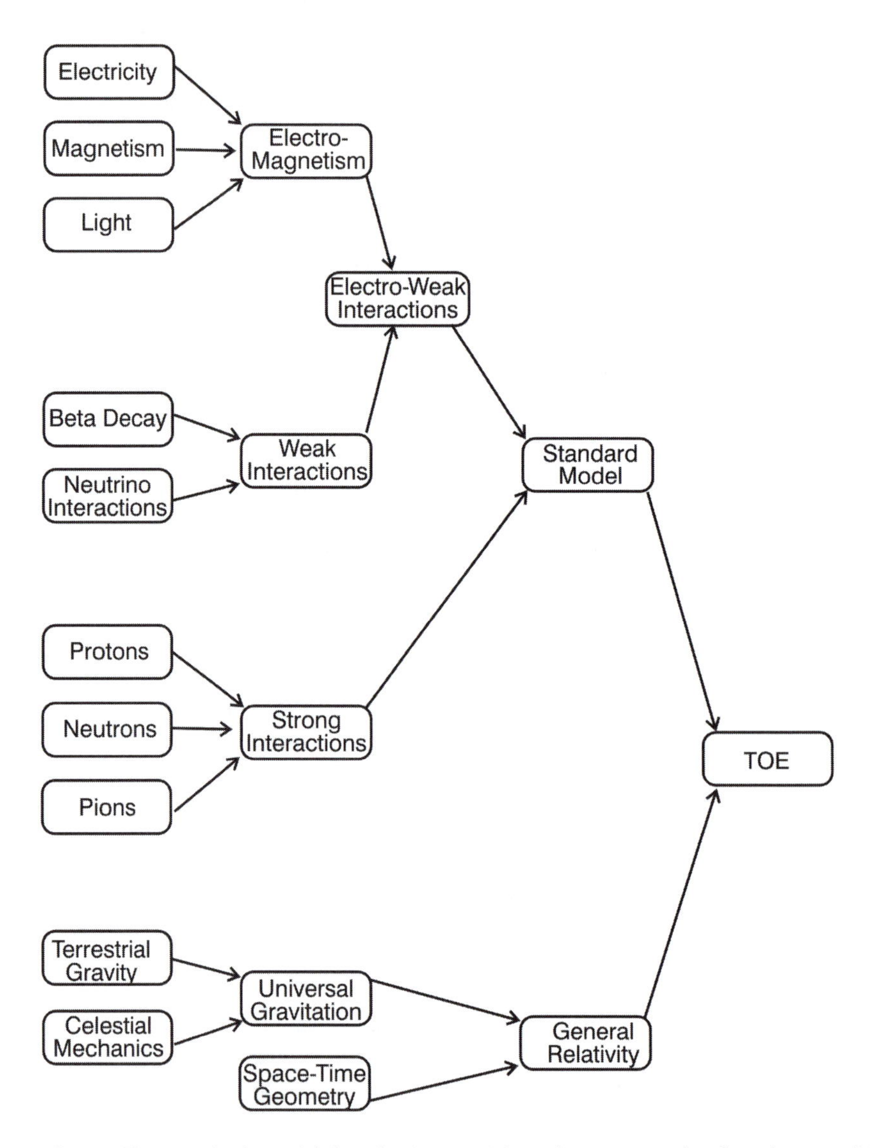

Figure G11. The standard model for physics particles. If a way can be found to combine this theory with the theories of universal gravity and the quantum field theory, it could result in a "Theory of Everything" (TOE) viz., a theory for all energy and matter.

research had gone a long way towards advancing this concept of a "theory of everything" (TOE) by predicting the "charm particle" as another type of quark (*see* Gell-Mann). Possibly more important was his contribution to establishing the unification of the "weak" and "electromagnetic" forces as "electroweak interactions" that are part of the standard model for particle physics (*see* Figure G11).

If the standard model for particle physics can be successfully combined with the concepts of universal gravity, relativity, and quantum field theory a grand unification theory (GUT) or theory of everything (TOE) may become a reality.

There are four basic or fundamental forces (sometimes referred to as fields) that naturally exist in the universe. Following is a brief description of these forces listed from the weakest to the strongest.

1. *Gravitation force*: Gravity exists throughout the universe and is experienced everywhere where there is matter (there is no negative gravity). Gravity acts on all particles no matter where located in the universe. Scientists have theorized that the force of gravity itself is composed of particles called "gravitons."
2. The *electromagnetic force* (field): This force acts on all electrically charged particles. The "particle" of electromagnetic force is called the "photon," which has particle and wave-like characteristics.
3. *Weak nuclear force:* This force causes radioactivity to act on electrons, neutrinos, and quarks. The particle's repulsive nature of the weak nuclear force is known as the "W boson."
4. *Strong nuclear force:* This force is responsible for the quarks that form neutron and protons in atoms' nuclei. Quarks also are responsible for "holding" together the neutrons and protons of the atoms' nuclei. The particle related to this force is called the "gluon."

GODEL'S INCOMPLETENESS THEOREM: Mathematics: *Kurt Godel* (1906–1978), United States.

> *Using axioms that are true within that system cannot prove the consistency for any formal system, including formal arithmetic and logic systems. A stronger system is required that has an assumed consistency.*

The explanation of Kurt Godel's incompleteness theorem is the opposite of a complete theorem, which involves a system where all logical statements made by a formal system can be proved by **axioms** of that very system. Godel claimed no formal system could meet that criterion. The incompleteness theorem may also be stated for mathematics as: *No finite set of axioms is adequate to form the basis for all true statements concerning integers, and there will always be statements about integers that cannot be proved to be either true or false.* The Godel incompleteness theorem also holds for nonmathematically based systems of logic, where a proposition can neither be proved nor disproved based on axioms of that particular system. Godel's incompleteness theorem disturbs many scientists because a noncontradictory system of mathematics may never be constructed, and the means to understand the physical nature of the universe fully may never be found. This theory has implications for mathematics in the sense that, if Godel's theory holds up, there are fundamental limits within the field of mathematics, as well as limitations on the acquisition of scientific knowledge about the universe.

Godel worked closely with Albert Einstein on the mathematical and practical aspects of the general relativity theory. He was particularly interested in concepts of time and at one time proposed that, theoretically, it was possible to travel back in time. Toward the end of his life he was plagued by mental and physical problems. Godel admitted that he could neither keep up with nor understand the newer theories and works of younger mathematicians. As his illness progressed, he believed he was being poisoned. Thus, he stopped eating and starved himself to death.

GOLD'S COSMOLOGICAL THEORIES: Astronomy: *Thomas Gold* (1920–2004), United States.

Gold's theory for the nature of the universe: *The universe exists in a steady state. It is unchanging in space and time, with no beginning or ending.*

Thomas Gold and some other astronomers rejected the big bang concept for the origin of an ever-expanding universe. In 1948 Gold referred to his theory as the *perfect cosmological principle*: a universe with no beginning and no ending, and thus the steady-state universe. In addition, his universe appears the same regardless of where it may be viewed, which was just the opposite of Margaret Geller's concept of a nonhomogeneous universe (*see* Geller). In other words, Gold's theoretical universe has a density that is not only constant, but all matter and energy will always be maintained in the same relative proportions. Because this theory conflicts with the laws for the conservation of matter and energy (laws of thermodynamics), the proponents of the steady-state universe needed to produce an idea for the continuous creation of matter and contended that this is just what occurs. Gold postulated that a continuous production of one hydrogen atom per cubic kilometer of space occurs every ten years. Although this amount of "new" hydrogen is undetectable and the vacuum of space is better than any vacuum that can be produced mechanically on Earth, it is an adequate "regeneration" of hydrogen to confirm the laws of conservation and maintain a steady-state universe. Currently, it is estimated that "empty" space contains one proton (hydrogen nucleus) per cubic centimeter (*see also* Geller).

Gold's theory for neutron stars: *The detectable high-frequency, periodic radio signals originating in pulsars are caused by rapidly rotating neutron stars.*

In 1968 British astronomers Jocelyn Bell Burnell (1943–) and Antony Hewish discovered that very short bursts of radio waves were being emitted every second from a new type of star they referred to as *pulsars* because the signals seemed to "pulse" on a very regular basis. Gold contended these signals originated from rapidly rotating neutron stars, which are extremely dense, spinning bodies that underwent collapse on a gradated basis. Neutron stars, through nuclear fusion, have exhausted their nuclear fuel. The result is that gravity becomes so extreme that the stars collapse into themselves, to the point where their protons and electrons combine to form neutrons. At the end of their "lives", these neutron stars collapse into a *black hole*. Because these stars are very small and very dense, their rotation period can be short enough to produce the detected high-energy radio signals. Gold's theory was substantiated when a new pulsar neutron star with a more rapid rotation was located in the Crab Nebula, even though this pulsar is slowing down about 3.5 seconds every trillion years.

Gold's theory of life on Earth and Mars: *Conditions on and below the surfaces of Mars and Earth are very similar and both may support life.*

Thomas Gold made many comparisons between the structure and conditions existing on Mars and Earth—for example:

- Several **meteorites** found in Antarctica appear to have come from Mars. Some contain several rare noble gases (e.g., neon and xenon), as well as nitrogen isotopes, unoxidized carbon, and petroleum-like hydrocarbon molecules.
- Heat on Mars and Earth increases with depth, and both must have water at some depth below surface levels. This deep liquid water came to the surface on Earth, but not on Mars, where the subsurface temperature keeps the water frozen.

- Although the surface chemicals on Mars and Earth are different, their subsurface chemistry seems similar. Both planets exhibit leftover debris, including meteorites, from their formation billions of years ago.
- Stable hydrocarbon molecules under great pressure are found at great depths on most planets. Earth's carbon-containing liquid and gas petroleum help maintain the carbon cycle by seeping to the surface where carbon dioxide from hydrocarbons is transported to the atmospheric-ocean-rock environments and then recycled by green plants. A similar process occurs for many planets but without the green plants, as we know them. All the major outer planets (Jupiter, Saturn, Uranus, and Neptune) have enormous amounts of hydrocarbons, mainly methane and ethane, plus ammonia in their atmospheres that are recycled as *petroleum rain*. Different forms of hydrocarbons have also been detected on the surfaces of asteroids.
- Evidence suggests that living organisms did not form these deep hydrocarbons. Rather, these primordial hydrocarbons were mixed with biological molecules and are a prerequisite for life. A dilemma exists because primitive microbes must in some way oxidize the hydrocarbon molecules to obtain energy, but the deep interior of planets does not contain such "free" oxygen. A possible means of oxidizing hydrocarbons would be for the deep microorganisms to use sulfur and iron sulfate compounds to oxidize and dine on the hydrocarbons. Organisms that thrive on sulfur have been found at the site of sulfur "vents" discovered at great depths on Earth's ocean floor.
- Life on Earth most likely started internally, not from transport of "organic seeds" to the surface from space or by other surface phenomenon. Therefore, life may be found at a primitive stage below the surface of Mars and some other planets.

In conclusion, without photosynthesis as a source of energy it is very unlikely that Mars has any surface life. But there is a good possibility that at some depth, there is chemically fed life. Physical conditions, such as temperature, radiation, lack of stable liquids, and so forth make surface life less likely than subsurface life on Mars.

See also Gamow, Guth; Lemaître

GOLDSTEIN'S THEORY FOR THE METABOLISM OF CHOLESTEROL, FATS, AND LIPIDS: Biology and Medicine: *Joseph Leonard Goldstein* (1940–), United States. Joseph Goldstein shared the 1985 Nobel Prize in Physiology or Medicine with Michael S. Brown.

> *Cholesterol is found in two main sources: the liver, and in foods. As particles, cholesterol is carried by blood and lymphatic fluid; thus they are vital products of metabolism of human body cells.*

Joseph Goldstein received his MD degree in 1966 from the University of Texas Southwestern Medical School. He did his internship at the Massachusetts General Hospital in Boston where he met his friend and long-time collaborator, the geneticist, Michael S. Brown (1941–). They shared an interest in genetics related to biochemistry, and in particular the study of cholesterol and other lipoproteins.

Cholesterol is formed from two different sources, that is, within the liver and from the fat in food we eat. Cholesterol is transported from the liver by the blood as spherical particles called lipoproteins that are a combination of fat and proteins. There are two types of lipoproteins: LDL (low density lipoproteins), and HDL (high density lipoproteins). The LDL particles are transported in the blood and are considered the "bad" cholesterol because an excess of it can clog the walls of arteries and restrict the flow of blood to the heart, thus increasing the risk of a heart attack or stroke. A healthy person has approximately 2 grams of cholesterol per liter of blood in their system. A person with 10 or more grams of LDL per liter of blood suffers from the metabolic disease called *familial hypercholesterolemia* (FH).

It should be mentioned that cholesterol is a very important chemical required for the proper metabolism of food and should not be eliminated completely from one's diet. It has two main functions in the body. First, it is a constituent of normal cell wall membranes, and two, it is converted to important steroids, hormones, and, bile acid secretions. About 90% of all cholesterol found in the body is incorporated in cell membranes. The deposit of cholesterol particles on the interior walls of arteries is a slow process, taking many years to accumulate enough particles to block the flow of blood. The rate of accumulation may be a factor of genetics (e.g., your parents may have had high cholesterol or died young of a heart attack or stroke). It can also be the result of a long-term diet of foods containing saturated fat (trans fats) along with, smoking, stress, and genetic factors.

Goldstein and Brown conducted extensive research on the nature of cholesterol and its effects on the human body as well as how to reduce the risks associated with familial hypercholesterolemia.

GOULD'S HYPOTHESIS OF "PUNCTUATED EQUILIBRIUM": Biology: *Stephen Jay Gould* (1941–2002), United States.

The evolution of a species is a series of episodic changes within relatively isolated populations.

In 1972 Stephen Jay Gould and Niles Eldredge arrived at their theory of catastrophic evolution they called "punctuated equilibrium" by examining fossil records. They disagreed with Darwin's concept of a slow but continuous organic evolution involving natural selection, which is a ladder-like, smooth progression of changes from one species to variations of species. Gould and Eldredge detected many gaps in the fossil records, which they claimed disputed this smooth progression. Gould believed such a smooth transition was very rare in nature due to a continuous "pruning" of the branching tree of evolution, which results in the extinction of species. He admitted slower transitional evolution might exist for larger populations or groups of species, but not necessarily for individual species. On the other hand, even though patterns found in fossil records indicate that many species are stable, there is evidence that catastrophic ecological events resulted in the abrupt appearance of new species. These new species then adapted to their new environmental conditions. One problem with the Gould–Eldredge punctuated equilibria theory is that fossils by their nature are never complete, and therefore recognizing species from their fossil remains can only be speculative. One commonly accepted example of the catastrophic theory is the extinction of the dinosaurs approximately sixty-five million years ago, about the same time that a massive asteroid struck Earth. It is assumed this collision created a world-covering cloud of dust and smoke that blocked out the sun for several years, resulting in the elimination of plant life and great numbers of animal life as well, including dinosaurs. Obviously, some plant and animal life survived—most likely simple plant life and small mammals and some sea life.

In the 1970s and 1980s Stephen Jay Gould reexplored the concept of recapitulation that had been out of favor with biologists for many years. It is based on the idea that for a variety of animals, the maturation of embryos in the early fetal stages advance through similar stages of basic structural development. At certain stages of development, the embryos of fish, reptiles, birds, mammals, and humans appear very similar in structure. In other words, very early growth for animals represents a recapitulation of eons of evolutionary development from one species to another. Another way to say this is that *ontogeny* (individual development) follows *phylogeny* (evolutionary history).

See also Eldredge; Haeckel

GRAHAM'S LAWS OF DIFFUSION AND EFFUSION: Chemistry: *Thomas Graham* (1805–1869), Scotland.

> *The identity of a gas is based on its physical properties that affect the gas's rate of diffusion and/or effusion.*

Diffusion may be thought of as an increase in the degree of disorder (entropy) that causes a dispersing of one substance throughout a different substance. The force that causes this dispersion is thc kinetic energy of the particles in the substance. The molecules in liquid and gaseous substances are much further apart than the particles of solids. For example, if a drop or two of colored dye is placed into a glass of clear water and is not disturbed, in time the dye will be dispersed evenly through the water molecules. The rate of diffusion is much more rapid in gases and liquids than in solids. Even so, if a smooth bar of gold is placed onto a smooth bar of lead and their surfaces are in tight contact, over time there will be some atoms of gold diffused into the surface of the lead. Because molecules are much farther apart in gases than solids, light can pass through the gas without much interference and gases are thus mostly transparent. At the same time, light passing through a clear liquid (with molecules closer together than gases) is more easily deflected than through a gas. Most people have experienced the sensation of knowing when a person who is wearing too much perfume or cologne walks passed them. The reason: molecules of the perfume or cologne are diffused into the moving air as they walk by.

Graham concluded from his investigations of the relationship between the rate of diffusion of a gas and its molecular mass that the rate of diffusion can be expressed either by knowing the gas's density or its molecular mass. Graham's *law of diffusion* states that the rate at which different types of gases diffuse is proportional to the square root of their densities, as expressed by the following equation:

$$\text{Rate of Diffusion} \propto 1/\sqrt{\text{density}}$$

Or, because the same volume of different gases (at the same temperatures) have the same number of particles (*see* Avogadro's Hypothesis), the rate of diffusion can be expressed as proportional to the molecular mass (MM) of the gas, as expressed in the following equation:

During his research on the rate of diffusion Graham observed that some substances pass through a membrane slower than others. He also observed that organic substances (blood, starch, sugar, etc.) diffused more slowly through a porous membrane than did some inorganic chemicals, such as salts (potassium and sodium chloride). He called the substances that passed more rapidly through the membranes *crystalloids* and the slower ones *colloids. Colloid* means "glue" in Greek. Since the time of the simple explanation for a colloid in Graham's day, colloidal chemistry has become an advanced science with techniques and procedures used in a great variety of industries, including ceramics, paints, agriculture, cleaning agents, soil treatments, and food preparations. One of the most important uses of Graham's discoveries is in the medical field that developed a "dialysis apparatus" used to assist patients who are experiencing some form of renal failure or kidney disease. Small and large particles of undesirable solutes are removed from the blood and are cleansed of impurities. Thomas Graham is recognized as the founder of the expanding discipline of colloidal chemistry.

$$\text{Rate of Diffusion} \propto 1/\sqrt{MM}$$

Graham also studied *effusion* that can be described as the movement (or escape) of the molecules of a particular gas through an opening that is small compared to the average distance the gases' molecules travel between collisions. The equation for the rate of effusion is, in essence, a combination of the above two equations for diffusion.

An interesting application of Graham's laws for the rates of diffusion and molecular mass was the system developed early in World War II to separate the lighter Uranium-235 which is the fissionable isotope from the heavier Uranium-238. A long diffusion chamber over several hundred yards in length was used to separate U-235 as a gaseous version of uranium hexafluoride (UF-6). The concept was that the lighter UF-6 that contained the isotope U-235 would diffuse to the end of the long tunnel faster than the heavy U-238. By using a long tunnel (chamber) the gas collected was enough of the fissionable U-235 to complete the research that led to the atomic bomb.

See also Avogadro; Ideal Gas Law

GUTH'S THEORY OF AN INFLATIONARY UNIVERSE:

Astronomy (Cosmology): *Alan Harvey Guth* (1947–), United States.

> *The "newborne" universe at the instant of the big bang passed through a phase of exponential expansion. The negative pressure of the vacuum of space drove this inflation.*

Much of Guth's inflation theory and ideas that are additions to the big bang theory have recently been confirmed. In 2001 the "Wilkinson Microwave Anisotropy Probe" (WMAP) satellite was placed in orbit by a Delta II rocket from Cape Canaveral, Florida. Three years worth of data from WMAP was provided to the public on March 17, 2006. The data includes temperature readings from two points in the sky at the same time, readings of cosmic background radiation, polarization readings, and other important data. As WMAP scans the heavens, it covers about 30% of the sky each day. The data gathered by this probe and other instruments strengthen Guth's theory for the inflationary universe. Note: for those interested NASA has a website where pictures of the "baby universe" can be accessed. It is: http://www.gsfc.nasa.gov/topstory/2003/0206mapresults.

In 1948 the big bang theory for the origin of the universe proposed by George Gamow and others explained a great deal about the universe as it is today, including background microwave radiation and an ever-expanding universe. But this theory did not explain all observable events. For one, why was the present universe so uniform and similar in all directions of space, while on such a large scale it is more likely to exhibit some major irregularities? To overcome this and other problems perceived with the big bang, Guth first proposed some refinements to the theory in 1980. His inflation theory stated that the new universe underwent an exponential expansion of 10^{30} in the size of its radius within the first microsecond (10^{-43} to 10^{-35} seconds). This first instant of beginning is sometimes referred to as the start of the period of grand unification theory (GUT) where all four physical forces were united into one force. Guth also suggested that the GUT theory provides evidence that it is possible that the universe started from nothing. In other words, there was nothing before the big bang. Scientists today are looking for a similar simple GUT theory that is sometimes called the theory of everything (TOE). This microsecond is also the beginning of the period of the second law of thermodynamics where everything started to become more disorganized and the cooling of the universe began (and continues to cool). This is explained by the laws of thermodynamics.

HABER'S THEORIES: Chemistry: *Fritz Haber* (1868–1934), Germany. Fritz Haber was awarded the 1918 Nobel Prize for Chemistry.

The Haber process: *By using high temperatures and a metallic catalyst, atmospheric nitrogen can be chemically combined with hydrogen to form synthesized ammonia* ($N_2 + 3H_2 \leftrightarrow 2NH_3$).

During World War I Germany was cut off from its supply of nitrate salts from Chile, an important source of the chemical used for fertilizers and explosives. Fritz Haber developed the process of using "free" nitrogen from the atmosphere and converting it into "fixed' nitrogen. This synthesis, under normal conditions, produced very limited amounts of useful ammonia. To increase the yield to industrial proportions, in the early 1900s Haber ran hot steam over hot **coke** at 250°C under high pressure of 250 atmospheres (one atmosphere is normal air pressure) while using a catalyst to increase the rate of the reaction, thus producing more ammonia than would be possible during the normal chemical reaction. This is known as the Haber water gas process, where the nitrogen of the air and the hydrogen from the steam are "fixed" as ammonia. Haber was awarded the 1918 Nobel Prize in Chemistry for this achievement. The Haber process was adapted as an industrial process by Carl Bosch in 1913. Bosch's industrial processes resulted in significant amounts of ammonia that could be used for commercial purposes. Ammonia, in its various forms, is one of the major industrial chemicals produced worldwide. Most living plants, with the exception of legumes, such as beans, cannot take in free nitrogen from the atmosphere, but must rely on nitrogen that is found in soil. Without the Haber/Bosch process for the industrial production of fertilizer, the agricultural producers worldwide would be incapable of producing an adequate food supply for an ever-increasing population. In addition to its use in the manufacture of fertilizers, ammonia is used as a refrigerant, and also in the manufacture of synthetic fabrics, in photography, and in the steel and petroleum industries. Regrettably, a form of ammonia fertilizer, when mixed with petroleum, has become a cheap, but effective, type of explosive for a variety of terrorists.

Haber's theory for extracting gold from the oceans: *Gold can be removed from seawater through the use of high pressures and temperatures and proper electrochemical catalysts.*

During World War II Haber assisted the German government in the development of poison gases. After the war, he tried to help Germany pay off its war debts by extracting gold from seawater. In addition to using very high pressures and temperatures with catalysts to increase the speed of extraction of the gold, Haber used an electrochemical reaction to extract the large amount of gold present in seawater—without much economic success. An estimated eight thousand million tons of gold is dissolved in all the oceans' waters. However, to date, no economical process has been devised that would extract a significant amount of gold from seawater. Although his efforts were unsuccessful, Haber's techniques did lead to the current process of extracting bromine from seawater.

HADAMARD'S THEORY OF PRIME NUMBERS: Mathematics: *Jacques Salomon Hadamard* (1865–1963). France.

> *Prime numbers between 1 and n are distributed in an orderly fashion when considered in the formula:* $\pi(n) = n/\log_e n$*, which only works when n approaches infinity.*

In the above formula πn represents the number of the primes between 1 and *n* (it does not work for small numbers). Although others developed this theorem, Hadamard proved it in 1896.

Although Hadamard is best known for his proof of the prime number, he was interested in a number of other areas of mathematics. Educated at the Ecole Normale Superieure in Paris, he taught in several French universities before accepting the chair of mathematics at the College of France, also in Paris, where he remained until his death. He published over three hundred papers during his career, including one in which he defined a singularity as a point at which the function is no longer regular but was rather a set of singular points that will accommodate continuous functions. In modern mathematics it is referred to as "lacunary space." Hadamard's lacunary space exists as a set of singular points similar to a continuity of a function. The study of this concept by mathematicians, physicists, and cosmologists continues today. Hadamard also advanced the concept of solutions for differential equations. An example is his "well-poised problem" as a problem whose solution exists for a given set of data, but the solution also depends on the continuous use of the given data. In

> In his book *Psychology of Invention in the Mathematical Field* Hadamard investigated how scientists, particularly mathematicians and physicists, did their thinking. He surveyed about one hundred leading scientists of his day and asked how they accomplished the mental aspects of their work. Hadamard used introspection (a means of contemplation of one's own thoughts) to describe his mathematical thought processes and was interested in how others did their work. Some responded that they saw their mathematical concepts in color. Einstein is well known for his "thought" experiments and said he felt sensations in his forearms as he contemplated his ideas. Nikola Tesla was able to formulate his theories in his mind as "models" related to potential inventions based on his theories. He could then, without physically diagramming the models, instruct his workers how to construct working models of his ideas. Others described similar ways of forming mental images that did not use language or cognition to describe their thoughts.

1896 Hadamard proved the prime number theorem that was proposed by Gauss and Riemann, and which has proved to be the most important discovery in number theory.

A prime number is any positive integer (number) that has no divisors except itself and the integer 1. A prime number has just two factors. Therefore, 1 cannot be a prime number. If a number has more than two factors, it is called a composite number. Some examples of prime numbers are: 2, 3, 5, 7, 11, 13, 17, 19, 23, 29, and so on.

See also Gauss; Riemann

HADLEY'S HYPOTHESIS FOR THE CAUSE OF THE TRADE WINDS: Meteorology: *George Hadley* (1685–1768) England.

> *Earth rotating on its axis causes vertical rising hot air from the tropical circulation cell at the equator to form the trade winds that blow westward near the equator and eastward at more northern latitudes.*

Even before the time of Christopher Columbus, sailors knew that when sailing west toward the islands off the coast of Africa they must go south to pick up the westerly winds. They also knew that if they wanted to return to northern Europe and England, they needed to go farther north to pick up the easterly winds. George Hadley attended Oxford University and upon graduation became a barrister. He was more interested in physics than law and became an amateur meteorologist who directed the meteorological research observatory for the Royal Society. He was intrigued with the movements of cold air (that is denser and thus heavier than warm air) that replaced less dense hot

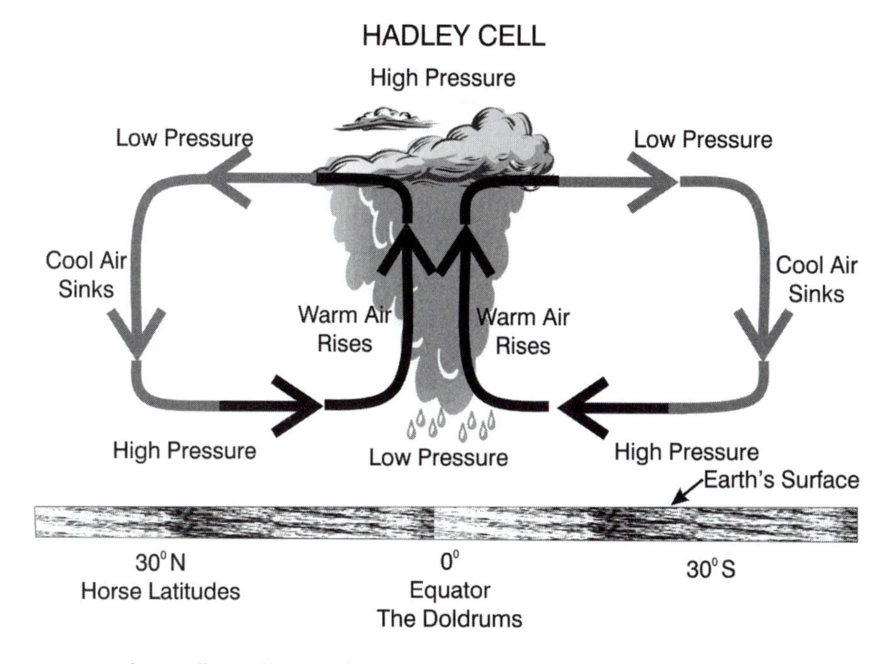

Figure H1. The Hadley cell's circulation is based on the physical principle that warm air is less dense, thus it will rise until it becomes cooler and more dense and therefore will sink toward the surface of the Earth.

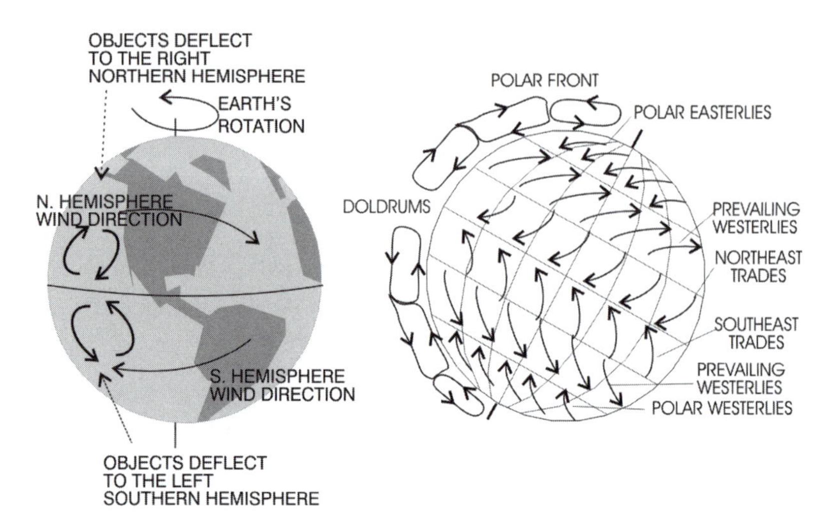

Figure H2. The trade winds were used by sailing ships to head west carried by the prevailing winds at a lower latitude in the northern hemisphere and return westward at a higher latitude.

air thus causing a circulation that was continuous with the cooler air moving under the less dense hotter air. The cooler air replaces the warmer air causing masses of air moving in giant patterns of circulation. This circulation, regardless if it is in a room where the air is heated with a stove or for giant air masses in the atmosphere, the air will continue to circulate as long as there is a difference in the temperatures of two masses of air. At warmer climate regions the heated air is "pushed" up by the cooler heavier air, heating up the air as it rises. This circulation of cool air from northern latitudes moving close to the surface of Earth toward warmer southern latitudes near the equator forms a pattern known as a "cell." The warm air mass then moves toward the higher cooler latitudes where they are again cooled as the circulation continues as long as Earth is warmed by the sun. This movement of air masses creates a circulation that is now called a "Hadley cell" (*see* Figure H1). The Hadley cell carries heat and moisture from the tropics to the more northern latitudes. Hadley was perplexed as to why this moving air did not take a direct path to the poles from the equator. Instead the air moved westward as it proceeded north.

He proposed a theory for the cause of the westward motion in the northern hemisphere of the winds known as the *northeasterly (NE) trade winds*. Hadley believed that Earth's rotation played a crucial role in determining the direction of the large mass of heated air from the tropics (*see* Coriolis' Effect). This circulation of tropical air mainly occurs between 30°N and 30°S latitudes as the hot, moist air from the tropics sinks at higher latitudes as it becomes cooler. The cycle is completed as the cooler air returns to the equatorial tropics, is again heated, and continues the cycle. The Hadley cell explained the westward flow of air at these low latitudes, but not why the wind changed directions at latitudes greater than 30°N. The 30° latitude in the northern hemisphere became known as the "horse latitude." A similar cell to the Hadley cell called the *Ferrel cell* is formed at latitudes greater than 30° north. The wind changes direction in the Ferrel cell and flows from north to south. The Coriolis effect causes the winds north of the horse latitudes to flow eastward. It might help to remember that when meteorologists refer to the "NE trade winds" they are referring to the direction from which the winds are flowing. This means the NE trade winds are flowing from east to west. The opposite is true for the westerly winds north of the horse latitudes that are flowing from west to east. At warmer climate regions the heated air is then "pushed" up by the cooler heavier air, heating up the air as it rises.

See also Coriolis

> The horse latitudes are regions in the subtropics between 30° and 35° in the Northern and Southern Hemispheres. The air in these regions is relatively dry and thus exhibits high air pressures that produce weak or no winds. (Note: the lower the air pressure, the more severe are storms with winds.) Sailors, particularly in the Northern Hemisphere, called this region around 30° latitude the "horse latitudes" because sailing ships would be stuck in a calm area for days. If the calm lasted for many days and water and food for the horses and cattle on board were running low, the sailors would throw the animals overboard so they would have enough water to survive until the ship moved out of the calm area. There is no explanation of why they just didn't eat the animals. Interestingly, many of the dry and desert regions of the world, including the great Sahara desert, also lie in the horse latitudes. Because this latitude is also near the Tropic of Cancer (in the Northern Hemisphere and Tropic of Capricorn in the Southern Hemisphere), it is also known as the Calms of Cancer or Calms of Capricorn.
>
> This information on the direction of air movements was extremely important during the days of sailing ships. Sailors now knew the reason they must take a more southerly route to the New World and a more northerly route to return to Europe. Hadley, a well-known and respected lawyer and amateur meteorologist was elected a fellow in the Royal Society in 1745. The Hadley Centre for Climate Prediction and Research in England, as well as a crater on the planet Mars, are named for him.

HAECKEL'S BIOLOGICAL THEORIES: Biology: *Ernst Heinrich Haeckel* (1934–1919), Germany.

Haeckel's law of embryology: *Ontogeny recapitulates phylogeny.*

Ernst Haeckel spent his professional adult life attempting to use the concept of evolution as a means to develop a unifying theory for the field of biology. He contended that the embryological stages of all animals were a recapitulation of their evolutionary history. *Ontogeny* is the embryonic development of organisms, while *phylogeny* is the evolutionary history of a species. Haeckel also believed that at one time there were animals on Earth that resembled all the developmental stages of the embryos of animals presently found on Earth. In other words, he incorrectly believed there were prehistoric mature animals that resembled the embryos of modern animals. Haeckel hypothesized that the mechanism for inheritance of traits was present in the nuclei of cells, even though he was unaware of chromosomes, genes, and DNA. There is some relationship to the similarities of early stages of development of animal embryos in that the cell nuclei of all animals contain a very large proportion of the same DNA. Mammals, including humans, share approximately 95% to 98% of the same DNA, and chimpanzees and humans share about 99% of the same DNA. Although it was partially supported by Stephen Jay Gould, most biologists no longer accept Haeckel's recapitulation theory (*See also* Gould).

Haeckel's theory of evolution: *All animals are derived from inanimate matter.*

Haeckel developed a hierarchy evolution design based on the premise that all animal matter was derived from inanimate matter and then progressed upward to humans.

Haeckel observed that there was a direct relationship between and among plants and animals and the environments in which they lived. Like many before him, he observed that there seemed to be a symbiotic relationship between animals and their environment. In 1868 Ernst Haeckel coined the word "ecology" to describe the study of plants and animals and their relationship to each other and their environment: the soil, water, atmosphere, light, heat, oxygen, carbon dioxide, and all the other elements in their biological surroundings. Ecology is, in essence, a study of interrelated "systems" of living organisms and their inorganic environments. The term *ecology,* from the Greek work *oikos,* meaning "household," is from the same root word as *economics.* Ecology was not a recognized science until the early part of the twentieth century. Many people confuse environmentalism (an ideological political movement) with ecology (a science based on systems).

He came to this conclusion from the scientific concept of symmetry and proposed his inorganic origin of animal life by relating the symmetry of the simplest animals to the symmetry of inorganic crystals. Although this theory as related to evolution is not now accepted, there is some acceptance of the idea that, at some point in history, inorganic chemical elements combined to become self-replicating, prebiotic organic molecules. These reproducing molecules are assumed to have developed into primitive DNA and cells. This is one theory for the origin of life.

Haeckel also contended all higher multicellular animals (chordates) with three layers of cells (ectoderm, mesoderm, and endoderm) were derived from animals, such as jellyfish and sponges and other marine invertebrates (gastrula), which are composed of only two layers of cells (ectoderm and endoderm). He was the first to distinguish protozoa (single-celled animals) from metazoa (multicellular animals).

Haeckel's concept of social Darwinism: *Human culture and society conform to the laws of evolution.*

Ernst Haeckel, an ardent supporter of Darwin's concept of organic evolution, predated current social Darwinists, such as Edward O. Wilson, by several decades. Haeckel's concept of social Darwinism asserts that human society is as much a product of nature as is human anatomy and physiology, and thus is no different from other organisms. In other words, natural selection applies as much to human behavior, cultures, societies, and religions as it does to the nature of the organism. This concept is not fully accepted by all biologists, partly because it is seen as antireligious and places humans in the same category as other animals and partly because it is used as a rationale for **eugenics**.

See also Aristotle; Darwin; Dawkins; Gould; Wallace; Wilson (Edward Osborne)

HAHN'S THEORIES OF NUCLEAR TRANSMUTATIONS: Chemistry: *Otto Hahn* (1879–1968), Germany. Otto Hahn received the 1944 Nobel Prize for Chemistry.

Hahn's and Meitner's nuclear isomerism: *Nuclei of different radioactive elements exhibit identical properties.*

Otto Hahn discovered a new radioactive form of the element thorium (Th) which he called "radio-thorium," and Lise Meitner discovered protactinium (Pa). Radioactive isomerism is defined as the phenomenon that occurs when some radioactive elements with the same atomic number (number of protons) and the same number of neutrons in their nuclei possess different properties and behave in different ways from the nonisometric forms of the same element. Interest in the concept of isomerism resulted in the collaboration between Hahn and Meitner. In other words, an isomer of an element has exactly the same atomic number and mass but not the same characteristics. This discovery for radioactive elements led to their more important discovery.

Hahn–Meitner–Strassman theory of nuclear fission: *Nuclear transmutation occurs in heavy as well as lighter elements.*

Ernest Rutherford first discovered the transmutation of the nucleus of one element into the nucleus of another element—one that has a nucleus with a different atomic number and mass—when he observed oxygen being transformed into nitrogen. Marie and Pierre Curie discovered similar reactions with uranium and radium. Hahn and his collaborators theorized there are two important aspects to nuclear transmutation. First, these types of nuclear reactions always emit light beta particles, high-energy electrons, and/or heavy alpha particles (helium nuclei). Second, the transmutation of one element's nuclei to form different nuclei occurs only between elements that are more than two places apart on the **Periodic Table of the Chemical Elements.** Hahn noticed these rules did not apply to heavy elements when he used heavy atomic nuclei (neutrons and alpha particles) to bombard uranium, which turned into lead. He also bombarded the heavy element thorium with neutrons. He then theorized there must be an intermediate emission of particles and lighter elements involved in stages between the uranium and lead. In essence, Hahn split the uranium atom by bombarding it with neutrons, thus opening the door to research in developing a fissionable chain reaction and later the atomic bomb and the generation of electricity and useful isotopes by atomic reactors.

See also Curie; Frisch; Meitner; Rutherford

HALDANE'S THEORIES OF GENETICS, EVOLUTION, AND ORIGINS OF LIFE: Biology (Genetics): *John Burdon Sanderson Haldane* (1892–1964), England.

> *Genes follow a rule for sex determination in hybrids, meaning that for the first generation of offspring between two different species, the sex genes are recessive and the hybrid will not develop as an embryo or, if born, will be sterile.*

Haldane was the founder of what is known as population genetics, but his main discovery was a rule that determines the sex of a hybrid animal. Haldane's rule described the well-known example of hybrid inviability and sterility. He noticed that an offspring between a male and female of different species was either stillborn, or when a live birth occurred, the sterile offspring displayed only one sex, usually heterogametic (XY) rather than homogametic sex (XX or YY). In essence, the rule states that if one particular sex is born of mating between a male and a female of different species, their offspring, if living, will be a hybrid whose sex is heterogametic. This means the offspring will most likely be male because males provide the sexual differentiation, which is the evolutionary path to differentiation and formation of new species. Haldane was intrigued by human genetics. He even prepared a map of the X chromosome with the genes that cause color blindness.

Mathematics can be used to explain population genetics and evolution.

Although there were other biologists interested in the premise that mathematics can be used to develop a theory of population genetics, it was Haldane who published a series of papers indicating that there is a direction and rate of change of the genes related to mutation and migration. One paper was titled "Mathematical Theory of Natural and Artificial Selection." Another paper titled "The Causes of Evolution" was a compilation of his research on population genetics that strongly supported and strengthened Charles Darwin's theory of natural selection as the primary operator of evolution. He was the first to explain evolution and natural selection in mathematical terms as related to Mendelian genetics.

John Haldane was interested in science from an early age and assisted his father, a physiologist, in setting up and conducting experiments in his father's home laboratory. While in the military in World War I, he was twice wounded. He became interested in research with respiration and its related chemistry after the Germans used poison gases on Allied forces. This led him to work on the development of an improved gas mask. He held positions in several British universities where he established his reputation in human genetics. He also was a member of the British Communist party and made political speeches. He moved to India permanently in 1957, ostensibly in protest of Britain's participation in the invasion of the Suez Canal.

During his extensive research studies he experimented on himself, acting as a guinea pig. One experiment involved exposing himself to extreme levels of oxygen. This caused a severe reaction that led to a seizure that damaged his spinal cord. In another experiment he used a pressure chamber to rapidly decompress his body and ruptured both eardrums. In yet another experiment he inhaled carbon dioxide to see how it affected muscles related to respiration. He also drank a mixture of $NaHCO_3$ (sodium carbonate) and NH_4Cl (ammonium chloride) that produced hydrochloric acid in his blood, the object of which was to observe the changes in the levels of sugar and phosphate in his blood and urine.

Haldane's theory of the origin of life: *Assuming Earth's original atmosphere was a mixture of ammonia, hydrogen, methane, and water vapor, these factors were sufficient to cause the formation of organic compounds, which were then deposited in the oceans.*

These early oceans were more like a hot, dilute soup that further evolved into complex organisms. This theory and research is related to the later studies and experiments carried out by Harold Urey and Stanley Miller at the University of Chicago where they used such a mixture and a high-energy spark to simulate lightning. They did produce some amino acids that are the building blocks of life, but not life.

After Haldane's move to India in the late 1950s, he was able to do extensive research on genetics. He wrote over three hundred papers and articles for scientific and political publications including the communist *Daily Worker*. He became an Indian citizen in 1961. He died of cancer in 1964, but not before writing a series of humorous poems that chronicled the challenges of the disease that killed him.

See also Darwin; Kimura (Motoo); Mendel; Miller; Urey

HALE'S SOLAR THEORIES: Astronomy: *George Ellery Hale* (1868–1938), United States.

Hale's sunspot theory: *The spectra of sunspots can be used to identify the elements in the outer reaches of the sun.*

George Hale spent most of his life seeking funds to build larger and larger telescopes. His efforts resulted in the forty-inch telescope at the Yerkes Observatory on Geneva Lake in Williams Bay, Wisconsin; the sixty-inch reflector telescope and the

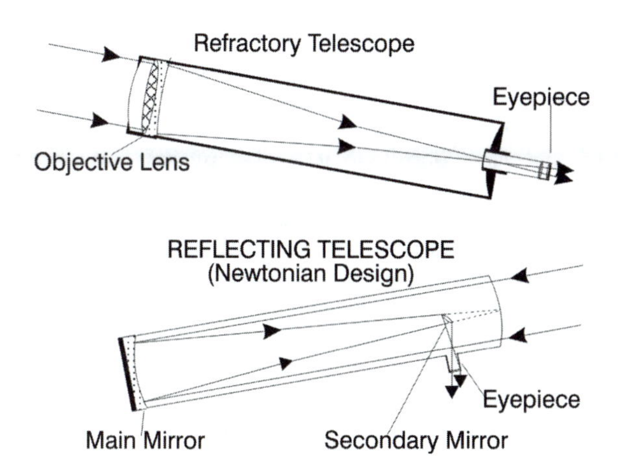

Figure H3. Diagrams of the refractory and the reflecting telescopes. The size of the refractory telescopes is limited by the distortions in large glass lenses. Reflecting telescopes collect more light from objects by use a parabolic mirror, which can be made larger than a glass lens.

100-inch Hooker telescope, both installed at the Mount Wilson Observatory located in the San Gabriel Mountains of Southern California; and finally the 200-inch Pyrex mirror reflecting telescope installed at the Palomar Observatory in north San Diego County, California. Hale spent the last years of his life supervising the construction of the 200-inch telescope at Palomar and thc buildings associated with it. He was one of the few astronomers who realized that bigger and more powerful telescopes were needed to advance the study of astronomy. Some of these telescopes were of the *refractor* type that uses a series of lenses, while the larger instruments were of the *reflector* type, which use mirrors (*see* Figure H3).

An instrument he invented, called the *spectroheliograph*, enabled him to photograph and make measurements of sunspots. He viewed and photographed small bands of wavelengths of the sun and identified the double spectral lines emitted by the element calcium, which led to the study of the chemical makeup of the outer layers of the sun. Hale used his progressively more powerful telescopes to effectively prove that the Martian "canali" (canals) reported by Schiaparelli are nonexistent.

Hale's hypothesis for extraterrestrial magnetic fields: *Sunspots exhibit strong electromagnetic fields.*

About ten years after using his spectroheliograph to describe the chemical elements in sunspots, George Hale identified the Zeeman effect, which occurs when the magnetic field splits the spectral lines of light from the sun, enabling the viewer to detect specific elements in the sun. This was the first time that a magnetic field was detected as originating from another object beyond Earth. The so-called sunspot cycle is an irregular cycle of about eleven years between the high and low portions of the frequencies of the cycles.

See also Lorentz; Schiaparelli; Zeeman

The story of the conception, funding, and construction of the 200-inch Hale Telescope located in the Mount Palomar Observatory is fascinating. Hale had secured funding for earlier smaller telescopes located in various institutions. He sought funds for the construction of his dream of a giant reflecting telescope from the Rockefeller Foundation. In 1929 the foundation agreed to donate $6 million to the California Institute of Technology. Hale was appointed chairman of the group responsible for its completion. The task proved challenging. The first 200-inch mirror that cost $600,000 was a failure due to imperfections and distortions in the glass. Hale suggested they use a new type of glass called Pyrex and the first mirror cast with this new type of heat resistant glass was successful in 1934. It weighed sixty-five tons and required several months to cool under controlled conditions. The giant disk was shipped from the Corning Glass Works in New York to California on a special truck that never accelerated more than twenty-five miles per hour. The grinding of the parabolic Pyrex disk to form the reflecting mirror was delayed during World War II. By the time it was completed and commissioned as the Hale telescope in 1948, George Hale had been dead for nearly a decade. Although other larger, as well as satellite telescopes have been developed, the Hale continues to be a very impressive and productive instrument.

HALLEY'S THEORIES FOR COMETS AND STARS: Astronomy: *Edmond Halley* (1656–1742), England.

Halley's theory for the return visits of comets: *Gravity that affects the planets also affects the paths of comets.*

Edmond Halley not only accepted Newton's laws of motion and gravity, but he was such a staunch advocate of these laws that he provided funds for the publication of Newton's famous publication *Principia.* Halley managed to upset John Flamsteed (1646–1719), his predecessor as royal astronomer at the Greenwich Observatory, by allowing Newton to publish Flamsteed's astronomical observations before Flamsteed, Britain's first Astronomer Royal and founder of the Greenwich Observatory, had completed them.

Halley theorized that comets must react similarly to planets because comets are also affected by the gravitational attraction of the sun. He collected data for several past comet sightings and noted there appeared to be repetition in the number of years for the sightings. He then calculated that the comets of the years 1456, 1531, 1607, and 1682 were actually the same comet that traveled in a large, elongated ellipse around the sun. Using this seventy-six-year cycle, he predicted the same comet would reappear in the year 1758. Halley died before he could see his comet return, but it was named after him and ever since has been known as Halley's Comet. The same comet was first viewed and recorded in the year 240 BCE. The most recent visit was in 1986, and the next is expected in 2061 or 2062.

Halley's theory of stellar motion: *Stars are not fixed, and their movements can be detected over long periods of time.*

Ancient stargazers, as well as the Greek astronomers, recorded the positions of the brightest stars. Many years later, Edmond Halley also catalogued the positions of three of the brightest stars: Sirius, Procyon, and Arcturus. He noted their positions had changed, and thus stars must not be in fixed positions. Due to the great distances of stars from Earth, any movement of a star is difficult to observe. Halley felt the great astronomer Tycho Brahe had made the best measurements of these same stars' positions, but at that time Tycho was unaware of their slow apparent movement due to the tremendous distances of the stars from Earth. Comparing his own records with Tycho's data, Halley established the fact that the real motion of the stars was detectable only after long periods of viewing. Later astronomers used the concept of parallax, where the position of a star is measured at six-month periods as Earth revolves around the sun, but these measurements made over a short period of time did not detect star motion. Thus, it was determined the movement of stars was random. It was not until the early 1900s that the Dutch astronomer Cornelius Kapteyn observed that two different streams of stars were moving in opposite directions. This proved that the movement of stars was not random but rather proceeded with some order, as was discovered in galaxies that continue to move away from us and from each other at increasing velocities.

Edmond Halley, the first to study the stars in the Southern Hemisphere, also identified two "cloudy" areas, one named the Magellan Cloud, and the other a star group referred to as the Southern Cross. He developed meteorological maps and found that the subtropical trade winds blow from the southeast toward the equator in the Southern Hemisphere and from the northeast in the Northern Hemisphere. (Early sailors, including Christopher Columbus, understood this occurrence and made use of the trade winds when crossing the Atlantic Ocean eastward below the Tropic of Capricorn and returning to Europe north of the Tropic of Capricorn.) Halley correctly related this phenomenon to a major circulation in the atmosphere due to the difference in the heat produced by the sun in the areas above and below the equator as affected by the rotation of Earth.

See also Coriolis; Hadley

HALL EFFECT OF ELECTRICAL FLOW: Physics: *Edwin Herbert Hall* (1855–1938), United States.

An electric force is generated when a magnetic field is perpendicular to the direction of an electric current. This force is perpendicular to both the magnetic field and the current.

A more technical explanation is that a transverse electric field will develop in a conductor that is carrying a current while it is in a magnetic field. The Hall effect occurs when a magnetic field is at a right angle to the direction of the current flow. This modulates and multiplies voltages and can act as a magnetic switch for semiconductors, which is important for electronic applications. These devices, also referred to as *Hall generators*, are used in brushless induction motors, tachometers, compasses, thickness gauges, and magnetic switches that do not require contact

Edwin Hall received his PhD degree from the Johns Hopkins University in Baltimore, Maryland, in 1880. Soon after, he joined the Harvard faculty and was appointed professor of physics in 1895. He remained in this position for twenty-six years. While working on his doctorate related to electric currents within a magnetic field, there was a controversy as to whether there was action on the electricity itself or just on conductors carrying the electricity. He set up a unique experiment using thin sheets of gold foil to detect that the electric potential was acting perpendicular to both the current and magnetic field. The peculiar effect was soon named the Hall effect and is explained as: When an electric charge is moving in a conductor, it will experience a transverse force that tends to move the flow of current to one side of the conductor. The Hall effect has many applications in modern electronics industries, as well as in electric and magnetic equipment. Almost one hundred years after Hall's experimental evidence for this effect, the 1985 Nobel Prize for Physics was awarded to Klaus von Klitzing for his additional work on this phenomenon.

points. Devices employing the Hall effect can detect minute hairline cracks in the wings and bodies of airplanes, as well as in very sensitive instruments that can measure weak magnetic fields.

See also Ampère; Faraday; von Klitzing

HAMILTON'S MATHEMATICAL THEORIES: Mathematics: *Sir William Rowan Hamilton* (1805–1865), Ireland.

Hamiltonian functions: *A special function of the coordinates and the moments of a system is equal to the rate of change of the coordinate with time.*

Sir William Hamilton worked out the mathematics expressing the sum of kinetic and potential energies of any dynamic (changing) system. It is related to the Hamiltonian principle, which states that the time integral of the kinetic energy minus the potential energy of a system is always at a minimum for any process. The equation involves both the motion and time for the system. The Hamiltonian function was an important step in developing the theory of quantum mechanics.

Hamilton's quaternions: *By ignoring the products of multiplication, a system of internal algebra for "hypercomplex" numbers can exist for "quaternions" (Latin for "four").*

In developing his theory, Sir William Hamilton applied algebraic functions to non-Euclidean geometry involving *n*-dimensional (multiple dimensions) analytic geometry, which involves more than the three dimensions $(x,y,z \ldots ,n)$ of space. The quaternion is a hypercomplex number in algebra. It has always been accepted that $A \times B = B \times A$, but for quaternions, the commutative law of multiplication does not hold. Therefore, $A \times B$ does not necessarily equal $B \times A$. The quaternion theory is important in the field of abstract algebra, but it has little use in physics, where vectors are more applicable. However, it did lead to the development of mathematical matrices, such as matrix mechanics as applied to Werner Karl Heisenberg's quantum theory. Hamilton made other contributions to optics, mathematics, and abstract algebra.

See also Heisenberg; Lagrange

Sir William Rowan Hamilton had an interesting, if somewhat tragic life. He was introduced to mathematics at the age of thirteen when he studied algebra. At age fifteen he found an error in a work published by Laplace. At age eighteen he entered Trinity College in Dublin, Ireland, where he showed interest in the classics and mathematics. While at Trinity College, he was introduced to a friend's daughter, Catherine, with whom he carried on a love letter romance for most of his life. When her mother decided that she should marry a much older and well-to-do man of the clergy who could provide for Catherine, an extremely upset Hamilton became ill, read and wrote poetry, and considered suicide.

In 1826 during his senior year at Trinity he received an unusual honor in both the classics and mathematics. This was somewhat unusual because he seldom attended classes. Soon after graduation he was appointed professor of astronomy at Trinity College even though he lacked experience as an observer of the heavens. Although still smitten with Catherine, he married a local woman, Helen Maria Bayly, who was often ill. He continued his work because he had two children with Helen, but his failed marriage soon led to heavy drinking. It is reported that in 1843 as he was walking on his way to the Royal Irish Academy he was thinking about his concepts of noncommutative algebra called "quaternions" (four). He is said to have carved the formula for quaternions in a stone on the Brougham Bridge. It was: $i^2 = j^2 = k^2 = i\ h\ j = -1$. Hamilton and Catherine kept in touch with each other which neither spouse seemed to mind. Hamilton presented Catherine with his book of *Lectures on Quarternions* just two weeks before she died.

HARDY'S MATHEMATICAL THEORIES: Mathematics: *Godfrey Harold Hardy* (1877–1947) England.

The Hardy–Weinberg law: *The allelic frequencies in a population remain the same in subsequent generations.*

Note: An **allele** is a particular pair of genes (or group of genes) located at the same position on a chromosome.

This law is related to population genetics that states mathematically that inherited characteristics are of more importance to a population of organisms than inheritance is to individuals. It is well known that populations evolve from environmental changes by natural selection. But it is an allele of a group of genes that determines the specific characteristics of an organism, such as height, color, and so forth. When a gene mutates (alters), it may pass from parent to offspring. If these new (mutated) genes increase the frequency of successful change in future generations, there will be evolution of the population. The Hardy–Weinberg law is, in essence, a statement for an *equilibrium equation* that states that if no factor interferes with the allele frequencies in a population, evolution will not occur. However, there are forces that change allele frequencies such as mutation, mate selection, migration, and drift. The Hardy–Weinberg law is named in honor of Godfrey H. Hardy, called "Harold" by friends and colleagues, and Wilhelm Weinberg (1862–1937), a German physician, who independently and in the same year (1908) formulated the identical principle of population genetics.

Godfrey Hardy was educated at Cambridge University, Hampshire, England, and at Trinity College, one of the constituent colleges of the University of Cambridge. After graduation he became a fellow in mathematics in Trinity College but left in 1919 to take the Savilian Chair of Geometry at Oxford University. He returned to Cambridge n 1931 to become the Sadleirian Professor of Pure Mathematics, a position he held until 1942. (The Savilian Chair was established in 1619 by Sir Henry Savile; the Sadleirian Chair was established by Lady Sadlier in 1701.) Hardy's main interests were in the theory of numbers, inequalities, integrals, complex functions, and the Riemann zeta-function. In addition to his work in pure mathematics, he was considered an excellent teacher who introduced modern mathematics to several generations of British students. He sponsored a brilliant Indian mathematics student Srinivasa Ramanujan (1887–1920) for a scholarship at Cambridge in 1914. They became close friends as Hardy mentored and tutored Ramanujan. In an interview, Hardy is reported to have said that his discovery of and collaboration with Ramanujan was the one romantic period of his life. He was known as an eccentric who enjoyed cricket games, was an outspoken critic of Christianity, and refused to have his picture taken—there are only five photos of him in existence.

Hardy investigated many areas of pure mathematics and published multiple papers in collaboration with many other mathematicians in areas such as the Riemann zeta-function, prime numbers, the Fourier series, divergent series, Diophantine analysis, and complex analysis. Explanations of his mathematical studies are beyond the scope of this book.

See also Fermat; Fourier; Hamilton; Lagrange; Laplace

HARKINS' NUCLEAR THEORIES: Chemistry: *William Draper Harkins* (1873–1951), United States.

Harkins' rule for isotopes: *The isotopes of elements that have odd-mass numbers are less abundant than are the isotopes of elements with even-mass numbers. Both are represented by whole number ratios for their atomic numbers and weights.*

William Harkins proposed the whole number rule about the same time as Francis William Aston. The rule determined that the mass and the occurrence of isotopes of stable elements are related. When isotopes of stable elements are considered, the numbers of neutrons in the nucleus are related as whole numbers—never fractions of neutrons. Harkins predicted both the existence of the neutron and the element deuterium (heavy hydrogen whose nucleus contains a neutron as well as a proton (*see* Figure O1 under Oliphant). Using a mass spectrometer, which he invented, Aston determined this whole number concept was not always correct. Sometimes the mass of isotopes was slightly more or less than a whole number. Aston called this the *packing fraction*, which occurs when the hydrogen nuclei join to form helium nuclei (fusion) and a small amount of energy is produced. Therefore, Harkins' whole number rule was modified to conform to new experimental data.

Harkins' theory for the hydrogen-helium-energy reactions: $4H_1 \rightarrow He_4 + energy$.

Marie Curie demonstrated the concept of radioactivity produced by the fission of nuclei of radioactive elements, resulting in the production of enormous amounts of energy. However, it was William Harkins who applied his knowledge of isotopes and neutrons to address Aston's "packing fraction" principle to determine how hydrogen could be converted into helium with a tiny fraction of matter converted to energy. This confirmed the basic laws for the conservation of matter and energy. It was later determined that such a conversion reaction maintains the energy output of the sun and other stars. It was not until several decades later that the concept was applied beyond the laboratory, resulting in nuclear energy to produce electricity, isotopes, and the atom (nuclear) bomb.

See also Aston; Curie; Fermi; Rutherford; Urey

HARVEY'S THEORY FOR THE CIRCULATION OF THE BLOOD: Biology: *William Harvey* (1578–1657), England.

> *The beating heart propels blood through the arteries, where, after it reaches the body's extremities, it is then circulated back to the heart through the veins.*

From the time of Galen (second century CE), medical practitioners taught that blood was produced in the liver. Then in the seventeenth century, William Harvey concluded that for this to be true, a tremendous amount of new blood would have to be produced continuously by the liver to maintain a constant flow. Thus, he theorized the same blood must circulate through the blood vessels. Harvey dissected cold-blooded animals to observe the pumping action of their hearts. Because circulation is slower in frogs and snakes, he observed and traced the blood passing from the right to the left side of the heart through the lungs, not through "pores" in the septum (tissue that divides the two sides of the heart), as was taught and believed from the days of Galen. Harvey observed that as the heart beats, **systolic** contractions occurred, swelling the arteries with blood as it was pushed throughout the arterial system of the body. At the **diastolic** phase, the heart was again filled with blood that was returned by the veins. Harvey declared the heart to be a pump. This established his theory that the heart caused the blood to circulate throughout the body via the arteries and then back to the heart through the veins. Nonetheless, no reasonable explanation existed for how the blood was transferred from the ends of the arteries to the veins to complete the circuit back to the heart. Harvey correctly theorized the presence of very small blood vessels, too small to be seen, that provided the passage of blood from the ends of arteries to veins. The validation for Harvey's discovery had to wait until the invention of the microscope and Marcello Malpighi's observance of the tiny web of blood vessels in the thin skin of a bat's wing, which he named *capillaries* (Latin for "hair-like").

See also Galen

HAÜY'S GEOMETRIC LAW OF CRYSTALLIZATION: Geology and Mineralogy: *René Just Haüy* (1743–1822), France.

All minerals exhibit six different basic forms dependent upon how the molecules are joined to produce crystals.

Haüy's law of crystallography was a more-or-less serendipitous discovery that happened when he dropped a piece of calcite spar (calcium carbonate $CaCO_3$) crystal that shattered in many smaller pieces. Being a mineralologist, he recognized the odd patterns of the fragments with smooth surfaces that met at similar angles as being rhombohedral in shape (a rhombohedron figure may be thought of as a "stretched" out cube with six faces). He was also familiar with the work of Nicolaus Steno in the late 1600s who discovered that the angle between faces of a particular crystal are similar, as well as the works of Robert Hook, Christiaan Huygens, and Sir Isaac Newton who all considered that crystals are compiled of stacks of particles, similar to a stack of bricks. Therefore, Haüy hypothesized that crystals were formed in layers that are now called *unit cells*, and that, as a whole, the crystal was formed by different geometric shapes (*see* Figure H4).

After more investigation, he decided that chemical changes were responsible for the differences in shapes of crystal minerals. Haüy expressed his work as a mathematical theory in his writings, which impressed the scientific community. In addition, he experimented with *pyroelectricity* (to heat) that is the generation of an electric current by heating a crystal (this is related to, but not the same as, *piezoelectricity*, that means "to press"). Both of these processes are used today in the form of thin films of semiconducting compounds. One example is a heat-detecting device to identify the heat from a human or animal body just a few feet away and convert this into an electrical signal. Another is to use $LiTaO_3$ (Lithium tantalite), that has both piezoelectric and pyroelectric properties and is used to create an experimental small-scale nuclear fusion known as "pyroelectric fusion."

Haüy was arrested during the French Revolution and almost lost his life in prison. However, later during Napoleon's rule he was released, whereupon he became professor of mineralogy at the French National Museum of Natural History in Paris. But again, disaster overtook him

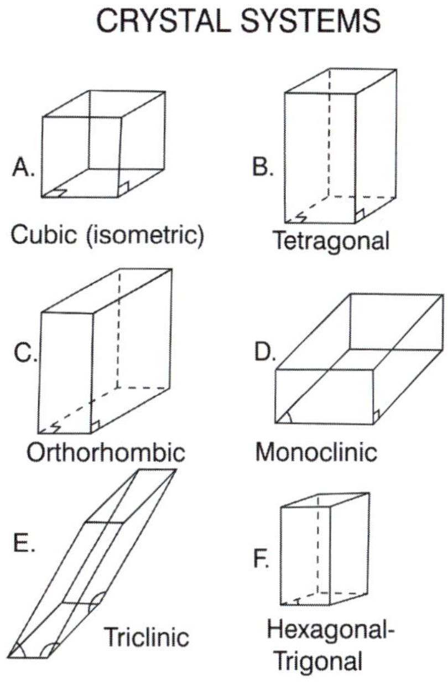

Figure H4. Various chemical changes result in different arrangements of the particles of crystals, resulting in different structures and thus a variety of basic shapes.

during the period of the Restoration government. He ended his life in poverty, but he will always be known as the father of mineral crystallography and for maintaining his self-respect right up to the end of his life.

HAWKING'S THEORIES OF THE COSMOS: Astronomy: *Stephen William Hawking* (1942–), England.

Hawking's theory for the nature and origin of the universe: *Applying the submicroscopic theory of quantum mechanics to the macroscopic universe produces a wave function similar to a wave function for elementary particles, thus shaping the nature of the universe.*

Stephen Hawking theorized that according to the quantum wave function theory, as related to Einstein's theory of relativity, there is no past or future because the time function exhibits characteristics of special dimensions (relativity). The universe can, by its own accord, be both zero in size or infinite in size. It can have no beginning or no ending. This led Hawking to a new version of "open inflation" that spontaneously created a "bubble" out of nothingness. In other words, the big bang, which is generally accepted as how our universe began, had to have something with which to start. Hawking's use of quantum mechanics provides this answer by showing how the wave function starts as a *singularity* (a region of space-time where one or more components of curved spaces become infinite.)

These ideas preceded the three different geometries or shapes for the universe. The first concept maintains the universe is flat space as related to the rules for Euclidean geometry, in which parallel lines never meet or cross, a triangle always has exactly 180° as the sum of its angles, and the length of a circle is its circumference (two times pi × the circle's radius or $2 \pi r$). Although the space in the universe generally is considered flat, space-time is not. For most of our science and measurements on Earth, three-dimensional space involves L for length, L^2 for area (length × height), and L^3 for volume (length × height × width, or depth). Astrophysicists add time (T) as the fourth dimension to the three dimensions of space to describe matter in relativistic terms. In other words, the structure of matter changes in space over long periods of time—thus, the term "space-time." Contrary to older theories, the universe is not static but rather is a dynamic, changing physical entity. This also means that over great distances and long periods of time, the extended universe will not remain flat. At its outer reaches, the total curvature of space-time is determined by the density of matter as well as the time involved for the continued expansion of the universe. An analogy: When standing on the surface of Earth and looking toward the horizon, for about a mile the surface seems flat, just as the space near us that we can see seems flat. But when viewed from a spacecraft, earth looks curved—just as the distant sectors of the universe also appear to be curved.

The second theory for the geometry of space is that space is spherical similar to a globe. This represents a closed universe, which is similar to Einstein's original static universe theory. Curved surfaces are positive in the sense that space curves back on itself similar to the surface of Earth. Because a closed universe is finite, parallel lines would always converge (e.g., line of longitude for Earth); the sum of the angles for a triangle laid out on the surface of a sphere is always greater than 180° (the sum of the angles for a triangle can total up to 540° when projected on the surface of a sphere). And the circumference of a circle inscribed on the globe's surface is always *less* than $2\pi r$.

The third geometry is a hyperbolic "open" universe whose negative surface would look something like a horse saddle. In an open universe, parallel lines would diverge and never meet, the sum of the angles for a triangle would be less than 180°, and the circumference of a circle inscribed on its surface would be *greater* than $2\pi r$. Before astronomy became a science, the closed geometry theory was usually accepted, despite the fact that this implied a finite, static, unchanging universe, which is inconsistent with the observable evidence. According to the latest research, the flat universe theory is the most unlikely geometry for the actual universe.

Hawking's black hole theory: *Mini black holes, also known as primordial black holes, were formed soon after the beginning of the big bang.*

Stephen Hawking was not the first to advance the black hole theory. John Archibald Wheeler named this phenomenon, which describes the aftermath resulting when a massive star has "burned up" its nuclear fuel. The dying star cools down to almost absolute zero and shrinks below the critical size, meaning its radiation can no longer overcome its internal gravity, resulting in a "singularity" as a bottomless hole represented as a point in space-time. This concept was first known as the *collapsed star phenomenon*. No nearby light or matter can escape the strong gravitational attraction from this bottomless hole—thus, the name "black hole." Einstein's theory of general relativity asserts that gravity is related to the curvature of space-time, and massive objects, such as giant dead stars, distort space-time. This distortion causes an *event horizon* as the boundary for a black hole. Once an object moves close to the ever-expanding boundary, extreme gravity pulls it into the hole for which there is no escape, not even for light. Most black holes seem to be located near the center of galaxies. The more massive a black hole, the larger its size. It can capture and compress many massive stars and might weigh as much as 10^{36} kilograms (10 followed by 36 zeros). On the other hand, there might be smaller black holes (it should be remembered that this concept does not refer to a hole as in an empty hole in the ground; rather, it is an area of compressed mass where light cannot travel fast enough to escape the overwhelming gravity created by one or more collapsed stars).

Hawking expanded the concept of black holes by considering thermodynamics and quantum gravity. Ordinary space is not really empty. It is filled with pairs of particles. Some are positively charged particles that annihilate partners that have negative charges (in space there are pairs of particles and **antiparticles**). Hawking theorized that for black holes, the partner particles become separated, with the negative particle falling into black holes just as all other matter does and the positive antiparticle escaping and giving off energy that can become a real particle or even approach infinity. These particles and antiparticles can exist because of *vacuum fluctuations*, which are fields similar to light, magnetic, and gravity fields. They are also referred to as *virtual particles*. Hawking also theorized that soon after the big bang, many mini black holes existed. They were no larger than the nuclei of hydrogen atoms (protons) and yet weighed billions of tons and produced tremendous energy. Hawking claims that black holes are not 100% black because quantum mechanics caused them to produce particles and forms of radiation at a regular rate. John Wheeler, a theoretical physicist, stated that what falls into a black hole, including ordinary matter such as buildings, cars, or people, is independent of the radiation and thus will come out the same as it went in. Hawking stated that the emission of particles and radiation from a black hole would reduce its mass to the point where it will have zero mass. At this point, the buildings, cars, and people, including information that fell into the hole cannot come out because there is no longer adequate mass for this to happen.

It might be mentioned that in July 2004 Stephen Hawking revised his three decades' old theory of black holes. His new model indicates that black holes do not just "swallow" up objects of matter and radiation, but also send information out of the hole. It seems there are two kinds of black holes: 1) stellar size (the size of stars) that are the leftover stuff of massive dead stars that have imploded and 2) supermassive black holes that have mass greater that a million times the mass of the sun. These are the types of black holes that are found in the center of galaxies and are considered to be related to the processes involved with the way galaxies evolve. There are some smaller black holes that fall between these two categories. It is now also believed that there may be other types of objects that are just the opposite of black holes that are sending matter and radiation into the entire universe. It appears that much more theoretical research is needed before we can completely understand our universe.

Hawking's theories for inflation and singularity: *The original expansion of the big bang was at first rapid, slowing down as the universe aged. The densities of black holes are singularities.*

Before Einstein's theory of relativity, the general concept was that the universe remained the same: it was static. In 1929 Edwin P. Hubble published his measurements that indicated galaxies were receding from Earth and from each other at an ever-increasing rate. Later, Roger Penrose theorized that a collapsing star is a singularity, as are all black holes. *Singularity* means a single point where the density of matter is infinite, just as the curvature of space-time is infinite. It is a region in space where one or more parts of the space-time curvature becomes infinite. It is also known as a single point of a specific function. The concept of singularity is consistent with and conforms to physical laws. On the other hand, space-time, being infinite with no boundaries, is relative; thus there is no beginning or ending to space-time.

Hawking combined quantum theory with gravitation as expressed by Einstein to propose his concept of *quantum gravity*. He then reversed Penrose's singularity theory to explain how the big bang came to be. In other words, there was a singularity—a point in nonlinear time when a single massive point started the expansion of the universe. This expanding universe is explained by the inflation theory, which states that this three-dimensional "spreading" was extremely rapid, causing great heat, radiation, rapidly expanding gases, and energy that overcame gravity at this point. There are several concepts that explain what occurred after this singularity event created the nascent universe. One states that the early rapid inflation of gases slowed and cooled down, and in time these gases (mostly hydrogen) formed helium and the galaxies, stars, planets, and other objects in space. The leftover radiation and matter, galaxies, and stars are still overcoming gravity and thus expanding but possibly slowing down. The quantum gravity theory states that many billions of years from now, there will be a period of deflation that will be caused by gravity taking over again. This scenario is called the *big crunch*, which will end in a new singularity. It somewhat represents the geometry for a closed universe, where gravity finally wins and a collapse occurs. Another theory states that inflation will continue at an ever-increasing rate of expansion, resulting in a lower density of matter in the universe, or new matter will be created continually to maintain the current density of the universe. This scenario represents the geometry for the open universe. The geometry for a flat universe is somewhere between these two extremes, but it best fits the *standard inflationary theory*, which suggests there is much more matter in the universe than we can see or even realize. This so-called dark matter may not be adequate to justify a flat universe theory, but research and speculation continue to find other types of matter or energy in the universe. During the first decade of the twenty-first century, several high-resolution instruments will be placed in orbit to make measurements that will address the many questions and theories related to the origin, nature, and fate of the cosmos.

See also Einstein; Feynman; Hubble; Penrose; Wheeler

HAWORTH'S FORMULA: Chemistry: *Sir Walter Norman Haworth* (1882–1950), England. Sir Norman Haworth and Paul Karrer shared the 1937 Nobel Prize for Chemistry.

> *Carbon and oxygen atoms are linked in specific ratios to form carbohydrate rings of sugar molecules.*

Organic chemistry had succeeded in synthesizing a number of different forms of sugar molecules, mostly linear open-chain structures. Sir Norman Haworth was the first to demonstrate his *prospective formula*, which was his hypothesis leading to closed chains of sugar molecules. He proposed two types of closed rings for molecular sugars. One was the sugar molecule that formed a ring composed of five carbon atoms and just one oxygen atom (glucose), whereas the other consists of the molecular structure for a sugar ring with only four carbon atoms and one oxygen atom (fructose) (*see* Figure H5). By demonstrating that a chain of carbon-oxygen atoms for several types of sugars could be synthesized into rings, Haworth synthesized a variety of polysaccharides (a sugar composed of many monosaccharides) that are produced today.

In addition to his success with synthesizing what is known as the Haworth formula for carbohydrate molecules, Sir Norman Haworth and Tadeus Reichstein independently succeeded in analyzing and synthesizing a variety of vitamins. Their first success identified the substance found in orange juice, which was first called "hexuranic acid." Later, it was identified as vitamin C and named "citric acid" by Haworth. Their discovery enabled the production of synthetic vitamins with the same chemical makeup as natural vitamins. Their work led to a decrease in malnutrition and diseases caused by vitamin deficiencies.

See also Krebs

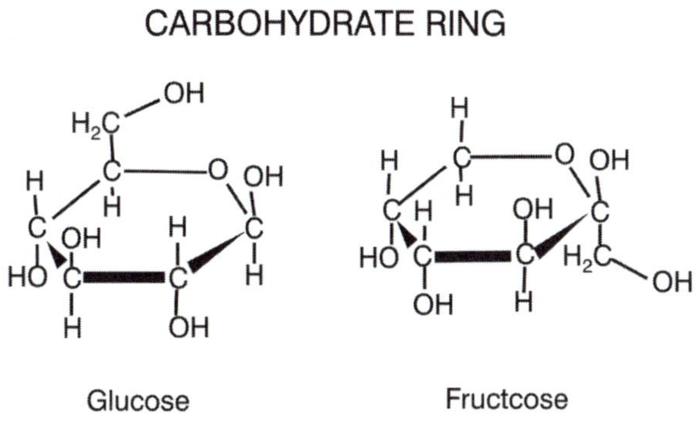

CARBOHYDRATE RING

Glucose

Fructcose

Figure H5. Glucose has 5 carbon atoms in its closed ring sugar molecule, while fructose has just 4 carbon atoms in its closed ring sugar molecule.

HEISENBERG'S UNCERTAINTY PRINCIPLE AND THEORY OF NUCLEONS: Physics: *Werner Karl Heisenberg* (1901–1976), Germany. Werner Heisenberg received the Nobel Prize in Physics in 1932.

Heisenberg's uncertainty principle: *The simultaneous measurement of the position of an electron affects the measurement of its momentum, leading to the uncertainty of either its observed position or movement.*

Werner Heisenberg used matrix mathematics to develop a formulation of quantum mechanics to explain the uncertainty principle as a statistical means to measure the probability of the position and momentum of submicroscopic particles. This principle went a long way toward the development of a formulation of quantum mechanics that was equivalent to the Schrödinger formulation but differed in form (this is analogous to saying the same thing in English and in French). The Heisenberg formulation perhaps emphasizes the uncertainty principle more strongly than does the Schrödinger formulation. Nature at the very small scale has a wavy conformation that can be described only by quantum mechanics. Natural physical laws hold true for everyday objects in the universe, including us. However, in the submicroscopic universe, these same natural laws do not always apply. It is possible for observers to measure accurately the position of a minute, submicroscopic, subatomic particle independently, and it is possible to measure a particle's *momentum* (a vector quality for a particle's velocity × mass) accurately. But both cannot be accurately measured at the same time, thus creating uncertainty. Rather, the small size of their joint uncertainties must be larger than the number referred to as Planck's constant ($\hbar$). Heisenberg insisted the only way to interpret the atom and the subatomic particles that compose them was to observe the radiation (light) they emit. Heisenberg and other physicists developed systems and equations to accomplish this feat. Heisenberg's equation for the uncertainty principle is: $\Delta x \bullet \Delta p \geq \hbar/4\pi$, where Δx is the uncertainty of the particle's position in any direction, Δp is the uncertainty in determining the momentum related to that position, and $\hbar$ is Planck's constant. Statistical probability was used to measure the degree of uncertainty (of a particle's position and momentum) that is related to the development of matrix mechanics that was the precursor to quantum mechanics.

Heisenberg's nucleon theory: *Nucleons are isomers of protons and neutrons.*

Werner Heisenberg also developed the theory that nucleons were composed of neutrons and protons (baryons), which make up the nuclei of atoms. He proposed that both the proton and neutron had approximately the same mass but different spins and thus were forms of **isomers**. Based on his theories of matter, Heisenberg attempted to devise a unified field theory for all elementary

Werner Heisenberg studied under three excellent physicists, Wolfgang Pauli, Max Born, and Niels Bohr. He did his best work in moving fluids. At twenty-five years of age he published an important paper describing the concept of the uncertainty principle for which he is best known. During these early years he published several important papers and at the age thirty-one he received the Nobel Prize in Physics.

During World War II Heisenberg worked with the nuclear physicist Otto Hahn who had shared in the discovery of nuclear fission. Heisenberg became head of Germany's nuclear weapons program. For reasons not completely understood today, the German atom bomb program was not a success. Some say it was more or less sabotaged by the scientists working on the project who may, or may not, have belonged to the secret "Wednesday Society"—the underground resistance group that attempted to assassinate Hitler. Heisenberg's membership in this group has never been proven—one way or the other.

particles that related their characteristics and energies. Neither he, Albert Einstein, nor several other scientists were able to finalize such a grand unification theory (GUT) or a theory of everything (TOE) that would include all submicroscopic particles from the quark up to large molecules, ordinary matter, and all forms of energy, gravity, and the endless universe. (Most theoretical physicists believe that when such a theory is produced, it will be expressed as a simple equation.)

See also Bohr: Einstein; Maxwell

HELMHOLTZ'S THEORIES AND CONCEPTS: Physics: *Hermann Ludwig Ferdinand von Helmholtz* (1821–1894), Germany.

Helmholtz's theory for heat and work: *Heat and work are equivalent, and one can be converted to the other with no loss of energy.*

Hermann Helmholtz was familiar with James Joule's concept of work and heat, and Lord Kelvin's work with thermodynamics. Lord Kelvin later proposed a concept for the conservation of energy. However, Helmholtz was not the first to relate mechanical work to heat. That honor goes to the German physician and physicist Julius Mayer (1814–1878). It might be helpful to define some terms as used by scientists. *Heat* is the result of molecules bumping into each other in random motion. The greater their total motion, the greater the amount of heat. The mechanical equivalent (molecular motion) of heat is measured in calories: 1 calorie = 4.185 joules, while 1 joule = 10,000,000 ergs (the joule is a unit of energy or work). Heat results from mechanical *work*. Try rubbing the palms of your hands together very fast. Note the heat generated from the mechanical motion (work) of your hands. Neither the heat produced by this rubbing friction nor the heat generated by muscle energy that is used to produce this work is lost. Rather, the heat soon disperses into the air around your hands, slightly increasing the motion of air molecules (warming up the air), and is thus conserved. Heat is a form of energy transferred from a body with a higher temperature to a body with a lower temperature—never in the opposite direction (*entropy*). Work is accomplished when a force on a body transfers energy to that body. Work (W) = applying a force (F) over a distance (d) ($W = F \times d$). Work, in the scientific sense, is not exactly the same as in the common sense, which usually refers to some physical or mental effort—although both involve energy and heat. *Energy* is the capacity to do work. Heat and work are forms of energy and are interchangeable. There are many forms of energy, such as heat, light, sound, radioactive, mechanical, work, and so forth; and they are all conserved, but

Helmholtz also explained his theory on a physical basis by speculating that the total energy of a large group of interacting particles is constant, basing his ideas on the perceptions of Robert Hooke. Using a microscope, Hooke observed the action and interaction of minute particles in a fluid. He related the motion of the particles to heat resulting from what we now know as the molecular action within a body, even before molecular motion was known. Nonetheless, it provided Helmholtz with the idea that work and heat are related. As the law of conservation of energy was refined, it stated that the total amount of energy in the universe is constant. Energy can neither be created nor destroyed, but it can be changed from one form to another. For example, mechanical, magnetic, electrical, chemical, light, sound, kinetic energy, and others are all interconvertible. Also known as the first law of thermodynamics, it is one of the most basic universal physical laws of nature. As did many other scientists, Helmholtz tried unsuccessfully to develop a unified field theory based on his concepts of work/heat/energy, electrodynamics, and thermodynamics.

not in the sense that environmentalists use the word "conservation," usually meaning "saved" and not to be used. In scientific terms, *conservation of energy* or *mass* means that it can be changed from one form to another without any total loss of the original energy/mass ($E = mc^2$).

As a physician, Helmholtz believed animal heat generated by muscle contractions was related to his theory that heat and work are convertible from one to the other in a *quantitative relationship*, with no loss of total energy. This led to the law of conservation of energy, one of the fundamental laws of physics. Imagine yourself raising a baseball from the top of a table to the height of your head. In doing so, you have converted mechanical work to potential energy in a gravitational field. However, you have done the mechanical work at the expense of the energy stored in your body. One might say you did 1 joule of *work*, which is equal to about 0.25 calorie of *heat* because work and heat are forms of energy and interchangeable. You also increased the potential energy of the baseball, which would be released as kinetic energy if the baseball were dropped, resulting in a conservation of energy (none was lost or gained).

See also Carnot; Clausius; Hooke; Joule; Kelvin; Maxwell

HELMONT'S THEORY OF MATTER AND GROWTH: Chemistry: *Jan Baptista van Helmont* (1579–1644), France.

Plants grow, increasing their weight, but without reducing the weight of the soil in which they grow.

Helmont was one of the early scientists who carefully observed and measured well-planned experiments, the most famous of which was the observation of the growth of a specific plant over a five-year period. He noted the specific weight of soil in which it was planted, the plant's growth weight, and the weight of the amount of water it used. At the end of five years, Helmont weighed the plant as well as the soil in which it grew and noted the plant weighed about 165 pounds, but the soil lost less than one-fourth pound in weight. However, he incorrectly attributed all the growth weight to water, which led to a mistaken concept of matter. In another experiment, he burned about sixty pounds of wood, and then weighed the remaining ashes, which equaled less than one pound. He concluded that the lost weight was composed of water and four "gases,"

> Helmont was educated at a Catholic university that specialized in medicine, mysticism, and chemistry, but he refused to accept a degree from the university because the church's teaching conflicted with his beliefs. Helmont established a private research lab in his home and conducted some early experiments on matter and energy. Although many of his conclusions were off the mark, his writings led to more research by others. For instance, the Roman Catholic Church taught, in medicine, that it was possible to treat and heal a wound caused by a weapon by treating the weapon rather than the person's wound. Contrary to church dogma, Helmont stated that this was not a supernatural phenomenon but rather it was a natural phenomenon. He was placed under house arrest for this and other heretical ideas and was prevented from publishing his papers. He also had some odd ideas, such as a belief in the alchemists' philosopher's stone, and that all matter was composed of water (fish were nourished by water). He was one of the first "scientist/philosophers" to record careful observations of nature before drawing conclusions. After his death, his son published most of his works in a book called *Ortus medicinae* in 1648.

which he named *pingue* (methane), *carbonum* (carbon dioxide), and two types of *sylvester* gas (nitrous oxide and carbon monoxide). He was correct that carbon dioxide resulted from combustion as well as the chemical process of fermentation. He also correctly identified the other gases (but not by their current names). Helmont also believed there were several different types of air, just as there were different forms of solids and liquids. He related these "airs" to chaos because they had no specific volume as did liquids and solids, and he coined the term "gas," which soon was accepted along with the common terms "solid" and "liquid," providing a general concept for the three forms or states of matter.

HENRY'S PRINCIPLES OF ELECTROMAGNETISM: Physics: *Joseph Henry* (1797–1878), United States.

Electromagnetic induction: *By moving a conductor through a magnetic field, an electric field is induced in the conductor* (*see* Figure F1 under Faraday).

Solenoid-type magnets were developed and improved in the early part of the nineteenth century. These devices consisted of coils of insulated wire, which produced a magnetic field within the coil when electricity passed through the wire. The English physicist and inventor William Sturgeon (1783–1850) made further improvements when he placed an iron rod inside a coil and observed that, while electricity flowed through the loops of wire forming the coil, the rod became magnetized. However, when the current was turned off, the rod no longer retained its magnetism. Thus, by turning the current on and off, the rod became an on-and-off magnet with the ability to control various mechanical devices. His source of electricity was generated by a battery of simple voltaic wet cells that produce direct current electricity of low voltage (about 1.5 volts for each cell). However, this amount of electricity was adequate to produce electromagnetism, but more current is produced if several cells are connected in series ($8 \times 1.5 = 12$ volts).

The number of turns of wire loops used to form a coil was limited because if any of the bare wires came in contact with each other, they would short out, breaking the electrical circuit. Henry solved this problem by using insulated wires, which could be wrapped close together and overlapped without creating a short circuit, thus producing much stronger electromagnets. He lifted as much as a ton of iron by using a small battery of voltaic cells to supply the electric current. Michael Faraday and Joseph Henry independently proposed the idea that magnetic properties from this coil could be induced to an iron bar. Joseph Henry, who is said to have developed this theory, did not receive credit at first for the principle of magnetic induction because Michael Faraday's findings were published a few months before Henry's. Joseph Henry is immortalized inasmuch as the universal SI constant for inductance is named after him. The Henry SI unit (H) for induction occurs when an electromotive force of one volt is produced when the electric current in the circuit varies uniformly at a rate of one ampere per second.

Henry's concept of an electric motor: *A moving wire that cuts across a magnetic field induces an electric current. Therefore, the process should be reversible.*

Faraday's device used a copper wheel to cut through a magnetic field and thus induced a current to flow through the wire. This later became the concept for the dynamo or electric generator. Joseph Henry questioned what would happen if this process were reversed. He devised a machine with a wheel that would turn when an electric current was sent through the electromagnet. The result was the first electric motor, which over many years of development has become one of the major technological developments of all times. It might be said that Faraday, Henry, and other scientists who worked with electricity and electromagnetism created the laborsaving conveniences of the modern world, that is, using motors that were powered by electricity produced by dynamos (generators).

Henry's concept of "boosting" electricity over long wires: *If a method for maintaining the strength of an electric current over long wires can be found, an electromagnetic communication device could be developed.*

In addition to Joseph Henry, several others imagined an electromagnet solenoid that could open and close a "clicker" circuit using a "key" to send signals through wires. The problem was the natural resistance to the flow of electricity in copper wires. Henry solved the problem by developing the electric relay (similar to a solenoid) that uses a small coil as an electromagnetic switch. This turns large amounts of electric current on and off, thus making unnecessary the circuit's continuous connection. This relay enabled electricity to be sent over long stretches of wires at intervals, and thus the success of the national system of telegraph communication. The problems related to the early electromagnetic telegraph were money and support for implementation, not the scientific problems. Joseph Henry was so admired in the United States that in 1846 he was named the first secretary of the new Smithsonian Institution located in Washington, D.C.

See also Ampère; Faraday; Tesla

HERSCHEL'S STELLAR THEORIES AND DISCOVERIES: Astronomy: *William Herschel* 1738–1822) and *Caroline Lucretia Herschel* (1750–1848). England.

Herschel's theory of the sun's movement: *The sun (thus the entire solar system) is moving in the direction of the constellation of Hercules.*

Although he was not able to calculate the sun's speed of movement, William Herschel based his theory on observations to determine the direction of movement.

Herschel's theory for the structure of galaxies: *Based on his observations while using a telescope of his own design, William Herschel was the first to explain the structure of galaxies. He theorized that the Milky Way has more stars in its center and fewer toward its celestial poles. He also proposed that the Milky Way and many other galaxies are shaped like a flat, circular "grindstone."*

William Herschel is credited with the discovery of infrared radiation by passing a beam of sunlight through a prism and then placing a thermometer at the end of the spectrum just beyond the visible deep red end of the light spectrum.

Caroline Herschel, the sister of William Herschel, made a number of contributions to the field of astronomy in her own right. Both were born in Germany, but in 1757 during the Seven Years War, William moved to England where he made a living as a music teacher and performer. In 1772 Caroline joined her brother in Bath, England, where he became an organist at the Octagon Chapel and she became a successful soprano singer. During this time William became interested in astronomy and began making telescopes to observe the heavens. Caroline joined him in his observation. While using a large reflector telescope, they soon discovered a celestial phenomenon in 1781 that William believed was a comet. In the meantime, their work impressed King George III of England who gave William a pension in 1782 to serve as his astronomer and commissioned him to build a large reflector telescope. Caroline then abandoned her singing career to become William's full-time assistant. Other astronomers soon identified what William and Caroline thought was a comet as a new planet. William wanted to name it "*Georgium Sidus*" meaning George's Star, after his patron George III, but another astronomer suggested that the name "Uranus" be given the new planet.

Much of their work extended and improved upon the observations of other astronomers using this improved telescope. Caroline recorded William's observations as they extended the one hundred nebulae listed in the catalog compiled by the French astronomer Charles Messier (1730–1817). The list soon expanded to over two thousand nebulae that the Herschels discovered. As William became involved in other activities, Caroline used their telescopes to do more and more observations on her own. In 1783 she discovered her first new object, a cluster, now known as NGC2360. Later that year she discovered her second object, a galaxy known as NGC205. Other astronomers as well as King George III who recognized Caroline's discoveries paid her as the assistant to her brother for her contributions to astronomy. In addition to discovering many galaxies, Caroline had discovered eight comets by 1797. She became a well-known observational astronomer in her own right. Before her death she prepared a catalog of twenty-five hundred nebulae. Although it was never published, she did receive the Gold Medal from the Royal Astronomical Society in recognition of her contributions to the field.

In 1887 William discovered two satellites of Uranus named Titania and Oberon. Soon after, he found two satellites of Saturn (Mimas and Enceladus). He designed and built a new type of telescope that was forty feet long with a twelve-meter mirror. It eliminated the need for a secondary mirror that was installed in Newtonian-style scopes. This improved the viewing resolution but required agile observers, as they had to climb up the forty-foot scope.

See also Bode

An interesting mnemonic device for remembering the nine planets is "My Very Educated Mother Just Served Us Nine Pizzas." A problem with this sequence is that in August 2006 at a meeting of the International Astronomical Union in Prague, Czechoslovakia, and after years of debate, astronomers who had remained for the last day of the conference voted to "downsize" Pluto to a minor planet that no longer fits their classification as a true planet.

The controversy: Only 424 astronomers participated in the vote, representing less than 5% of the world's astronomers. As a result, there will be three main categories of objects in the solar system, as follows: 1) *planets,* the remaining eight (Mercury, Venus, Earth, Mars, Jupiter, Saturn, Uranus, and Neptune); 2) *dwarf planets,* Pluto and any other round object that "has not cleared the neighborhood around it, and is not a satellite"; and 3) *small solar system bodies,* that orbit the sun.

Many astronomers expect the decision to be overturned sometime in the future, and to that end, are circulating a petition in support of a reversal.

HERTZSPRUNG'S THEORY OF STAR LUMINOSITY: Astronomy: *Ejnar Hertzsprung* (1873–1967), Denmark.

The luminosity of stars decreases as their color changes from white to yellow to red due to a decrease in the star's temperature.

Previous to Ejnar Hertzsprung's theory, a star's distance from Earth was determined by the Doppler effect, the result of sound or light waves lengthening (lower

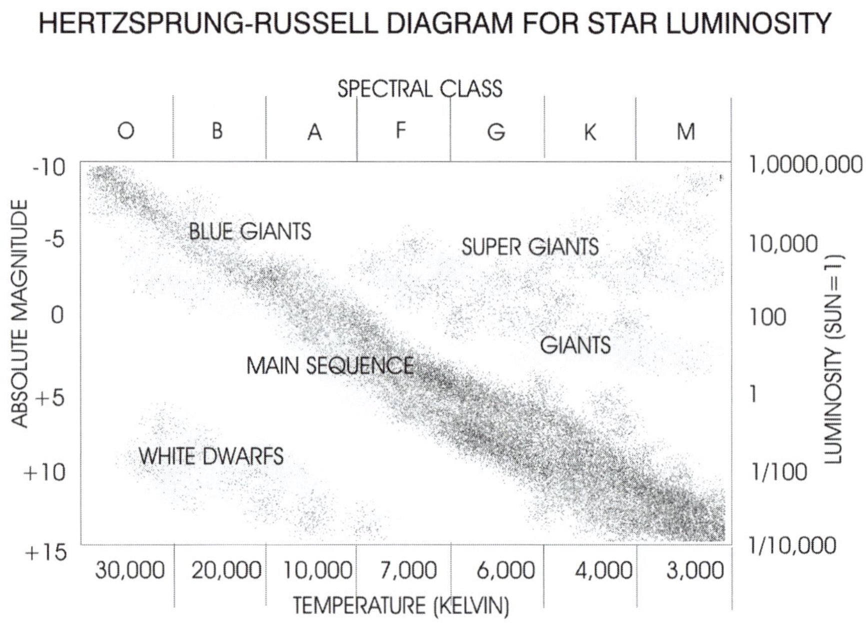

Figure H6. The Hertzsprung–Russell diagram depicting the color and magnitude of stars. It indicates how stars vary according to size, temperature, and brightness. Our sun is an average size star located about in the middle of the main sequence.

frequencies) as the object producing the sound or light waves moves away from the observer. Vice versa, the waves shorten and become higher in frequency as the object producing the waves approaches and moves closer to an observer. Hertzsprung improved this technique for determining the distance of stars by using photographic spectroscopy to measure the inherent brightness or luminosity of stars. He determined there were two main classes of stars: the supergiant stars that were very bright and the much more common and fainter stars referred to as the main *sequence type stars*. Henry Russell, who charted the stars according to Hertzsprung's classification, confirmed Hertzsprung's theory in 1913, and both are credited with developing the star diagram. The Hertzsprung–Russell diagram (referred to as the H-R diagram) arranged stars on the x-axis according to a classification using temperature and color, and also on the y-axis according to their magnitudes and luminosities (*see* Figure H6).

The diagram reveals several clusters or groupings of stars:

1. The most massive types of stars are the *supergiants*—the largest and brightest stars that show a large spread in intrinsic brightness. Although a few supergiants are larger than the orbits of Jupiter, most of them are smaller. All supergiants are comparatively rare and short-lived, but even at their great distance from Earth, many are redder than the main sequence of stars, but larger. They can be viewed easily because of their great brightness. Many supergiants are binary star systems (two stars circling each other as they are caught in each other's gravitational systems). Binary stars are somewhat easier to locate due to their brightness.

2. The *giants* are not quite as large and bright as the supergiants. They are yellow to orange to red and about one hundred times brighter than the sun with radii ten to twenty times that of the sun. Several examples of giants are Capella, Arcturus, and Aldebaran.
3. The *main sequence of stars*, ranging from very large dim stars to small, bright stars. The sun is near the middle of the main sequence.
4. The comparatively small *white dwarfs* are about 10 magnitudes less than the main sequence of stars in the diagram and are in their last stages of evolution.

The large bright and cooler stars are found in the upper right of the diagram, while the smaller white dwarfs, which are somewhat dimmer and hotter, are located in the lower left of the x-y coordinates of the diagram. The sun is found among a group of small stars known as red dwarfs, which are located near the middle of the main sequence. It is somewhat more brilliant than its neighbors.

Hertzsprung's work and the H-R diagram led to a better method for determining the distance of stars and galaxies from Earth, as well as aiding in the study of the evolution of stars. The H-R diagrams are also called "color-magnitude" diagrams because the spectra used to place stars into categories is dependent on a star's color which is determined by its temperature.

See also Hewish; Russell

HERTZ'S THEORY FOR ELECTROMAGNETIC WAVES: Physics: *Heinrich Rudolf Hertz* (157–1894), Germany.

Electromagnetic waves, produced by electric sparks, behave the same as light waves.

Heinrich Hertz expanded the electromagnetic theories of James Clerk Maxwell and Hermann Helmholtz by erecting an apparatus consisting of a metal rod with a small gap cut in the center to detect different wavelengths of the electromagnetic spectrum. When electricity was sent through the rod, a rapidly vibrating spark was produced at the gap, generating high frequencies in the rod. Not only was Hertz able to detect these high frequencies (electromagnetic waves), but he also determined that these electromagnetic waves would reflect and refract off surfaces just the same as light waves. More important, he discovered that electricity is transmitted as electromagnetic waves that travel at the same speed as light. He also showed the wavelengths of electricity were quite a bit longer than gamma radiation or even light rays. They were named *Hertzian waves*. Today we know this range of electromagnetic waves with very long wavelengths as radio waves. They have wavelengths not only much longer than light waves, but longer than most other waves of the electromagnetic spectrum. Hertz's work not only confirmed Maxwell's electromagnetic spectrum theory, but also paved the way for radio communications (*see* Figure M4 under Maxwell). Hertz experimental work more-or-less established the field of electrodynamics. He also published two papers that analyzed and confirmed Maxwell's theory in 1890. His theory and experimentation (based on the theories and experiments of Nikola Tesla) led to the development of wireless communication proposed by Tesla. Their work led to wireless telegraph, radio, television, cell phones, etc. The SI metric unit for expressing frequency is named for Heinrich Hertz. It is known as the Hz.

The terms "wavelength" and "frequency" are often confused—the *velocity* of a wave, its *frequency*, and its *length* are all interrelated components of waves. Waves, some very short, others very long, can be formed in all kinds of substances in any of the three states of matter. Waves occur in water and other liquids, gases, earthquakes, and electromagnetism (light, radio, X-ray, etc.). The frequency of a particular wave is given in terms of the number of wave crests that pass a given point in a second, whereas the wavelength is the distance between the successive crests of each wave passing a point. This wavelength is directly proportional to the wave's velocity but inversely proportional to its frequency (i.e., how many crests of the wave pass a given point over time). This can be expressed in the following equation: $V = \lambda f$, where V is the velocity of the wave, f is its frequency, and λ (lambda) is its wavelength. If two of these facts concerning a wave are known, the unknown fact can be calculated, that is, to determine the wavelength, use the variation of the equation: $\lambda = V/f$ (by dividing the velocity of the wave by its frequency will give you the unknown wavelength).

It is reported that when experimenting with various frequencies Hertz also accidentally detected the photoelectric effect. Others already knew this effect, but Hertz used ultraviolet light (wavelengths just shorter than visible light) to "knock" out electrons (photons) from the surface of particular types of metal plates.

See also Curie (Pierre); Helmholtz; Maxwell; Tesla

HESS' SEA-FLOOR SPREADING HYPOTHESIS: Geology: *Harry Hammond Hess* (1906–1969), United States.

The sea floor "split" near the middle of the Atlantic Ocean provides an opening for deeper magma to protrude, thus renewing and spreading the ocean floor under landmasses, which eventually separated the continents.

At one time Earth's entire land mass was connected. Over millions of years, this mass separated to become distinct continents. How this occurred has perplexed geologists because it did not seem possible that landmasses could pass over layers of solid rock. Hess's hypothesis provides a possible answer. Hess based his hypothesis on evidence that fossils found in the "newer" ocean beds are much younger than those found on continental landmasses. As the ocean floor spread out, it approached the landmasses and dipped under the continents, while at the same time forming a ridge at the origin of this expansion (*see* Figure H7).

This movement of ocean floors resulted in the development of rifts or breaks near the center of the Atlantic and Pacific Ocean beds as the seafloor was "pulled" apart along a rather narrow crack down a central ridge in the ocean floor over seventy-five thousand miles long. This produced weak spots where the magma (molten rock) protrudes through what is known as the midocean ridge, which is a deep canyon running down the middle of the North and South Atlantic Oceans, across to the Pacific and Indian Oceans to form the center of what is known as the Great Global Rift. As volcanic material rose up from Earth's mantle and out of the great fissure, it constantly fills up the crack, creates new ocean floors, and spreads out toward the continents. This spreading of the sea floor eventually was forced underneath land masses and is responsible for the North and South American continents moving farther apart and westward, while moving Europe and Asia eastward. The continents did not actually drift or float but were rather fixed to plates that were forced apart and, in some places, forced together.

Hess' research and hypothesis led to the science of plate tectonics, from the Greek work *tektonikos*, meaning "builder." Plate tectonics, an important geological theory, is

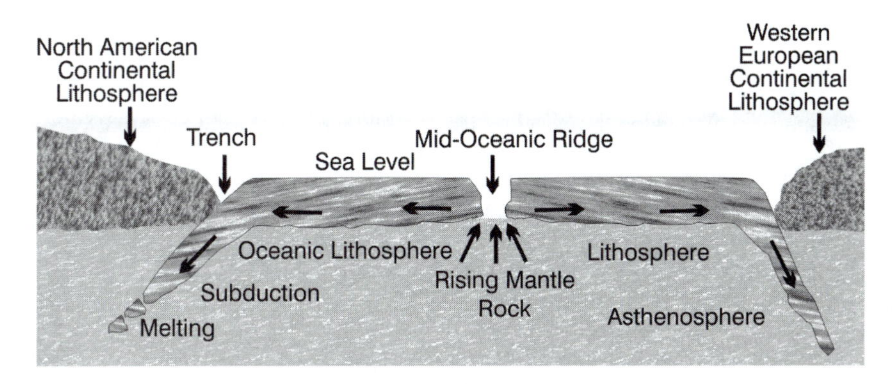

Figure H7. The mid-oceanic rift is a deep canyon running down the ocean sea floor as it spreads both east and west

as important to geologists as evolution is to biologists. In essence, it states that worldwide, there are six large plates and some smaller ones that were moved by the spreading of the oceans' floors. These ocean floors and plates move apart about two inches each year as new magma rises into the midocean rifts. Earthquakes and volcanoes are in evidence where some of the major plates clash, such as on the West Coast of the United States and South America and in the plates bordering Japan and much of the Pacific Ocean. This movement produced, and still produces, mountains and island chains. Hess' hypothesis was later confirmed by studies of the magnetic properties of the new magma material compared to the older magnetic properties of the ocean floor.

See also Suess; Wegener

HESS' THEORY FOR THE IONIZATION OF GASES: Physics: *Victor Francis Hess* (1883–1964), United States. Victor Hess shared the 1936 Nobel Prize for Physics with Carl Anderson.

As altitude increases, so does the level of ionizing radiation that affects gases.

Victor Hess noted that, over time, **electroscopes** that had received a charge when exposed to air lost their charge. He theorized that the upper atmosphere emitted some kind of radiation. First, in 1910 he timed the discharge of the leaves in his electroscope at different levels as he ascended to the top of the Eiffel Tower in France that is about three hundred meters high. Not much difference was noted. Next he tried his experiment in a balloon. At low altitudes, the rate of discharge was much the same. But as he increased the altitude from several thousand feet to over sixteen thousand feet, a much more rapid discharge of the electroscope became evident. He made ten balloon trips, five at night, over the period from 1911 to 1913. His original theory proved

Victor Hess was born in Austria where his father was a forester. He attended the local gymnasium and later in 1901 entered the University of Graz where in 1906 he received his PhD in physics. He held several teaching positions before and after his famous "cosmic" ray experiments for which he shared the 1936 Nobel Prize with Carl D. Anderson who discovered the positron. He moved to United States when he was offered a professorship at Fordham University in New York City. He became a citizen in 1944. After the atom (nuclear) bomb was dropped, he studied atmospheric radioactive fallout from the top floors of the Empire State Building in New York. Later he studied radiation levels emitted by granite rock deep in the subway station at 190th Street. In 1948 Hess and another investigator developed a method of detecting and measuring minute amounts of radium in human bodies. Following World War II this proved to be an important means of detecting radiation poisoning before it became life threatening. He could also measure any radiation in the breath of people who worked around radioactive materials. He determined that different people have different levels of tolerance to radiation. However, the effects of exposure to radiation by the human body are cumulative and may take many years before radiation sickness becomes evident. Victor Hess opposed the use of all nuclear weapons and their testing, even underground.

correct. There was some form of radiation coming from outside Earth's atmosphere that was absorbed by the more dense gases at low altitudes but was not impeded at higher altitudes. He concluded that very strong radiation entered Earth's atmosphere far from Earth. At first he believed the ionizing radiation came from the sun. During a total eclipse of the sun, he noticed that the rate of discharge at great heights was always the same as when there was no eclipse, and that the level of radiation was the same during the daylight hours as it was at night. Hess then theorized that the radiation came from outer space. The scientific community remained unconvinced. At the time, most people continued to believe this form of radiation came from either the sun or some unexplained phenomenon on Earth. His work led to the discovery of cosmic rays by Robert Millikan a few years later.

See also Millikan

HEWISH'S THEORY OF PULSARS: Astronomy: *Antony Hewish* (1924–), England. Antony Hewish shared the 1974 Nobel Prize for Physics with Martin Ryle.

Pulsars are rotating neutron stars that send out electromagnetic signals with great precision that can be detected on Earth.

Antony Hewish was born in Cornwall, England, and entered Cambridge University in 1942. However, in 1944 he was soon involved with war-related research with several government agencies. When he returned to Cambridge in 1946, Hewish joined the research team at the Cavendish Laboratory headed by the British radio-astronomer Sir Martin Ryle. He continued teaching along with his research on radio astronomy. Jocelyn Bell Burnell, his graduate research assistant, was assigned the job of using the main telescope and then analyzing the results on charts produced by the telescope. She noticed variability in the scintillation of one of the pulsating radio signals from a small star in the Milky Way galaxy. She correctly figured that the rate of pulsation was much too fast to be produced by a regular quasar. At that time, Bell Burnell had adequate training to realize that this was evidence of a new type of radio star that was later referred to as a pulsar. When Hewish and Ryle received the 1974 Nobel Prize, Jocelyn Bell Burnell was excluded. Both men were extremely upset about this exclusion, as were many other astronomers, because it was she who had actually discovered the data that led to the identification of pulsars (Jocelyn Bell Burnell went on to a distinguished career in astrophysics before retiring from the University of Bath, England, in 2004).

The word "pulsar" is the abbreviation for "pulsating radio star." Hewish recognized the importance of Bell Burnell's discovery and named it PSR 1919+21. Later, it was discovered that pulsars act as "lighthouses" because they not only rapidly emit radio pulses, but also pulses of light and X-rays. There are three classes of pulsars: 1) pulsars whose rotations are powered by the internal energy derived from that rotation; 2) X-ray pulsars that emit X-rays that can be detected on Earth that rotate from potential gravitational energy; and 3) magnetic pulsars whose energy of rotation is derived by the decay of strong magnetic forces as internal heat dissipates toward their poles. Although pulsars are very small in comparison to other bodies in the universe (only about ten to fifteen miles in diameter), their mass is several times the mass of our sun. Their speed and period of rotation also varies.

Several astronomers, including the late American astronomer Thomas Gold, hypothesized that pulsars are rapidly rotating neutron stars. Due to the mysterious nature of pulsars, some astronomers considered that they might be a form of communication from extraterrestrials (ET). One wag even named their pulsar LGM-1 for "little green men." At one time several astronomers in England and the United States seriously considered this possibility based on the fact that the signals were so regular. Today, even though much more still needs to be uncovered about pulsars, astronomers no longer consider the possibility of ET signals a viable explanation.

See also Gold; Hertzsprung

HIGGINS' LAW OF DEFINITE COMPOSITION: Chemistry: *William Higgins* (1763–1825), Ireland.

> *When atoms of specific elements combine to form molecules of a compound, they do so in a ratio of small, whole numbers.*

William Higgins claimed to be the first to conceive that simple and multiple compounds composed of the same elements are combined in a ratio of small, whole numbers. Despite having no experimental evidence for his proposition, he based his law of definite or constant composition on the concept that for any molecular compound that contains atoms of the same element, these atoms combine in the same fixed and constant proportions (ratio) by weight. Higgins' theory was speculative. Shortly after, John Dalton expanded Higgins' atomic theory by adding the statement that atoms are neither created nor destroyed in chemical reactions and that the molecules of compounds are composed of one or more atoms of the elements forming the compound. Further, these elements always combine to form compounds (molecules) in ratios of small, whole numbers of the atoms of the constituent elements. Higgins' law is best understood when considering molecules of oxygen: O_2 (oxygen) and O_3 (ozone). Dalton's contribution to the law is better understood when comparing the five oxides of nitrogen—for example, for NO, NO_2, N_2O, N_2O_3, and N_2O_5, each compound molecule has a different ratio of small, whole numbers by weight combining the two elements, nitrogen and oxygen. Higgins' law of definite composition relates to the ratio of the small, whole numbers of atoms of the same element combining to form a molecule (e.g., O + O + O_2), whereas Dalton's law of multiple proportions relates to the ratio, in small, whole numbers, of the weights (atomic mass) of the atoms of *different* elements combining to form a compound (e.g., 2H + O = H_2O). Today, most scientists consider Higgins and Dalton among the founders of modern chemistry.

See also Atomism; Dalton

HIGGS' FIELD AND BOSON THEORIES: Physics: *Peter Ware Higgs* (1929–), England.

> *At very low temperatures, the symmetry of the electromagnetic force breaks down, producing massive particles (Higgs bosons) from formerly massless particles.*

Peter Higgs endeavored to unify electromagnetic waves and the "weak" force into a single "electroweak theory" (*see* Figure G11 under Glashow). He and other physicists observed that at high temperatures, electromagnetic photons and weak force W and Z bosons were indistinguishable from each other. Both seemed massless. Bosons are elementary particles that obey a particular type of statistical mathematics. Some examples are photons, pi mesons, and particles with whole number "spin," as well as nuclei of atoms composed of an even number of particles. Higgs subjected these particles to extremely low temperatures, at which point their symmetry no longer existed, and therefore photons could be distinguished from the W and Z bosons. A mathematical expression of this phenomenon became known as the *Higgs field*, which attributed the weightless bosons with mass. This formerly weightless boson, which now had theoretical mass, became known as the *Higgs boson*. An elementary particle, the Higgs boson, derived from the combined theory of electromagnetic and weak interactions, has not yet been detected. It is more massive and has a much higher level of electron volts, meaning that one of the newer, high-energy superconducting supercollider particle accelerators will be needed for its detection.

Scientists are still not sure if the Higgs boson really exists with a mass less than the Z particle. However, they hope to discover it and then fit the boson into the scheme for the "electroweak theory." For over ten years they have been using the Large Electron Positron (LEP) collider located in Geneva, Switzerland, that accelerates electrons and positrons to very high energies and guides them into collisions. Since then, the Large Hadron Collider (LHC) at CERN has been placed into position to continue research on the Higgs boson. Also the Fermi National Accelerator Laboratory in Batavia, Illinois, was used to smash particles together at very high speeds that produce energies high enough to convert the particles into smaller bits of matter. Hopefully, one form of these small bits of matter will be a Higgs boson. Because it lasts for only a small fraction of a second and then decays into something else, scientists look for what it might decay into. One of the purposes of the Superconducting Super Collider (SSC) that was being constructed in Texas in 1991 was to discover the Higgs boson. After the large underground circular accelerator was started, Congress canceled the funds for this project in 1993. The abandonment of the SSC devastated the United States' worldwide position and lead in the field of particle physics.

HODGKIN'S THEORY OF ORGANIC MOLECULAR STRUCTURE: Chemistry: *Dorothy Crowfoot Hodgkin* (1910–1994), England. Dorothy Hodgkin was awarded the 1964 Nobel Prize for Chemistry.

> *The structure of complex organic molecules can be determined by use of X-ray analysis.*

The use of X-ray diffraction to study subatomic particles too small to be seen by optical microscopes was developed by Max Von Laue. His concept stated that X-rays were electromagnetic waves similar to light waves but with much shorter wavelengths that could "see" particles invisible to ordinary optical microscopes because the submicroscopic particles were smaller than the length of a light wave. He developed a technique for passing X-rays through a crystal, forming a diffraction pattern of the crystal's structure. Dorothy Hodgkin perfected the technique to produce diffraction patterns that revealed the structures of comparatively large organic molecules. Working with her teacher, John Desmond (J.D.) Bernal (1901–1971), she made the first X-ray photographs of the large protein molecule pepsin. Hodgkin then produced a three-dimensional photograph of penicillin, which aided in understanding its molecular composition. Later Hodgkin used X-ray diffraction to determine the structure of the vitamin B_{12} molecule, which is constructed of over ninety atoms. This enabled the vitamin to be produced in quantities adequate for the virtual elimination of pernicious anemia. She received the 1964 Nobel Prize for Chemistry for her work with B_{12}. Subsequently, with the aid of computers, she successfully determined the structure of the insulin molecule that has over eight hundred atoms.

HOFFMANN'S THEORY OF ORBITAL SYMMETRY: Chemistry: *Roald Hoffmann* (1937–), United States. Roald Hoffmann shared the 1981 Nobel Prize for Chemistry with Kenichi Fukui.

> *Using quantum mechanics, it is possible to predict and explain the symmetry of chemical reactions.*

Roald Hoffmann and Robert Burns Woodward collaborated in the development of an orbital molecular theory, now referred to as the *Woodward–Hoffman rule*. Woodward is known for his work on synthesizing natural substances of very complex structures, whereas Hoffman was more concerned with how chemical bonds were formed and broken during chemical and cyclic reactions. Hoffmann advanced theories related to the electronic (orbital) structure of stable and unstable inorganic and organic molecules. His work on the transition states of organic reactions led to the concept of bonding and symmetry used for the analysis of complex reactions. The Woodward–Hoffmann rule outlined the **stoichiometry** taken during the total summing up of the many steps required to complete complex chemical reactions. This rule enables organic chemists to understand the structure of complex natural substances and synthesize them from simpler chemicals, resulting in chemical synthesizing (artificial production) in the laboratory of many substances found only in nature. A few examples are cholesterol, lysergic acid, reserpine, some antibiotics, and, most important, chlorophyll.

See also Woodward

HOOKE'S LAWS, THEORIES, AND IDEAS: Physics: *Robert Hooke* (1635–1703), England.

Hooke's law of elasticity: *The change in size of a material under strain is directly proportional to the amount of stress producing the strain.*

Robert Hooke was the first to apply mathematics to the concept of elasticity and relate this concept to the actions of springs. In essence, his law states that the distance over which a spring is stretched varies directly with its tension, as long as the spring does not exceed its limits of elasticity—that is, the strain is proportional to the stress. In other words, the more a spring is stretched, the greater becomes its internal tension. Once this was understood, springs could be designed for vehicles to provide smoother ground transportation. Hooke's law of elasticity applies to all kinds of situations and materials, from bouncing balls to the use of an elastic rod or fiber as a torsion balance. Hooke was the first to relate the simple harmonic motion concept for pendulums to the vibrations of micro "hair" springs used in the balance wheels of smaller watches. Hooke also improved the escapement movement in clocks first conceived by the ancient Chinese, by devising a method of cutting small but accurate cogs and gears. This special cog, called the *grasshopper escapement*, enabled pendulum-driving grandfather-style clocks and the pocket watch to be constructed as very accurate timepieces for their time in history (*See also* Huygens).

Hooke's cell theory: *All plants are composed of cells surrounded by a defined cell wall.*

Hooke's concept of cells resulted from the construction and use of his practical compound microscope. He observed and drew elaborate diagrams of objects such as feathers, insects, and fossils. His most famous microscopic observation focused on the walls of individual cells in a thin slice of cork, which is nonliving plant tissue. Hooke based his concept of and name for "cell" on the tiny monks' rooms in monasteries, called cells. Other biologists then examined living plant cells, all of which had well-defined cell walls, whereas the walls of animal cells were shaped like irregular membranes. Hooke's cell concept led to the development of the modern cell theory, which states that all living plants are composed of cells derived from other cells, as well as the extension of the plant cell theory to animal cells (*See also* Schleiden; Schwann).

Hooke's theory of sound and light waves: *Sound is transmitted by simple harmonic motion of elastic air particles, thus, light must also be transmitted by a similar wave motion.*

Robert Hooke applied his law of elasticity to air particles that are compressed (squeezed tighter together) and rarefied (spread farther apart) as these particles proceed from the source of the sound to a person's ears (the Doppler effect). He concluded that light, both colored and white, was transmitted in a similar wavelike motion through air. (Actually, light waves are not exactly similar to compression-rarefaction-type sound waves. Light has the dual properties of both particles and electromagnetic radiation or waves.) Hooke also extended his law of elasticity to form his own theory of gravity, which he based on mathematics related to the harmonics of planetary motion (a backward approach to the law of gravity as proposed by Sir Isaac Newton). In addition, Hooke claimed that he, not Newton, first conceived the concept that gravity obeyed the universal concept of the inverse square law. A disagreement developed in which Newton disputed Hooke's theories and claims.

See also Newton; Stark

HOYLE'S THEORIES OF THE UNIVERSE: Astronomy: *Sir Fred Hoyle* (1915–2001), England.

Hoyle's steady-state theory: *The universe did not originate from the big bang, but rather exists in a steady state.*

Sir Fred Hoyle disagreed with the big bang theory proposed by George Gamow and other cosmologists. The big bang theory, a term that Hoyle originated, some say disparagingly, states that the universe started as an incredibly dense point or tiny ball that contained all the matter and energy now existing in the universe. Steven Weinberg of the University of Texas and others assert that the beginning occurred as a singularity event where, in the first three minutes of creation, a small but rapidly expanding universe was so hot and dense that only subatomic elementary particles and energy existed. This was followed by the production of hydrogen and later helium as stars and galaxies. Other matter rapidly evolved and expanded and continues to expand today. As the stars evolved, they produced the heavier elements now found on the planets. Hoyle, a respected astronomer, rejected this theory of an expanding universe and proposed a continual creation of atoms and other matter to the extent that for a volume in space the size of a house (about thirty thousand to fifty thousand cubic feet), only one atom is created each year. He further claimed this constant creation of matter explains the formation of new galaxies. Hoyle believed the universe is closed and thus exists as a steady-state universe. Cosmological research, observational evidence, and mathematics over the past forty years or so seem to discredit the concept of a closed steady-state universe, but the debate continues (*See also* Weinberg).

Hoyle's theory for the origin of the solar system: *The original sun was a binary star, one of which separated and exploded. Over time, the force of gravity coalesced the exploded matter of the second star that, due to mutual gravity, formed the planets, comets, asteroids, and so forth, of the present solar system.*

Fred Hoyle made a number of contributions to astronomy and cosmology and provided a great deal of mathematical support for various theories. His theory for the formation of the planets in our solar system states that our sun was, at one time, one twin of a two-star (binary) system. One star exploded, but the gravitational attraction of the remaining star (our sun) maintained the pieces of the exploded star in orbits around our sun. In time, these chunks of matter attracted each other and piled up into great masses that became the existing planets, still revolving around our star as they are captured by the gravity of its great mass. This theory is considered a viable account for the formation of the planets as well as comets, asteroids, and meteors.

Hoyle's theory for the formation of the elements: *Hydrogen is "fused" into helium inside stars, which also combined to form heavier elements.*

One of Fred Hoyle's most important theories explains how hydrogen is converted into helium inside the sun by the reaction of atomic (nuclear) fusion, which is the same reaction that occurs when a nuclear fusion (hydrogen) bomb explodes. In addition to this reaction that creates helium plus all the energy output of the sun, Hoyle theorized that a similar process occurred inside the sun to form the heavier elements. One example is the formation of a carbon atom (atomic number 6, mass number 12) from three atoms of helium (atomic number 2, mass number 4): ${}_2\text{He-4} + {}_2\text{He-4} + {}_2\text{He-4} \rightarrow {}_6\text{C-12}$. A similar reaction formed the lighter elements as well as the heavier elements with which we are familiar on Earth. This relates to Hoyle's theory for the formation of the planets of the solar system when the chunks of the exploded twin of our sun agglomerated. These chunks of matter contained all of the known elements at the time the planets were formed.

Hoyle's disclaimer for the reptile/bird theory: *Reptiles did not evolve into birds.*

Hoyle believed there were interstellar grains similar to bacteria that brought life to Earth from outer space. Svante August Arrhenius first proposed this theory, called *panspermia*. However, the theory is no longer valid because it has been demonstrated that cosmic radiation would kill extraterrestrial life forms of this type. This idea not only influenced Hoyle's concept for the origin of life, but also affected his ideas related to the evolution of species and the use of fossils to explain evolution. He was unconvinced that fossils represented extinct species. Hoyle became involved in a dispute with a geologist over a type of fossil claimed to represent a species between a dinosaur-type reptile and a bird (Archaeopteryx). Hoyle claimed it was a fake because he insisted the feathers had been glued onto a reptile skeleton to make it seem part bird. The British Museum conducted many tests and found no evidence of glue or of any other deception. Recent studies of DNA, bone structure, and other anatomical comparison have established an evolutionary relationship between birds and reptiles.

See also Arrhenius; Redi

HUBBLE'S LAW AND CONSTANT: Astronomy: *Edwin Hubble* (1889–1953), United States.

Hubble's law: *The velocity of a galaxy that is receding from us is proportional to its distance from Earth.*

Even before Edwin Hubble could utilize the one-hundred-inch Mount Wilson telescope in California, he studied faint "clouds" of gas and dust that appeared as fuzzy images. He considered some of these areas as originating from our Milky Way galaxy. Other images seemed to originate from more distant areas of space, which were called *nebulae*. Once he was able to use a powerful telescope, he identified these more distant dense "clouds" of luminous gases as clusters composed of many millions, perhaps billions, of stars that are billions of light-years from Earth. He identified two types of these nebulae galaxies—one as spiral, the other as elliptical. He further classified elliptical galaxies as to their shapes approximating a circle. Although not all observed phenomena in deep space fall into these two classifications, Hubble's descriptions are still the basis for galactic classification. He also discovered several cepheids, which are stars that vary in their brightness (period-luminosity). These bright variable stars provide a means for measuring the distance of galaxies relatively near us—about one million light-years distant. From these data he proposed Hubble's law, v = Hd, where v is the recessional velocity of the galaxy, d is its distance from us, and H is known as Hubble's constant. To develop this law, Hubble measured the distance of about a dozen and a half galaxies of several different classifications and related their receding velocities to the degree of red shifts in their light. He then devised the Hubble diagram where the x-axis is the distance and the y-axis is the amount of red shift of the wavelength of the galaxies' light (*see* Figure H8).

Hubble's constant: *The original value of the Hubble constant (H) was 150 km/sec/ 1,000,000 light-years.*

Edwin Hubble overestimated the value for his constant by a factor of eight to ten. It has since been corrected as H = 15-30 km per second per Mpc. The symbol H for the constant is sometimes written as H_0, the range of fifteen to thirty kilometers is still not an exact known distance, and *Mpc* is a *megaparsec*, which is equal to 10^6 parsecs. The parsec is a unit used in astronomy to measure very large stellar distances. It is equal to 3.856×10^{13} km, or 3.2615 light-years. Hubble's constant is important for two reasons. First, it provides the factor necessary for relating the red shift from the light of stars to their distances, thus providing a means of calculating the observable size of the universe. And, second, the reciprocal of Hubble's constant provides a means to determine the age of the universe. It is possible to calculate how long it would take galaxies to backtrack (contract) their now-expanding movements to their state at the origin of the

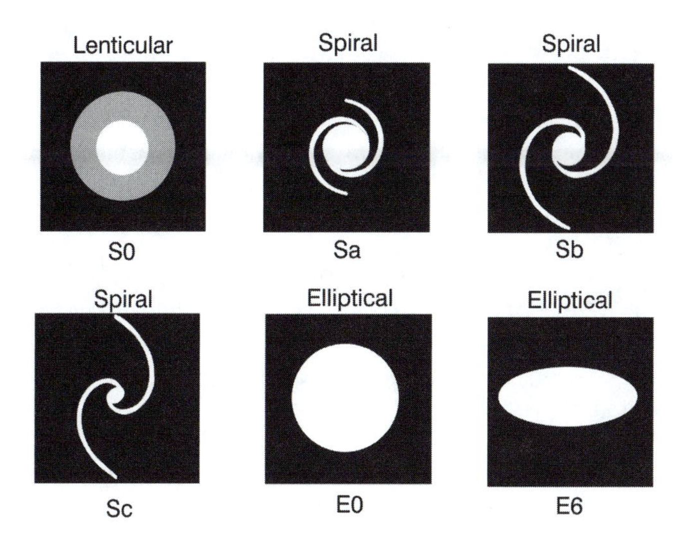

Figure H8. The apparent shapes of different galaxies are partly the angle of the viewing telescope.

universe. The current figure for the age of the universe, as calculated by the reciprocal of Hubble's constant, is between ten and twenty billion years, with a reasonable estimate of fourteen to fifteen billion years since the time of the big bang.

See also Gamow; Hale; Hertzsprung; Hoyle

HÜCKEL'S MO THEORY OR RULE AND THE DEBYE–HÜCKEL THEORY:

Chemistry: *Erich Armand Arthur Joseph Hückel* (1896–1980), Germany.

Hückel's molecular orbital theory: *Aromatic cyclic compounds similar to benzene are planar compounds that have 4n + 2 π-electrons (when* n *equals either a 0 or just a small whole number, it will determine how many aromatic rings a particular compound will have).*

Hückel developed his MO rule in 1930 to explain π-electron systems that limited rotation of ring structures similar to benzene-like organic ring compounds that contain double C bonds (C=C). The Hückel 4n+2 rule could be used to determine if ring molecules of aromatic compounds such as benzene are composed of C=C bonds (*See also* Kekule;Van't Hoff).

Hückel was born near Berlin, Germany, and was educated at the University of Göttingen. After receiving his doctorate in experimental physics in 1921, he became an assistant to the Dutch physicist, Peter Debye, at the Federal Institute of Technology in Zürich, Switzerland. While in Zürich, together they developed the Debye–Hückel theory for electrolytic solutions. However, Hückel's work was mostly ignored until his death, primarily because he had a difficult time communicating his ideas to other scientists. Although he never won a Nobel Prize, other scientists in his field have acknowledged his contributions to the field of chemistry.

Debye–Hückel theory of electrolyte solutions: *In concentrated solutions, as well as dilute solutions, ions of one charge will attract other ions of opposite charges.*

This theory assumes that strong electrolytes are fully dissociated into ions while in a solution state. This theory is used to calculate the properties of electrical conductivity for dilute solutions. This theory also accounts for the conductivity of ionic solutions by considering the forces between the ions within the solutions. Their theory takes into consideration the interactions of ions in a solution that explained the differences between dilute solutions of electrolytes and what is referred to as an ideal solution.

HUGGINS' THEORY OF SPECTROSOPIC ASTRONOMY: Astronomy: *Sir William Huggins* (1824–1910), England.

The chemical composition of stars and nebulae can be determined by the use of a spectroscope.

After receiving an early education at home, William Huggins was sent to a private school for further training. After graduating from the City of London School, he tried involving himself with his father's silk business but soon gave that up to devote himself to science. He became familiar with the work of Robert Bunsen, who invented the gas Bunsen burner, and Gustav Kirchhoff who, along with Bunsen, invented an instrument that used an optical prism device. They called it a spectroscope because it could be used to separate visible light emitted from chemicals that were vaporized by Bunsen's gas burner that produced a high-temperature nonluminous flame. Gas produced from the burning of any substance produces a unique pattern of spectral lines that can be distinguished when viewed through the spectroscope. The unique signature spectrum for each element is created by its atomic structure that is different for each specific element. As an atom's electrons jump between different orbitals (energy levels), electromagnetic energy in the form of light is generated. These "excited" electrons are what cause the unique spectrum of atoms of different chemical elements. Kirchhoff determined that the dark lines seen when viewing chemical substances were the same dark lines in the spectrum of light from the sun. From this data it was determined that the sun was composed of very hot gases.

Being familiar with this information, Huggins first examined the spectra of the stars. After acquiring a fine spectroscope, he examined the spectra of several chemicals found on Earth and then, in collaboration with

Although he was considered an "amateur" astronomer, the discoveries by William Huggins had a great impact on the science of astronomy. He made use of the discovery of the red shift by Christian Doppler and Armand Fizeau that proved that light waves from a source receding from an observer were of a lower frequency and thus of a longer wavelength than were the frequencies and wavelengths of waves approaching the observer. This shift known as the Doppler effect applies to light waves as well as sound waves. Huggins made use of this discovery of the red shift of light from stars by viewing a lower frequency (thus longer wavelength) that produced light near the red end of the light electromagnetic spectrum, which meant that that source of light was receding from earth. Huggins examined the spectrum of the star Sirius and determined that it had a noticeable red shift and, more important, that the degree of shift was proportional to the star's velocity away from Earth which was about twenty-five miles per second. This was evidence that the universe is expanding as distant stars recede from Earth as well as from each other.

a professor of chemistry at King's College in London, Huggins began a life-long investigation of the spectra of the stars and nebulae. In 1863 Huggins published his first findings that described the spectra of stars indicating that they are composed of the same chemical elements as the elements found on Earth. Following this discovery he resolved the mystery of the composition of nebulae that looked like interstellar clouds of gas or particles (e.g., Orion), which astronomers were unable to distinguish as stars. When Huggins looked at several nebulae with his spectroscope, to his surprise he saw only a single bright line. This proved that a nebula was not a clump of many stars, but rather a mass of luminous gases. After observing the spectra of several dozen nebulae he determined that about a third of the nebulae were of the gaseous type. Huggins also examined the spectra of the tails of comets and discovered that this gaseous "tail" is composed mostly of hydrogen. He also determined that the temporary star, Nova Coronae, was composed of mostly hydrogen.

As Huggins became older and his eyesight deteriorated, he had difficulty viewing objects through his spectroscope. His wife Margaret Lindsay Murray (Lady Huggins) then assisted him in viewing and recording observations. Together they prepared an *Atlas of Representative Stellar Spectra* in 1899. This atlas represented spectra from lambda 4870 to lambda 3300 along with an interpretation of their spectra and a detailed discussion of the evolutionary order of the stars.

See also Bunsen; Doppler; Fraunhofer; Kirchhoff

HUYGENS' THEORIES OF LIGHT AND GRAVITY: Physics: *Christiaan Huygens* (1629–1695), Netherlands.

Huygens' wave theory of light: *A primary light wave front acts as a spherical surface that propagates secondary wavelets.*

Christiaan Huygens was the first to conceive that light was propagated as waves and could support his theory by experimental observations. During Huygens' lifetime, most scientists supported the particle theory of light. Huygens, however, demonstrated that when two intersecting beams of light were aimed at each other, they did not bounce off one another, as would be expected if they were composed of minute particles with mass (conservation of momentum). He further theorized that light would travel more slowly when refracted through a denser medium than air. This was in direct opposition to Sir Isaac Newton's concept that light would maintain its speed when refracted toward the normal angle of light's motion. A more detailed statement of Huygens' wave theory is: *At every point on the main spherical wave front, there are secondary wavelets that at some time in their propagation are associated with the primary wave front.* Both the wavelets and wave front advance from one point in space to the next with the same speed and frequency. Since Huygens' time, the wave theory has been refined, and the wave–particle duality is now generally accepted. Electromagnetic radiation (light) waves are now also considered as particles, called *photons*, as explained by the quantum theory for the photoelectric effect and diffraction.

Huygens' concept of gravity: *Gravity has a mechanical nature.*

Huygens disagreed with Newton's law of gravity, which was based on Newton's laws of motion as well as the concept of force at a distance. Huygens considered Newton's theory of gravity as lacking the means to explain mechanical principles based on Cartesian concepts. Huygens used the term "motion" to mean "momentum" and considered the center of gravity to extend outward in a straight line similar to centrifugal force.

He based his theory on mechanics, which states there is no loss or gain of "motion" following a collision between bodies (conservation of "motion" or momentum). He further developed a mathematical explanation for perfect elastic bodies. Huygens' emphasis on the mechanistic nature of gravity was in direct conflict with his wave theory of light, based on secondary wave fronts he considered as massless waves that would not obey mechanistic rules. Over time, Newton's concept of gravity was accepted but revised to conform to the new theories of relativity and quantum mechanics by Einstein and others. Today's theory of gravity is still based on the relationship of bodies' masses and distances separating them. There is a modern concept for the wave nature of gravity based on yet-to-be discovered wave particles called *gravitons* (*see also* Einstein; Newton).

Huygens' concept for using the oscillations of a pendulum for timekeeping: *Based on the mathematics of curved surfaces and centrifugal force, the motion of an oscillating simple pendulum should maintain periodic time.*

Needing a device for accurate timekeeping for his work in astronomy and relying on the concepts developed by Galileo, Huygens worked out the mathematics of the pendulum, enabling him to develop the first accurate practical clock (*see* Figure G2 under Galileo). Much like today's grandfather clocks with a weighted bob suspended at the end of a long rod, it made use of slowly dropping weights and a crude escapement cog to maintain the swing of the pendulum at regular intervals. Huygens also developed a method for grinding lenses that improved the resolution of telescopes. Along with his pendulum clock and telescopes, Huygens made several discoveries, one of which was to identify a separation in the rings of the planet Saturn, another was the discovery of one of Saturn's moons.

See also Galileo; Hooke; Newton

I

IDEAL GAS LAW: Chemistry: *Robert Boyle*, England; *Jacques-Alexandre-Cesar Charles* and *Joseph-Louis Gay-Lussac*, France. This entry is a combination of several laws, theories, and hypotheses developed by several scientists.

The ideal gas law may be expressed as PV = nRT, or PV/T = nR, where P = pressure, V = volume, T = temperature, n is the number of moles of gas in the system, and R is the gas constant equal to 8.314 joules.

The French physicist and inventor Guillaume Amontons (1663–1705) was the first to measure and state the relationship between the temperature of a gas and its volume. He accomplished this by using an improved air thermometer similar to one devised by Galileo (*see* Figure G3 under Galileo). Amontons demonstrated that the volume of a gas increased at a regular rate when heated, and, conversely, its volume decreased when the gas was cooled. He also concluded that this relationship applied to all gases. His work was undiscovered, and Robert Boyle was later credited with recognizing the relationship between the volume and pressure of a gas at a constant temperature.

The ideal gas law is also referred to as the *law of perfect gases* or the *generalized gas law*. It is a combination of two laws and a constant to form a *universal gas equation*. The ideal gas law is expressed in the equation PV = RT, where P is the pressure exerted by 1 mole of gas in a volume V; R is sometimes expressed as nK, where n is Avogadro's number and K is Boltzmann's constant. It is the gas constant at absolute temperature $T = -273.2°C$ in the Kelvin scale, or R = 8.314 joules per gram-mole-Kelvin. The ideal gas law can also be determined in a physical sense from the kinetic theory of gases. There are four assumptions for the use of the kinetic theory to derive the law:

1. *The gas within a given volume contains a very large number of molecules, which follows Newton's laws of motion and maintains randomness* (*see* Avogadro's number).
2. *The actual volume of the total mass of the gas molecules is very small compared to the total volume the gas molecules occupy (the molecules are very small).*

3. *The only force acting on the molecules is the force resulting from short-term elastic collision (kinetic energy and conservation of momentum).*
4. *The gas molecules do not attract or repel each other.*

The ideal gas law is a generalized law that relates the temperature, pressure, and volume of a gas under ideal or perfect conditions, which do not exist in nature. It is a combination of Boyle's law (pressure × volume) and Charles' and Gay-Lussac's laws (volume/temperature). The equation describes the behavior of only "real gases," with some accuracy, at relatively high temperatures, and low pressures. For some gases, it is a good approximation at standard temperature and pressure. At extremely high or low temperatures or pressure, the ideal gas law no longer applies to any gas. Understanding the gas laws was the beginning of modern chemistry. These laws have proven invaluable for almost all areas of science.

See also Avogadro; Boyle; Charles; Gay-Lussac

I-HSING'S CONCEPTS OF ASTRONOMY: Astronomy and Mathematics: *I-Hsing* (c.681–727), China.

I-Hsing was an intelligent child who showed extraordinary ability to remember what he learned and observed. At the age of twenty he was known for his knowledge of mathematics related to astronomy. He was called to serve under Empress Wu-Zetian (625–705), the autocratic ruler of China during the Tang Dynasty, but refused and became a Buddhist monk where he received the name I-Hsing (his original birth name is unknown). He is credited with several major achievements.

In the years 723 to 726 he was part of an expedition to measure the length of a meridian line in degrees on Earth's surface by using astronomy. Over the distance of 2,500 kilometers (about 1,550 miles) on this imaginary line, a team erected nine different campsites to make measurements along this line at the same time on June 21st. At noon on this date each station measured the sun's solstice shadow which happens to be the 24 hours (day) with the longest period of daylight. (The summer solstice, for the Northern Hemisphere, occurs when Earth's North Pole is pointing more directly toward the sun at 23½ degrees.) From measurements taken, I-Hsing used this data to calculate the length of a meridian degree. His figure was much too large, but this value was not corrected until the late Middle Ages.

Using his knowledge about the movement of celestial bodies, I-Hsing devised what was known as the "*Dayan* calendar" which was more accurate than any other calendar of that time and was used for almost one hundred years.

I-Hsing may have been one of the first to design an escapement movement for use in "accurate" timekeeping devices. An escapement is a ratchet-like device, usually a cogwheel, that allows periodic movement in one direction. I-Hsing designed a wooden wheel that had scoop-like compartments spaced around the edge of the wheel. These compartments would catch and hold water until the wheel turned forward to a point where the water would spill out and more water would fill the next compartment. His water clock would make exactly one revolution in a twenty-four-hour period and was as accurate as other more elaborate water clocks that did not use an escapement device but rather depended on water dripping through a small orifice at a specific rate and collected in a small graduated tube to mark off time periods. It is difficult to determine when and who invented water clocks and later mechanical clocks. The challenge to invent an accurate timekeeping device was one that puzzled and occupied many scholars in many lands throughout ancient and medieval history. The goal of accurate timekeeping was not achieved until the twentieth century. It was not until the Middle Ages that metal escapement cogwheels were incorporated into mechanical clocks to produce a regular movement. Early mechanical clocks were no more accurate than water clocks, and all of the early types of clocks averaged an error of about fifteen minutes a day.

See also Huygens

INGENHOUSZ'S THEORY OF PHOTOSYNTHESIS: Biology: *Jan Ingenhousz* (1730–1799), Netherlands.

Plants not only produce oxygen in daylight, but also absorb carbon dioxide.

In 1771 Joseph Priestley, an English chemist, demonstrated that plants in a contained environment of carbon dioxide could make the air breathable in that environment. Jan Ingenhousz confirmed Priestley's hypothesis that green plants produce not only a breathable gas, but that the gas was oxygen. He went one step beyond Priestley's statement. Ingenhousz's theory stated that green plants absorb carbon dioxide and emit oxygen when exposed to light by a process, now known as *photosynthesis* (meaning combining or synthesizing by light), and that plants also reverse this process in the dark. To date, no one has been able to artificially replicate this process in the laboratory. Scientists are working to create artificial systems of photosynthesis that will convert solar energy to more useable sources of energy, such as hydrogen, for use as a clean fuel. Recent experiments indicate that an atmosphere rich in carbon dioxide increases the rate of plant growth in nature as well as in greenhouses. Thus, one solution for reducing the amount of carbon dioxide in the atmosphere would be to plant more trees and other green plants. Ingenhousz's theory set the foundation for the concept of the relationship between living things. He demonstrated that animal life is dependent on plant life (oxygen and food), whereas plants depend on animals for carbon dioxide and the products of decomposition of dead animals and the wastes deposited by animals, as well as dead plants, thus establishing an ecological balance between the animal and plant worlds.

See also Lavoisier; Priestley; Sachs

INGOLD'S THEORY FOR THE STRUCTURE OF ORGANIC MOLECULES: Chemistry: *Sir Christopher Kelk Ingold* (1893–1970), England.

If an organic molecule can exist in two different states other than its normal structure, then it can only exist in a hybrid form.

Sir Christopher Ingold called this idea *mesomerism,* which he used to describe a process similar to the resonance or oscillation of molecules in certain types of organic structures as described by Linus Pauling. Ingold was interested in the molecular structures of particular organic compounds that can have two or more molecular structures. The "mesomerism" molecules of these compounds have the same basic structure but with different arrangements of their valence electrons, which led to Pauling's concept of organic molecules existing in an intermediate (hybrid) form. Isomerization (mesomerism) that forms organic compounds does not fit the octet rule for the Periodic Table of the Chemical Elements. Ingold published a paper "Principles of an Electronic Theory of Organic Reactions" that explained his theory in 1934. Other scientists also validated it by measuring bond lengths in these types of organic molecules. His theory led to a better understanding of special types of organic molecules, which became important in the development of new drugs, particularly antibiotics.

See also Pauling

INGRAM'S SICKLE CELL THEORY: Biology: *Vernon Martin Ingram* (1924–2006), United States.

> *Hemoglobin S (HbS) varies from normal hemoglobin A (HbA) in only one amino acid.*

A Chicago physician named James B. Herrick (1861–1954) first described sickle cell disease in 1910. One of Dr. Herrick's patients who came from the West Indies suffered from a form of anemia in which some of the patient's blood cells were sickle shaped (something like an open letter C)—thus the name *sickle cell anemia.* Some years later it was discovered that these special cells were related to a low level of oxygen in the blood. Hemoglobin molecules present in each red blood cell carry oxygen from the lungs to tissues and organs throughout the body and then bring carbon dioxide back to the lungs. In patients with sickle cell disorder, the hemoglobin molecules are abnormal, and after giving up their oxygen, have a tendency to agglomerate, become stiff, and form into sickle shapes and thus are unable to pass through small blood vessels. This "blockage" causes oxygen deprivation, manifesting in pain and damage to tissues and vital organs. The phenomena was further verified in 1927 by E. Vernon Hahn and Elizabeth B. Gillespie, a surgeon and an intern, respectively, at the University of Indiana Medical School in Indianapolis, and in 1940 by Irving J. Sherman, an undergraduate participating in a genetics study at the Johns Hopkins University in Baltimore.

Using new techniques the American physical chemist Linus Pauling and his graduate students at the California Institute of Technology separated type S hemoglobin from normal type A hemoglobin in 1947. The question, still unanswered at that time, was how hemoglobin S (HbS), which caused the debilitating disease, was related to the proteins and amino acids of the blood.

In 1956 Vernon Ingram split hemoglobin into smaller units and separated them further by using electrophoresis. This electrochemical process uses a weak electric charge, causing large molecules to separate from each other into different "paths" or tracks according to their individual characteristics, which provides a means of identifying the components of the original substances. This procedure helped Ingram to determine that the sickle cell hemoglobin was caused by changes in only one of over five hundred amino acids in the human body. The HbS appeared when the *glutamic amino acid* was replaced by the *valine amino acid.* Ingram then determined this was a mutation of the blood cells. The sickle cell disorder consists of at least two varieties of sickle cells. One form is characterized by a severe slackening of the blood flow, resulting in a **reduction** of oxygen in the blood vessels, which causes more restriction of the flow of oxygenated blood to the body's organs. This reduction of oxygen is the cause of genetic sickle cell anemia. Another kind is called a trait, which is not as devastating as the anemia form. A test for the presence of sickle cell disorders exists, but it cannot distinguish between the two types. Sickle cell disease is the first-known genetic disease to be identified. So far, there is only one drug used to prevent complications of sickle cell disease, but because it is a genetic disorder no drug has yet been found that will cure the disease.

See also Pauling

IPATIEFF'S THEORY OF HIGH-PRESSURE CATALYTIC REACTIONS:

Chemistry: *Vladimir Nikolayevitch Ipatieff* (1867–1952), Russia and United States.

> *By using high temperatures to increase pressure, it is possible to catalyze liquid hydrocarbon molecules.*

Ipatieff extended the original research by the French Nobel Laureate in Chemistry Paul Sabatier (1854–1941) who used finely powdered nickel as the catalysis to increase organic catalytic reaction. Sabatier used hydrogen for the catalytic hydrogenation to convert liquid oils (fact) to form more solid hydrocarbons, for example, margarine. Ipatieff designed a type of heavy steel autoclave that he called the "Ipatieff bomb." This device was strong enough to contain a mixture of a liquid and a catalyst whose temperature was raised above its normal boiling point and thus at increased pressures. This led to a method that used high temperatures for catalytic reactions that was much more efficient in creating organic hydrocarbon polymer compounds.

Valdimir Ipatieff was born, raised, and educated in Russia and was responsible for establishing several chemical research facilities for explosives and synthetic rubber. Lenin made him a lieutenant general of all military research in Russia and head of the Soviet chemical industry. Due to the fact that he was a soldier and former follower of the deposed Russian Czar Nicholas, whose entire family was murdered by revolutionists, as well as being dissatisfied with the Communist government, he felt threatened. During a trip to Germany to visit a friend, he emigrated to the United States rather than return home. Although he had no visa when he arrived in Chicago, an oil company supported him as a professor at Northwestern University to continue his research in petroleum chemistry. He and his assistant, the Polish chemist Herman Pines (1902–1996), expanded their research that led to the use of an acid to act as a catalyst for low-temperature organic reactions. In turn, this research led to the development of a method of synthesizing petrochemicals to form iso-octane, as well as other petroleum derivatives, that greatly improved the quality of gasoline by increasing its octane level, thus greatly enhancing the level of performance of airplanes. During World War II, the improved 100-octane aviation gasoline greatly increased the speed of the American and British warplanes, and thus was a major contributing factor in the defeat of the German Luftwaffe (air force). The use of catalytic chemistry was considered a major contribution to the Allied victory over Germany. The use of catalysts in the petroleum industry is directly responsible for the abundance of products manufactured from crude oil today.

See also Adams; Wilkinson

ISAACS' THEORY OF PROTEINS ATTACKING VIRUSES: Biology: *Alick Isaacs* (1921–1967), England.

When under attack by a virus, animal cells are stimulated to produce a protein (interferon) that interrupts the growth of the virus.

Alick Isaacs' study of various genetic varieties of the influenza virus led to his investigation of how the human body responds to different variations of a particular virus. He discovered a low-molecular-weight protein that had some effect on the way a virus multiplied and mutated. He named this newly identified protein *interferon* (also known as IFN). Interferon is a natural protein produced by the cells of the immune systems of animals. There are several classes and types of interferon that attacked foreign agents such as virus, bacteria, parasites, and tumor cells that can damage the animal's tissues and organs. It enters the bloodstream automatically whenever a virus invades the body. Although the body produces only small amounts of interferon, it enables healthy cells to manufacture an **enzyme** that counters the viral infection. As a natural component of an animal's body, it is referred to as a *biopharmaceutical*. Each species of animal, including humans, produces its own type of interferon, which cannot be interchanged between species. The human body produces three types of interferon: 1) type I is divided into *alpha, beta, omega, kappa zeta, tau,* and *delta* sub types; 2) type II, or *gamma* is a single subtype; and 3) type III, *lambda* with three different subtypes. However, all interferon is produced by only one type of cell but can act as a "trigger" to help other cells produce more cells to fight a disease. When first discovered, interferon was very expensive to produce because the human body produces only minute amounts. However, it is available now in large quantities due to the genetic engineering of the protein molecule. It is used to treat a variety of diseases, including liver disorders, such as hepatitis C, hairy cell leukemia, Kaposi's sarcoma, multiple sclerosis, genital warts, and diseases of the gastrointestinal tract.

J

JACOB–MONOD THEORY OF REGULATOR GENES: Biology (Genetics): *François Jacob* 1920–), France. François Jacob shared the 1965 Nobel Prize for Physiology or Medicine with Jacques Monod and André Lwoff (1902–1994).

> *A group of genes, called the "operon" genes, is expressed as a single unit that is responsible for regulating proteins.*

After receiving his medical degree from the University of Paris in 1947, François Jacob began working with others at the Pasteur Institute on the genetic material found in viruses. He and a colleague at the Institute, the French geneticist Elie Wollman (1917–), introduced the term "episomes" to describe the genes that sometimes have a separate existence in cells of a host, and yet at other times are integrated into chromosomes of cells that replicate themselves along with the chromosomes. During their studies they found that episomes act as viruses that infect bacteria as they are transferred to the genetic material of the bacteria cells. Following this research Jacob joined forces with the French biologist Jacques Monod (1910–1976) and the American biochemist Arthur Pardee (1921–) to expand his research on the regulation of gene activity. They coined the term "operon" for a group of genes—the *promoter* and the *operator* genes as well as the *structural* genes that provide the RNA code for specific proteins through a process called "transcription" of RNA and DNA. These operons are linked to specific repressor or activator genes and act as regulators that either shut down or turn on other genes. They are related to enzymes required for biosynthetic processes and regulate the production of proteins. The first operon discovered was called the "lac operon" that consisted of structural genes, a promoter gene, a terminator gene, and an operator gene. Most operons that have been studied are found in *E. coli* bacteria.

See also Lederberg; Robbins; Tatum

JANSKY'S THEORY OF STELLAR RADIO INTERFERENCE: Physics (Engineering): *Karl Jansky* (1905–1950), United States.

Unexplained radio interference on Earth originates in space.

Karl Jansky was assigned the task of solving the problem of radio noise interference in shortwave transatlantic radio-telephone transmission. He devised a rotating linear directional antenna that he mounted on an automobile wheel, enabling him to turn it through 360 degrees. Using this antenna, he identified a number of sources of "noise" or static originating in the atmosphere, some of which came from industrial sources, as well as thunderstorms. However, he continued to believe some of these unwanted signals originated from the sun. After more study he realized this was not possible because the time of the peak noise was shifting as Earth revolved around the sun during a twelve-month period. He finally rejected this solar theory when he was unable to detect a signal from the sun during the partial solar eclipse of August 31, 1932. Jansky's theory, which he published in 1932, stated there were two sources in space from which these radio signals emanated. One source was the Milky Way galaxy. The strongest signals came from the direction of the constellation Sagittarius configuration of stars and galaxies. According to the astronomers Harlow Shapley and Jan H. Oort, this is in the direction of the center of the Milky Way. Jansky's work and theory led to the development of the important fields of radio and X-ray astronomy. Currently, manmade satellites use his concept to explore radiation from deep space. X-ray astronomy measures the leftover radiation from the big bang at the time of the origin of the universe. These observations are expected to explain the age and nature of our universe more accurately. Unfortunately Jansky developed liver disease and died at the age of 45 before he could continue his promising research in the field of radio astronomy. The unit for the strength of transmitted radio waves was named the *Jansky* in his honor.

JANSSEN'S THEORY OF SPECTRAL LINES OF SUNLIGHT: Astronomy: *Pierre Jules Cesar Janssen* (1824–1907), France.

The bright yellow lines observed in the spectrum during a solar eclipse are the same lines observed when the moon does not block the sun. This is evidence of a new element never detected on Earth.

Over a period of nearly fifty years Pierre Janssen traveled on many expeditions around the world to view total eclipses of the sun. He was aware that the bright line he observed in his spectroscope when viewing light from the unobstructed sun could be considered evidence of the element hydrogen in the sun itself. On a trip to Guntur, India, he used a spectroscope to observe a total eclipse on August 18, 1868. During this eclipse, even though the moon passing in front of the sun blocked it out and there was no direct sunlight, there were bursts of light appearing around the edges of the moon. These prominences of light are known as the **chromosphere**. Janssen noticed a bright yellow line in his instrument that had a wavelength of 587.49 mm during the eclipse when there was no direct sunlight. He was criticized because never before had an element been discovered in space. A few months later, Norman Lockyer, an English astronomer, observed the same yellow lines and named it "helium" that he thought was a new form of hydrogen. In 1895 William Ramsay of Scotland discovered an element on Earth with the same spectroscopic wavelength as Janssens' spectral lines. Ramsay is given credit for the discovery of helium, although it was Janssen who first detected helium in the sun.

Pierre Janssen was a well-traveled astronomer. On one trip to Peru in 1857 he was able to detect and establish the magnetic equator for Earth. He also viewed two transits of Venus and viewed many solar eclipses. During the early days of the Franco-Prussian conflict in 1870 he escaped from the siege of Paris by hot-air balloon to complete a tour to observe an eclipse. Due to cloud cover, his efforts proved unsuccessful.

JEANS' TIDAL HYPOTHESIS FOR THE ORIGIN OF THE PLANETS: Astronomy: *James Hopwood Jeans* (1877–1946), England.

> *A passing star pulled off a lump of sun matter, which later broke apart and solidified into the planets as they revolved around the more massive sun.*

Prior to Jeans' time, the accepted concept for the origin of the planets was Pierre Simon Laplace's **nebula** hypothesis, that a contracting cloud of dust and gas from the sun formed the planets. A different tidal hypothesis was proposed by James Jeans, which asserted a large star passed close to our sun, causing a cigar-shaped protrusion of gas and matter to be pulled off the sun. This oblong formation of gas and other sun chemicals coalesced to form planets as they were sent into orbits around the sun, as they were affected by the sun's great gravitational force. Some years later other astronomers rejected Jeans' theory. They resurrected and revised Laplace's condensation hypothesis, which is still accepted by some astrophysicists. Later, Jeans demonstrated that all the outer planets had very cold atmospheres, but residual internal heat keeps them relatively warm. In 1923 Jeans conducted research at the Mount Wilson Observatory in Pasadena, California. Sir James Hopwood Jeans, knighted in 1928, was professor of astronomy at the Royal Institute in London from 1935 until his death in 1946. He wrote several excellent books including some textbooks in science. Some of his early titles are *Dynamical Theory of Gases* and *Mathematical Theory of Electricity and Magnetism*, both in the early 1900s. Later he wrote important books on astronomy, including *Problems of Cosmogony and Stellar Dynamics*, *Astronomy and Cosmogony*, *The Universe Around Us*, and *The Mysterious Universe*. These last two were written for the general public and were his most popular.

See also Hale; Laplace

JEFFREYS' THEORY OF GENETIC (DNA) PROFILING: Biology (Genetics). *Alec John Jeffreys* (1950–), London.

> *Parts of the gene that code for the protein myoglobin consist of a short sequence of DNA that is repeated.*

After receiving his PhD degree at Oxford and spending two years there as a research assistant, Alec Jeffreys became a professor of genetics at the University of Leicester in England in 1977. While there, he and his colleague Richard Flavell (1945–) made the important discovery that developed into what is known today as "DNA fingerprinting."

Rather than referring to it as "DNA fingerprinting," the procedure might more properly be called "DNA testing," "DNA typing, or "DNA profiling" because prints from the fingertips are not part of the samples used in the collection process. Although there is only a one in fifty billion chance that two people will have the same DNA sequence, unlike fingerprints that are different for all people including identical twins, identical twins can have the same DNA. In the recent past DNA profiling has not always been considered "foolproof" in courts of law. For example, during the O.J. Simpson murder trial in 1995 the jury ignored the evidence of his DNA found at the scene of the crime. Since that time the process for collecting and testing the samples of DNA and related laboratory procedures have greatly improved to the point that the courts and juries are now more likely to accept DNA evidence related to crimes.

Some of the other uses for the technique are building a national or world data bank of individuals' DNA; diagnosing diseases such as cystic fibrosis, sickle cell anemia, hemophilia, Huntington's disease, and several other possible genetic diseases. It can also be used to test for paternity, as well as for personal identification in case of death where the body is unrecognizable, as in the 9/11 disaster. The process can be used on blood-stained clothing or semen that is decades old. A famous example: the blood from Abraham Lincoln's stained clothing was analyzed and used to diagnose him as having a genetic disorder known as Marfan's Syndrome.

The original idea of taking an ink print of individual fingertips was based on the fact that no two people (including identical twins) have the same swirl pattern (ridges) on the tips of their fingers. This system has been used successfully for many years to identify specific individuals for many purposes, but the patterns of ridges on the fingertips can be altered by surgery, so the method is not foolproof. As a geneticist, Jeffreys was aware that individuals within the same species have most of their DNA sequence in common. In other words, our genetic makeup is very similar to other humans. In fact, chimpanzees have about 98.5% of the same DNA as humans—but that 1.5% difference constitutes the distinction between *Homo sapiens* (hominids) and apes (hominoids).

DNA is the same for all human tissues, organs, and products of these tissues and organs, including blood, semen, and saliva. However, while working with the gene that codes the protein myoglobin, Jeffreys observed that a short part of the gene repeated itself in a repetitive sequence. More important, the number of times this sequence was repeated was different in each individual's DNA. The amount of repeats was named VNTRs which means "variable number tandem repeats." At first Jeffreys did not understand the real significance of this difference in the DNA of every individual. He soon realized that the marker sequences could be used to identify individuals, just as physical fingerprints do, by using an enzyme to split the DNA molecule that can then be treated and identified by a process called **electrophoresis**. This DNA "fingerprinting" (which is really a misnomer) requires only a very small sample of a person's DNA.

Jeffreys' 1984 "accidental" discovery was first used in Narborough, Leicestershire, England, to solve rape and murder cases involving two young girls, one in 1983 and the other in 1986. Blood samples that were taken from nearly 5,000 boys and men from the area of the crime were sent to Jeffreys for DNA analysis. None matched the semen recovered from the girls. Sometime later, another sample from a twenty-eight-year-old cake decorator, Colin Pitchfork, was also sent to Jeffreys for analysis. Pitchfork's DNA was screened, and it matched the semen recovered from both victims. As a result, Pitchfork was convicted in 1987 of these two crimes.

See also Crick; Franklin; Pauling; Watson (James)

JEFFREYS' SEISMOLOGICAL THEORIES: Astronomy (Geophysics): *Sir Harold Jeffreys* (1891–1989), England.

Earth's core is liquid.

Harold Jeffreys was not the first to theorize that the core of Earth is liquid ever since it was determined by the German-born seismologist Beno Gutenberg (1889–1960) that there was a central core inside Earth. Earth's internal structure was studied by several geologists including Richard D. Oldham (1858–1936), Herbert H. Turner (1861–1930), and Perry Byerly (1897–1978), as well as Keith E. Bullen (1906–1976) who jointly published with Jeffreys the seismological tables now known as the "JB tables" that give the travel times of seismic waves. These tables include a tabular list of all the main seismic phases for the focal points of earthquakes at different depths in Earth related to different locations on the surface of Earth. Jeffreys established travel times for the arrival of what are known as the P waves (the primary waves originating from the earthquake) from the travel time for S waves (the secondary waves of the earthquake). With the knowledge of the speed and travel times of P and S waves, the distance between the observers on Earth to the source of the earthquake could then be established. These tables also could be used to authenticate the depth of an internal "quake" and its distance from an observer on Earth. His data and theory also supported the concept of substantial differences in the geology and composition of the internal "spheres" of Earth.

Harold Jeffreys further developed James Jeans' planetesimal theory that hypothesized that a huge tidal wave on the surface of the sun caused by a passing or possibly a colliding star resulted in a long filament of matter being "pulled" out of the sun and in time through the effects of gravity and the matter's momentum on this filament of matter became the solar system's planets, comets, and meteors. This theory was first conceived by the American geologist Thomas C. Chamberlin (1843–1928) and the American astronomer Forest R. Moulton (1872–1952) in the early 1900s. Jeffreys was also a mathematician who further developed the concept of probability statistics that became useful in the fields of physics and astronomy. He published over three hundred scientific papers and published seven books including *The Earth: Its Origin, History, and Physical Constitution* and *Theory of Probability and Mathematical Physics*, as well as several important publications on geology and seismology.

See also Agricola; Chang; Laplace; Richter; Ulam

JENNER'S INOCULATION HYPOTHESIS: Biology: *Edward Jenner* (1749–1823), England.

Injecting humans with cowpox fluid will immunize them from smallpox.

At a young age, Edward Jenner was apprenticed to a surgeon in London where he studied medicine before returning to his home in the county of Gloucestershire as a country doctor. At that time smallpox killed many people in England as well as on the European continent. An old English wives' tale asserting that milkmaids who had contracted cowpox would not be susceptible to the more deadly smallpox was widely

Even today, there is still some criticism of Jenner's inoculation procedures due to our current laws restricting the use of humans as experimental subjects without proper consent, which did not exist before the mid-twentieth century. Regardless of the current laws restricting the use of humans as experimental subjects without informed consent, many states require preschool children to be vaccinated against several communicable diseases before entering school. A few parents refuse to have their children vaccinated for religious and other reasons, including the belief that many of the vaccines are unsafe. If large numbers of children do not receive the required vaccinations, there is a possibility of an epidemic of one or more of these communicable diseases.

accepted in this rural area. Jenner was a research-oriented physician who based his research on case studies and clinical observations, which proved to be the forerunner of modern medical research of disease-causing bacteria and viruses. Today, he is considered to have laid the groundwork of the modern science of immunology. This "old wives tale" belief led Edward Jenner to hypothesize that deliberately infecting people with cowpox germs would prevent them from contracting smallpox. To do this, he extracted some fluid from a cowpox blister on Sarah Nelmes, a milkmaid. Using a procedure Jenner later named *vaccination* (from two Latin words: *vacca* for "cow" and *vaccinia* for "cowpox"), he injected this fluid into an eight-year-old boy, James Phipps. Jenner is also credited with introducing the term "virus." Six weeks after the cowpox injection, Jenner injected the boy with fluid from a smallpox blister. The boy did not contract smallpox. After some additional trials (some not always successful and a few subjects did develop smallpox), Jenner published his results. The public reaction to his experiment was, and still is, mixed, and at one time his vaccinations were banned in England. Then soon after a serious outbreak of smallpox, all English children were required to be inoculated.

In the seventeenth century about 10% of all deaths in London were attributed to smallpox. During the eighteenth century smallpox vaccination became more acceptable. Due to the success of Jenner's controversial experiment, the death rate from smallpox during the nineteenth century dropped from about forty per ten thousand people to about one in every ten thousand. Currently, the World Health Organization (WHO) claims that smallpox has been eradicated. However, it can still flare up in rural areas of India, Asia, and Africa, where many children are not inoculated. Smallpox viruses are also being produced and stored for use as a biological weapon by some countries. Several industrialized nations are vulnerable, including the United States, because their citizens are no longer required to be vaccinated against the disease. However, travelers to "at-risk" countries are required to receive the smallpox vaccination before departing for their destination.

See also Lister

JERNE'S THEORY OF CLONAL SELECTION OF ANTIBODIES: Biology: *Niels Kaj Jerne* (1911–1994), Denmark. Niels Jerne shared the 1984 Nobel Prize for Physiology or Medicine with César Milstein and Georges Köhler.

Diverse antibodies are present in humans at birth and, when attacked by a virus, can produce additional antibodies.

An antibody is a protein found in the blood that responds to its complementary antigen. It is also known as an immune body.

Niels Jerne was aware of the concept that the body's lymphocytes (white blood cells) produce a wide range of various types of antibodies that attack specific bacterial and viral infections. Jerne based his theory on the belief that each cell that produces a specific **antibody** is present in the body from birth, possibly transferred by the baby's mother or the result of cell mutation in the newborn. Bacteria or viruses infect the body by releasing their particular set of chemicals. The infected person's antibodies cause the lymphocytes related to particular bacteria or viruses to divide, producing clone cells that greatly increase the number of antibodies available to fight that specific infection. This theory led to the question of how all this genetic information was included in these original, at-birth cells. Jerne developed the concept of *somatic mutation*. Somatic body cells are the many types of cells that make up the tissues in the body, with the exception of the reproductive germ cells (ova and sperm). This cell mutation concept was the forerunner to Susumu Tonegawa's more complex antibody interactive control mechanism referred to as the "jumping genes" theory. Niels Jerne proposed a theory on the functioning of the immune system, but he neglected to consider the multitude of chemical compounds involved in modulating the immune system. His work and theory are responsible in large part for the current study of the immune system.

See also Koch

JOHANSON'S THEORY FOR THE EVOLUTION OF HUMANS: Anthropology: *Donald Carl Johanson* (1943–), United States.

> *In the pre-*Homo sapiens *species,* Australopithecus afarensis, *the males were larger than the females, indicating they did the hunting while females gathered and cooked food.*

In Ethiopia in 1972, Donald Johanson discovered several bones of a fossilized skeleton, which he identified as a small, three and one-half foot female who was as much a bowlegged, upright-walking, chimpanzee-like creature as she was human. The bones are believed to be between 3.2 and 3.8 million years old. He named this small-brained fossilized female Lucy, after the Beatles' song "Lucy in the Sky with Diamonds," which was playing over and over on the camp's phonograph on the night he made the discovery. Johanson named his new species *Australopithecus afarensis*. Previous to Johanson's discovery, Raymond Dart (1893–1988), the Australian anatomist and anthropologist, discovered a skull in a box of fossil-ferrous rocks that was sent to him by the owner of a rock quarry in Africa. Dart later identified the find as a new species of fossil primate that predated various species of *Homo sapiens*. Dart named this new species, *Australopithecus africanus* (meaning southern African ape). Although both discoveries were made in Africa, they were discovered in different regions. Eventually and not unexpectedly, other anthropologists challenged Johanson's discovery and theories. Johanson stated that bipedalism (walking upright on two legs) preceded the development of the large brain capacity in humans. This contradicted existing theory that claimed that prehumans developed large brains before they became bipedal. He also claimed that females of this prehistoric species stayed "home," pregnant, caring for children, and cooking the bounty secured by the larger and stronger male hunters. Adrienne Zihlman, an anthropologist, primatologist, and self-proclaimed feminist from the University of

Several years ago anthropologists announced the discovery of one small skull plus a few other bones on the Indonesian island known as Flores. They claimed that this skull and bones were of a newly discovered species of miniature prehumans. The scientists who discovered these bones named the new species "hobbits" (*see* J. R. Tolkien's books) that lived on the island from as early as thirteen thousand and possibly as far back as ninety-five thousand years ago. The October 28, 2004 issue of the journal *Nature* reported that the discoverers gave the "hobbits" the official name of *Homo floresiensis* after the island where not only the bones were found but also rather sophisticated stone tools recovered from a cave named Liang Bua. These little prehumans were slightly more than three feet tall with heads the size of chimpanzees. Some scientists speculated that this prehuman species were the ancient ancestors of the later species known as *Homo erectus* that spread across Africa into Europe and Asia about a million years ago. This would also make them a distant past relative to *Homo sapiens* who emerged as recently as one hundred thousand years ago (some anthropologists believe that prehumans existed on Earth more three million years ago although there is scant fossil evidence to support this theory). More recent examinations by skeptical anthropologists challenge not only the authenticity of the hobbit fossils, but also the importance of this "new" species to the evolutionary processes leading to modern humans.

California, Santa Cruz, challenged this theory. Zihlman claimed that this is a typical male anthropologist's sexist interpretation, while Johanson claimed that Zihlman's interpretation of the fossil record was that of an anthropologist seeking recognition of the role of females in the discipline, whereas it is more acceptable to consider all fields of science, including anthropology as genderless. Anthropologist Richard Leakey, son of the famous anthropologist Louis Leakey, claimed the various human species could be traced back even further in history than the fossils of *A. afarensis* (Lucy). He also claimed there were probably two or more branches to the ancestral tree of modern humans, not one as Johanson claimed. Others declared that because Johanson gathered the bones from different sites, the partially completed skeleton was not of a single person or even a female. Additional pre-*H. sapiens* fossils found in Africa seem to confirm Johanson's theory. Although Johanson's discovery and theories significantly contributed to science, some doubt remains as to the history of the ancestors of early *H.sapiens*. In addition, it has been claimed that *Australopithecus* is not a separate distinct species in the continued evolution of humans but may be just another unsuccessful, extinct proto-human species similar to the Neanderthals.

See also Leakeys

JOLIOT-CURIES' THEORY OF ARTIFICIAL RADIOACTIVITY: Physics: *Frédéric Joliot-Curie* (1900–1958) and *Irène Joliot-Curie* (1897–1956), France. Frédéric and Irène Joliot-Curie jointly received the 1935 Nobel Prize for Chemistry.

Radioactive elements can be artificially manufactured from stable elements.

Frédéric and Irène Joliot-Curie's experiment entailed bombarding the element boron (B-10) with two alpha particles (α), which are the nuclei of the element helium generated from the decay of the element polonium. This nuclear bombardment resulted in the formation of radioactive nitrogen (N-14). The bombardment of the lighter element boron with alpha particles changed (transmuted) boron to nitrogen, thus making a radioactive form of that element (B-10 + $\alpha \rightarrow$ N-14). In another experiment, they bombarded aluminum with alpha particles. After a short period of time they removed the source of alpha particles and noticed that though the aluminum continued to be radioactive, it was only for a short period of time. The reason: the aluminum atoms absorbed alpha particles and transmutated into an isotope of the element silicon that had a half-life of just 3.5 minutes (Al-26 + $\alpha \rightarrow$ Si-28).

Otto Hahn recognized the importance of this discovery and realized that the reaction could lead to the fission of the nuclei of larger elements and possibly result in a nuclear chain reaction. These discoveries led to the production of controlled nuclear fission chain reactions that have produced not only the atom (nuclear fission) bomb, but also electricity, and just as important, the production of a great variety of radioactive isotopes that are used in the medical and other industries. Note: a nuclear chain reaction occurs when a fissionable heavy element produces a succession of nuclear divisions that set neutrons free that then interact with and split other nuclei. This ends in stable nuclei of lighter elements, or if uncontrolled fission occurs, the reaction will end as a nuclear (atomic) explosion. A typical nuclear fission reaction starts by a heavy element absorbing and then producing neutrons. For example:

$$^{0}n \rightarrow {}^{235}U \rightarrow {}^{236}U \rightarrow {}^{93}Rb + {}^{141}Cs \rightarrow 2^{0}n$$

Irène, the daughter of Pierre and Marie Curie, married Frédéric Joliot the son of a local tradesman in 1926. After the marriage, Frédéric and Irène combined their surnames as Joliot-Curie. Frédéric began his research career at the Radium Institute in Paris under the guidance of Marie Curie, his mother-in-law, where he received his doctorate. He served in the French Resistance during World War II. In 1956 he became head of the Radium Institute, which was founded in 1914 by the University of Paris primarily because of the research efforts of Madame Curie.

Irène Joliot-Curie received her doctorate of science in 1925, also from the Radium Institute, based on her research on alpha rays of polonium (discovered by her mother). During World War II she served as a nurse radiographer.

Together and individually, Frédéric and Irène did important work in both natural and artificial radioactivity, the transmutation of heavier elements into lighter elements (e.g., uranium into lead), and the production of new radioactive elements and isotopes of elements. For their work, they jointly received the Nobel Prize for their discovery of artificial radioactivity and anticipated Otto Hahn's discovery of nuclear fission.

See also Curie (Pierre and Marie); Hahn

JOSEPHSON'S THEORY OF SEMICONDUCTORS: Physics: *Brian David Josephson* (1940–), Wales. Brian Josephson shared the 1973 Nobel Prize for Physics with Leo Esaki and Ivar Giaevar.

A DC voltage applied across a thin insulator between two superconductors produces a small alternating current whose frequency varies inversely to the voltage.

The BCS theory (named after John Bardeen, Leon Cooper, and John Schrieffer) demonstrated the concept of superconductivity at super low temperatures. The BCS theory states that under conditions of near absolute zero, electrons travel in pairs rather than individually, as the result of vibrations of the atoms. Josephson demonstrated this

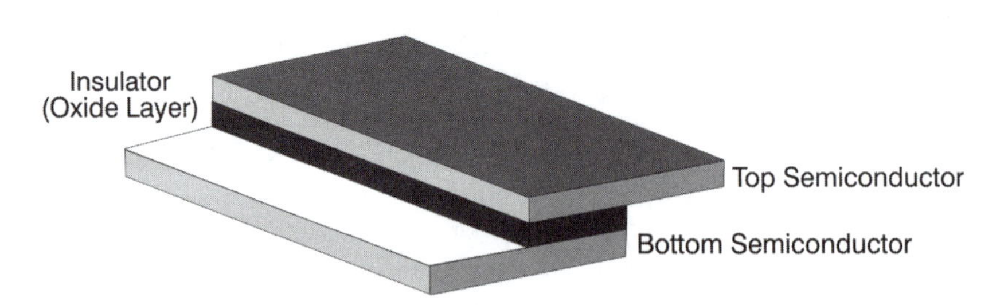

Figure J1. The Josephson Junction controls the operation of many devices that use semiconductors.

phenomenon by placing an insulator between two electron-conducting plates of metal known as a *Josephson junction*. The effects of these electrons flowing across this partially insulated junction produced a semiconducting flow of current known as the *Josephson effect*. A current can continue to flow across this junction for a short period of time even when the voltage is temporarily removed. In addition, a small current can produce an alternating current on the other side of the junction whose frequency varies inversely with the applied voltage. By using paired electrons Josephson maintained a tunneling effect that allowed the alternating super current to flow across the thin, insulating barrier of the **semiconductor.** Thus by changing the current's frequency the Josephson junction could be used as a means of controlling electronic devices somewhat like a switch.

This revolutionized the electronics industry and modern life. Today, the separator between the plates of semiconductors can be applied to a thickness of only one or a few atoms of material. Semiconductors are used in sensitive instruments to make accurate magnetic and electrical field measurements. Some applications of the Josephson junction are the detection of microwave frequencies, magnetometers, and thermometers to measure near absolute zero temperatures, and detection and amplification of electromagnetic signals. Of more importance, semiconductors are used to make high-speed (almost the speed of light) switching devices, which make modern computers possible.

See also Bardeen; Brattain; Shockley

JOULE'S LAW AND THEORIES: Physics: *James Prescott Joule* (1818–1889), Scotland.

Joule's law: *The relationship for heat produced by an electric current in a conductor is related to the resistance of the conductor times the square of the amount of current applied:* $H = RI^2$.

By experimentation, James Joule established the law that states that when a current of voltaic electricity is sent through a metal or other type of conductor, the heat given off over a specific time period is proportional to the resistance of the conductor multiplied by the square of the electric current. The equation for this law is: $H = RI^2$, where H is the rate of the heat given off as watts in joule units, R is the resistance in the conductor in ohms and I^2 is the amount of the current (amps) squared. The application of this law is important in all industries using electricity as a source of energy. The resistance to an electric current flowing through a conductor is analogous to the friction of air, the movement of engine parts, and tires on the road for a moving automobile. The electrical, as well as mechanical, energy is not just "lost," rather it is converted to heat, just as is friction. Joule was interested in improving the mechanical advantage of electric motors, but because they were very primitive during his lifetime, he devoted more of his work to improving the efficiency of steam engines. He accurately predicted that electric motors eventually would replace most other types of mechanical devices.

James Prescott Joule was one of five children in the family of a well-to-do brewery owner. Since he was a sickly child with a spinal deformity, both he and his brother were educated at home until the age of 15 and later by private tutors. The famous English chemist, John Dalton, taught them chemistry, physics, and the methods of scientific experimentation. Later in life Joule acknowledged that John Dalton encouraged him to increase his knowledge of science and of original research. When James' father died, he and his brother ran the brewery, which prevented him from attending a university. However, this did not deter him from setting up a laboratory in his home and continuing his interest in science after his day at the brewery. He became proficient in mathematics and learned how to make accurate measurements in the brewery. His home experiments resulted in his ability to measure slight increases in temperature under various conditions, which led to his theory for the equivalence of work and heat energy. The unit of work and energy was named after him (the Joule, or the symbol "J").

Law for the mechanical equivalent of heat: *A fixed amount of mechanical work (expenditure of energy) ends up in a fixed quantity of heat.*

Earlier in 1798 when Count Rumford was boring out the brass barrels of cannons, he noticed that large amounts of heat were generated. It became obvious that friction generated by the work of turning the bit in the metal resulted in heat. Julius von Mayer also was interested in this relationship and developed a figure for the mechanical equivalent of heat that was not very accurate. James Prescott Joule was the first to consider heat as a form of energy in his calculation. He conducted exacting experiments to determine the amount of heat generated not just by electricity but also by mechanical work. Joule calculated the amount of mechanical work needed to produce an equivalent amount of heat. He demonstrated that 41 million ergs of work produced 1 calorie of heat, which is now known as the *mechanical equivalent of heat*. Since 10 million ergs are equal to 1 joule, named after James Joule, 4.18 joules are then equal to 1 calorie of heat. Joule's work enabled others to perfect the law for the conservation of energy, which states that energy, like mass, cannot be created or destroyed but can be changed from one form to another.

Joule–Thomson effect: *When a gas expands, its internal energy decreases.*

James Joule collaborated with Lord William Thomson (later known as Lord Kelvin) to devise the Joule–Thomson effect, which is related to the kinetic theory of gases. They measured the change in energy involved when the pressure of a compressed gas is released and then expands. As a gas expands, the motion of molecules is reduced. In other words, its internal energy is decreased, and thus its temperature. It can be reheated if it can "consume" energy from its surrounding environment, thus providing a cooling effect to the area around it.

See also Ideal Gas Law; Kelvin

KAMERLINGH-ONNES' THEORY OF MATTER AT LOW TEMPERATURES: Physics: *Heike Kamerlingh-Onnes* (1853–1926), Netherlands. Heike Kamerlingh-Onnes was awarded the 1913 Nobel Prize for Physics.

Some metals lose their electrical resistance at super-low temperatures.

Heike Kamerlingh-Onnes experimented with the properties of matter at low temperatures and improved on the apparatus and procedures employed by Sir James Dewar in the late 1800s. This enabled him to use liquid hydrogen and the Joule–Thomson evaporation effect to cool helium to a temperature of 18 kelvin (which is 18° above absolute zero). After cooling the gas still more and by allowing it to expand through a nozzle, he determined liquid helium has a boiling point of 4.25 kelvin (it was known at the time that when a gas expands, its temperature is reduced). When the liquid helium is in an insulated container and the vapors are rapidly pumped away, the liquid helium is cooled still further to just 0.8 kelvin. This was as close to absolute zero as had so far been reached at that time in history. He was the first to study the nature of materials at this extremely low temperature and ascertain that molecular activity (kinetic energy) almost ceases at this temperature. This was the beginning of the science of **cryogenics**, which led to the observation of superconductivity, where metals lose their resistance to electricity, thus enabling electric currents to pass through wires without generating heat by internal resistance. In 1911 he determined that metals such as lead, tin, and mercury become superconductors at these very low temperatures. Today, scientists know that at least two dozen elements and many hundreds of compounds become superconductive at near absolute zero temperatures. In 1986 a ceramic-type compound exhibited superconductivity at about −196°C, which just happens to be the temperature at which nitrogen gas becomes a liquid. This is important because it is easier to reduce temperatures to near absolute zero using helium, but its source is limited and thus is very expensive. Although nitrogen can also be liquefied and used for low-temperature research, it is not as efficient for this purpose as is helium, but the supply of nitrogen gas is almost unlimited as four-fifths of air is nitrogen. Liquid nitrogen is now used for supercooled magnets in particle accelerators and magnetic resonance imaging (MRI) equipment in the medical industry. The size of the magnets required for this equipment can be greatly reduced through the use of supercooled magnets. Scientists have not found many other practical uses for superconductivity because of the difficulty of achieving and maintaining sufficiently low temperatures. Currently, scientists are at work attempting to achieve warm, ambient air superconductivity, which will provide the same low resistance to electricity but at temperatures much higher than absolute zero. Physicists in several countries claimed to have produced materials that become supercondutive at room temperatures. Most of these experiments can't be replicated even though a few exhibit some characteristics of superconductivity. It is hoped this research will succeed in achieving superconductivity at normal temperatures of room air. Such an accomplishment will result in the development of more cost-effective methods for transmitting electricity, as well as in the production of supermagnets to levitate high-speed trains.

See also Dewar; Joule; Kapitsa

KAPITSA'S THEORY OF SUPERFLUID FLOW: Physics: *Pyotr Leonidovich Kapitsa (or Pjotr L. Kapitza)* (1894–1984), Russia. Pyotr Leonidovich Kapitsa shared the 1978 Nobel Prize for Physics with Arno Allan Penzias and Robert Woodrow Wilson.

> *Thin film-vapor systems exhibit superflow properties at very low temperatures, with the resistance to flow increasing as the film's thickness increases.*

Pyotr Leonidovich Kapitsa developed an improved method for liquefying air, which enabled him to study the properties of liquid helium. He determined that liquid helium, known as He-II, behaves as a "superfluid" at near absolute zero and exhibits very unusual flow characteristics. At this temperature liquid helium appears to be in a perfect atomic macroscopic *quantum* state with perfect atomic order. He-II exhibits a super-thin film that manifests some novel forms of internal convection, including its ability to flow up the sides of its container, even when the container is closed. Kapitsa's methods for liquefying gases were used to facilitate the commercial production of liquid oxygen, nitrogen, hydrogen, and helium. Large-scale production of these gases enabled the development of very high magnetic fields used in many areas of research and technology, such as particle accelerators and nuclear magnetic resonance (NMR) instruments.

See also Kamerlingh-Onnes; Kusch

KAPTEYN'S THEORY OF GALACTIC ROTATION: Astronomy: *Jacobus Cornelius Kapteyn* (1851–1922), Netherlands.

> *Stars in the sky move in two different streams in two different directions.*

Kapteyn was an excellent observer of stars. In 1897 he discovered what became known as "Kapteyn's Star" that exhibited the greatest proportion of proper motion (8.73 seconds of annual motion) of any star known at that time. Later, Kapteyn's Star was relegated to second in its proper motion when Barnard's Star was discovered to exhibit 10.3 seconds of annual motion (Barnard's Star is named after the noted American astronomer Edward Emerson Barnard (1857–1923) who in 1916 found that it had the largest proportion of proper motion of all known stars). In 1904 Kapteyn cataloged over 454,000 stars of the Southern Hemisphere by using the photographic plates made by the Scottish astronomer David Gill (1843–1914). While studying these photographic plates, as well as doing his own observations, Kapteyn hypothesized that the stars in the heavens were moving and heading in two different directions. One stream consisting of about three-fifths of all the stars in galaxies was heading toward the constellation Orion and the other stream consisting of about two-fifths of all the known stars was heading in the direction of the constellation Scorpius. More important, the line between them led through the Milky Way galaxy that was originally discovered by William Herschel. Herschel calculated the size of the Milky Way galaxy as fifty-five thousand light-years across and eleven-thousand light-years thick. Using this model, Kapteyn proposed an arrangement and motion of the sidereal system, which he published in 1922, as a lens-shaped, rotating island universe that was denser at its center and less dense at the edges. This became known as "Kapteyn's Universe," which he estimated as forty-thousand light-years in size with the sun at its center. A more up-to-date figure for the size of the Milky Way is one hundred thousand light years across from edge to edge. At that time in astronomy's history, Kapteyn's and other astronomers' concept of a universe was limited to this single large galactic star system. Today astronomers believe that the universe is everything that can be observed and from which we can gain knowledge. This includes a multitude of "island universes" or galaxies.

See also Herschel; Oort

KARLE'S THEORY FOR DETERMINING MOLECULAR STRUCTURE: Physics: *Jerome Karle* (1918–), United States. Jerome Karle shared the 1985 Nobel Prize for Chemistry with Herbert A. Hauptman.

Mathematical methods can be used to deduct the molecular structure of chemical compounds and to explain the X-ray patterns formed by the compounds' crystals.

Jerome Karle began his academic and professional career at the University of Michigan where he met and married his wife, the former Isabella Lugoski (1921–), the renowned X-ray crystallographer and researcher. After completing the requirements for his doctorate at Michigan in 1943, he, along with Isabella, began work on the Manhattan Project at the University of Chicago. After the war, they went to work for the Naval Research Laboratory in Washington, D.C., where they began a life-long career and interest in the structure of crystals. Jerome Karle was concerned with the theoretical and mathematical aspects of crystallography, while Isabella Karle applied the practical application of her husband's research, as well as her own investigations. In every sense of the word, they were a "team."

At the time the Karles moved to the Naval Research Laboratory, they were joined by the American mathematician Herbert Hauptman (1917–). This was the beginning of three decades of collaborative research on crystallography by Karle and Hauptman that culminated in the 1985 Nobel Prize for Chemistry (after leaving the Naval Research Laboratory in 1970, Hauptman became research director of the Medical Foundation of Buffalo). In 1953 Jerome Karle and Herbert Hauptman published a monograph titled "Solution of the Phase Problem I: The Centosymmetric Crystal" in

Jerome Karle was born in New York City, received his undergraduate degree in biology from City College in New York and a master's degree from Harvard, followed by a PhD in physical chemistry from the University of Michigan in 1943. While at the University of Michigan he met and married Isabella Lugoski (1921–) the daughter of Polish immigrants in 1942. She did not hear any English spoken until she entered elementary school. At age 19 Isabella entered the University of Michigan where she received three degrees, including a PhD in physical chemistry in 1944. Jerome and Isabella worked together on the Manhattan Project in Chicago, Illinois, from 1943 to 1944 and as researchers at the Naval Research Laboratory (NRL) in Washington, D.C. Their work progressed from electron diffraction to improved methods of X-ray diffraction for the study of crystallography of matter. She applied her powerful techniques to directly calculate the diffraction patterns of many chemicals that were applied to improved medical procedures. In 1959 Isabella was appointed as head of the X-Ray Diffraction Section for the Structure of Matter at the National Research Laboratory. She has received many awards including the American Chemical Society's Garvan Medal in 1976 and was appointed president of the American Crystallographic Association, also in 1976. She received the 1985 Chemical Pioneer Award by the American Institute of Chemists. Isabella also received the $250,000 Bower Award and Prize for Achievement in Science by the Franklin Institute. However, she did not share in the 1985 Nobel Prize for which the work of her husband and Herbert Hauptman was recognized.

which they described a new method of forming X-ray diffraction patterns of crystals of chemical compounds. This monograph was the foundation for the solution to the "phase" problem of X-ray crystallography. The older, so-called heavy atom procedure that was used to enhance the crystal structure to more clearly indicate the crystal's diffraction pattern involved adding a heavy element to a particular area of the crystal to be studied. This technique was clumsy, and the resulting X-ray diffraction pattern could only be inferred. While at the Naval Research Laboratory, they arrived at a mathematical equation that could explain the arrangement of dots on films that showed the X-ray diffraction of crystals. In other words, using probabilistic methods for newly devised mathematical equation, the crystal's phase structure could be directly determined. This was a great improvement over the old "heavy atom" technique because the equation enabled the exact location of specific atoms within the crystal's molecules to be identified. Thus, the intensity of the "spots" on the film that depicted the types of atoms and their positions within the crystal's structure could be analyzed. It was mainly through the efforts of Isabella Karle, who called attention to the practical applications of crystal diffraction, that other scientists in the field were encouraged to adopt their methods for determining the three-dimensional structure of crystalline compounds. This research technique has applications in numerous fields, including pharmacology, molecular biology, chemistry, physics, metallurgy, geology, and genetics.

See also Franklin (Rosalind)

KEKULE'S THEORY OF CARBON COMPOUNDS: Chemistry: *Friedrich Kekule von Stradonitz* (1829–1896), Germany.

> *Carbon is a tetravalent atom (a valence of 4) capable of forming ring-type organic molecules as well as linear molecules.*

Friedrich Kekule von Stradonitz was the first to propose a structural formula that indicated atoms bonded with each other to form molecules. His study led to the concept of the carbon atom's structure consisting of four (tetravalent) bonds with a central nucleus by which it could form numerous types of molecules (*see* Figure V3 under Van't Hoff). He also related this unique atom to the basic structure of all organic (living) carbon compounds. However, one form of carbon puzzled the chemists of his day. Michael Faraday discovered that the molecule for the aromatic compound benzene contained six carbon atoms with a total of twenty-four shared (bonding) electrons, but benzene also had six hydrogen atoms; however, each hydrogen atom had only one bonding electron. When diagramed as a linear or even a branching molecule, this combination was impossible because each carbon atom had four bonds (valence) and each hydrogen atom had just one bond (valence), thus $6 \times 4 = 24$ for carbon, and $6 \times 1 = 6$ for hydrogen, totaling thirty bonding electrons in all for the molecule. Therefore, there existed an incorrect number of electrons to satisfy the octet rule for a linear structure such as C_6H_6. Reportedly, Kekule solved this problem one night as he dreamed. In his dream he saw different configurations of atoms forming various arrangements. One arrangement resembled a snake eating its own tail (*see* Figure K1).

BENZENE

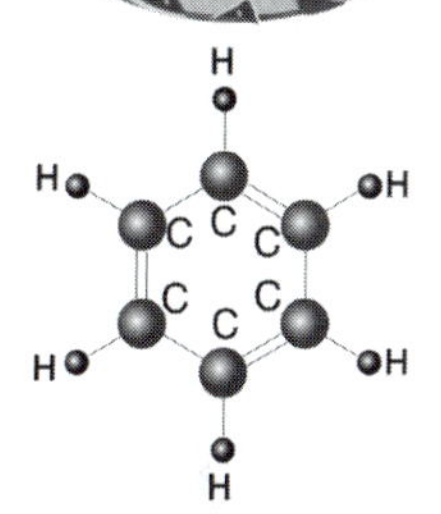

Figure K1. Artist's depiction of Friedrich Kekule's dream of a snake eating its own tail that aided him in solving the problem of the structure for the organic compound benzene, composed of a ring of carbon atoms.

He woke up energized, and, working the rest of the night, came up with the structure of the benzene ring. The ring consisted of each carbon atom sharing two of its four bonding electrons with another carbon atom, one valence electron with a partner on its other side on the ring, and one valence electron with a hydrogen atom outside the ring. The result is the classical benzene, hexagon ring. This answered many questions and was a revolution for organic chemistry. It was then possible to substitute another atom, a *radical*, or a molecule for one of the hydrogen atoms of the six in the ring. This resulted in what is known as a single substitution to form a derivative of the benzene ring (e.g., C_6H_5X). Specifically, if the free radical (a molecular fragment with a single unshared electron and no charge) such as NH_2 were substituted for one of the hydrogen atoms the compound aniline $C_6H_5 \bullet NH_2$ would be the result. In addition, it is possible to combine many of the hexagonal benzene rings to form more complex compounds containing many hexagonal structures. This answered the questions related to the great multiplicity of organic compounds. To some extent, this seems odd because Kekule considered the existence of atoms a metaphysical problem and claimed chemistry was concerned only with arriving at hypotheses that explained chemical structures and reactions. It might be noted that Kekule's rings were originally known as "Kekule Sausages" because he represented his hexagon molecule and its electrons as a somewhat difficult system of circles—not lines as shown in the diagram. The Scottish chemist Alexander Crum Brown (1838–1922) conceived the structure of the benzene ring as it is known today in 1865.

See also Couper

KELVIN'S CONCEPT OF ENERGY: Physics: *William Thomson Kelvin* (known as Lord Kelvin) (1924–1907), Scotland.

Kelvin's theory of thermodynamics: *When a gas cools, loss of volume is less crucial than loss of energy. Kinetic energy (molecular motion or heat) ceases at temperatures approaching absolute zero (zero Kelvin = −273.16°C).*

William Thomson Kelvin was knighted by Queen Victoria and given the title of Lord Kelvin of Largs for his work on electromagnetic fields. Kelvin was familiar with the works of Carnot, Joule, and Clausius, early explorers in the field of thermodynamics where mechanical heat is related to energy. The second law of thermodynamics deals with the concept of entropy, which states *that heat always flows from a warmer object to a cooler environment—and never in the opposite direction*. For example, a hot cup of coffee always cools and will never get warmer than its surroundings unless an external source of heat is applied to the cold coffee. This means that energy in a closed system is striving to reach a state where there is no transfer of energy—equilibrium. Eventually everything will be the same temperature. Another way of saying this is that everything becomes more disorganized (entropy) and "runs downhill" to a common level (equilibrium), unless more energy is pumped into the isolated system. But then it would no longer be a closed system. For instance, if Earth did not receive most of its energy from the sun, everything would run down, and it would soon be a very cold place. Kelvin provided a mathematical formulation of the second law of thermodynamics. Because entropy (disorganization) always increases, Lord Kelvin believed the universe would sometime in the future have maximum entropy and thus a uniform temperature. He called this the heat death of the universe. The first law of thermodynamics relates to the conservation of energy which states *that energy can be neither created nor destroyed but can be transferred from one form to another (e.g., mechanical energy to heat energy, as in rubbing your hands together vigorously, generating friction heat), or chemical energy transformed to light and heat (a burning candle)*. Heat itself is the manifestation of the kinetic energy of molecular motion, whereas the temperature of a substance is proportional to the average motion of the molecules when they are in thermal equilibrium, that is, the temperature is a measure of the average internal energy. Kelvin proposed a new scale for measuring the absolute temperature of matter, which at its zero point would be the lowest temperature possible. He started with what is known as the triple point of pure water, which is about 0.01°C, where equilibrium between the water, ice, and water vapor is established. At this temperature, water can exist in its three physical states at the same time: solid, liquid, and gas. The point was also used to set up the metric temperature scale originally referred to as the Centigrade scale because it was based on a scale of 100 but now referred to as the Celsius scale, where 0 is the freezing point of water (or melting point of ice) and 100 is the boiling point of water, where it attains its gaseous state. Selecting 100-degree units for this scale was arbitrary but assisted in metric calculations. Any units could be used, such as the units for the Fahrenheit temperature scale. Lord Kelvin used the same metric (based on 100) scale and, by extrapolation, arrived at −273°C degrees as the absolute zero point. This point was originally called A, for absolute. Therefore, water freezes at 273 kelvin and boils at 373 kelvin. Absolute zero was later refined to equal −273.16°C, or 0.0 kelvin. This absolute temperature scale, named after Lord Kelvin, provides for the measurement of very cold and very hot temperatures. Because no thermometer has been invented that can measure absolute zero, Lord Kelvin reached this point by theoretical consideration. Some people believe that all molecular motion ceases at this point. This is not quite correct because molecules of solids continue to "vibrate" but not move at random or exhibit any kinetic energy as molecules would in matter at higher temperatures. In other words, the energy has the lowest possible value at absolute zero and the entropy is zero.

Kelvin's theory of electromagnetic fields: *Electromagnetic fields travel through space as do light waves; the electric field vector and the magnetic field vector vibrate in a direction transverse to the direction of the wave propagation.*

Lord Kelvin's theory for electromagnetic fields stated the fields associated with alternating current (AC) electricity are waves that travel through space similar to light waves. His theory proved that both the types of waves not only are transverse waves, but also travel at the same speed. Kelvin's theory for electromagnetic fields was put to good use and provided the information necessary to lay the first successful transatlantic telegraph cable. Two previous attempts for an ocean-spanning cable had failed. In addition, his theory led to the electromagnetic theory of light developed by James Clerk Maxwell.

See also Carnot; Joule; Maxwell

KENDALL'S THEORY FOR ISOLATING ADRENAL STEROIDS: Biochemistry: *Edward Calvin Kendall* (1886–1972), United States. Edward Kendall shared the 1950 Nobel Prize for Physiology or Medicine with Philip S. Hench and Tadeus Reichstein.

> *The hormones and amino acids secreted by the adrenal glands are a number of steroids identified as* A, B, E, *and* F *that are responsible for many physiological activities of the body.*

Edward Kendall received his academic training at Columbia University in New York where he received his undergraduate and graduate degrees in chemistry. After initially working on research on the thyroid gland for a pharmaceutical company in Michigan and at St. Luke's Hospital in New York, he moved to the Mayo Foundation in 1914. He was appointed head of the Biochemistry Section in their graduate school. He subsequently became a director of the Division of Biochemistry and Professor of Physiological Chemistry. Even after his retirement in 1951, he remained a visiting professor at the Mayo Foundation. While at the Mayo Foundation, Kendall began the research that ultimately led to the discovery of a number of hormones produced by the cortex of the adrenal glands. He found that these adrenal secretions could be isolated and used to treat Addison's disease, as well as rheumatoid arthritis. As is often the case, similar research was being conducted independently in another country, in this case Tadeus Reichstein in Zurich, Switzerland.

Kendall's research on cortisone is related to the original work by Thomas Addison (1793–1860) of Scotland who in 1849 discovered a connection between the adrenal glands and a rare disease that later became known as Addison's disease (an insufficient secretion of hormones by the adrenal cortex). This discovery encouraged a number of medical researchers to further examine the secretions from this gland that is located near the kidneys. This new secretion was called the hormone "cortin." Further experimentation by Kendall in the 1930s led to the identification of more complex secretions. Four of these hormone compounds were named A, B, E, and F; other substances were found later in adrenal secretions. It was also determined that the adrenal secretion cortin was a steroid. Steroids are naturally occurring lipids (fatty components of living cells) that are derived from cholesterol produced by the body. In 1948 Kendall produced a few grams of the compound E, and with the assistance of Dr. Philip S. Hench (1896–1965), his colleague and fellow researcher at the Mayo Foundation, he used it to treat patients with rheumatoid arthritis. It then became known as a "wonder drug" and was renamed "cortisone" in 1949.

Cortisone and related forms of adrenal secretions can be artificially produced and used to treat (but not cure) a number of ailments, particularly those of an inflammatory nature, such as asthma and arthritis. This is somewhat similar to the use of insulin to treat patients with diabetes because both drugs must be taken daily to maximize the therapeutic effect. Many diseases are successfully treated with the use of cortisone and other steroids that have saved lives and mitigated the suffering of patients. Some examples are rheumatic and endocrine disorders; dermatological, collagen, and neoplastic diseases; allergic, edematous, and gastrointestinal diseases; and tuberculosis. Cortisone is also used when performing organ transplants to minimize the body's defense mechanisms to foreign substances in the implanted organ. Some of the possible side effects of using cortisone as well as some steroids are depression and other psychic disorders including physiological and/or personality changes, insomnia, mood swings, and possibly the development of glaucoma.

See also Reichstein

KEPLER'S THREE LAWS OF PLANETARY MOTION: Astronomy: *Johannes Kepler* (1571–1630), Germany.

Law I: *All the planets revolve around the sun in elliptical paths, and the sun occupies one of the two focal points for the ellipse (the other focal point is imaginary).*

Kepler's mathematical analysis of Tycho Brahe's data resulted in the concept of planetary orbits as being ovoid (egg shaped). However, after checking his data, he corrected an error in calculation and realized that all planetary orbits, including the orbit of Mars, are elliptical.

Law II: *An imaginary straight line joining the sun and a planet sweeps over equal areas in equal intervals of time.*

Kepler's second law follows directly from the first law. Also referred to as the law of areas, it is probably the most important and easiest to understand. Kepler measured the distance of a short path of a planet progressing along a segment of its elliptical perimeter. In addition, he measured the time elapsed for the planet to cover this short segment of its orbit. Using geometry, he determined the area for the pie-shaped wedge of space formed by the two sides of a triangle originating at the sun (the meeting point of these two lines), which extends to the perimeter. The area of this pie-shaped wedge was related to the distance covered by the planet along its elliptical perimeter in a given period of time. He then made similar measurements as the planet progressed to different segments (chords) of its orbit (*see* Figure K2). When the planet was at its closest to the sun, it traveled much faster in its orbit to cover an area equal to that area covered when it was farthest from the sun.

KEPLER'S LAW

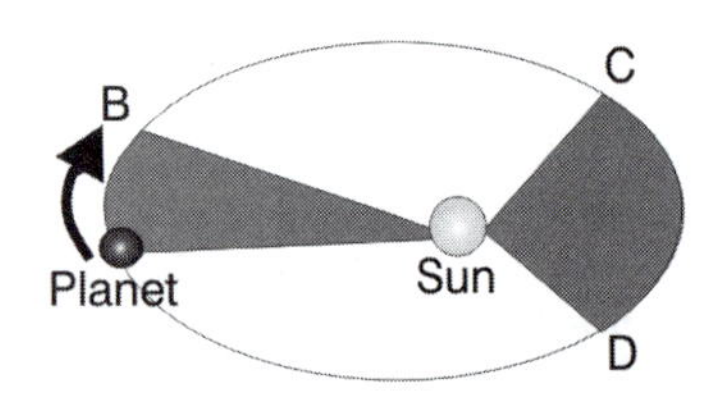

Figure K2. Kepler's second law states that planets revolve around the sun in an elliptical orbit and do so in a manner that equal areas are covered in equal times. (Figure not to scale.)

This law of areas was extremely important for Isaac Newton's formulation of his concept of gravity and his laws of motion. Kepler's second law also explains the theory for conservation of angular momentum for bodies in nonlinear motion. For instance, when an ice skater spins rapidly with arms extended and then pulls his or her arms in close to the body, increasing the rate of spin now conserves the momentum gained when the arms were extended. The same is true for an Earth-orbiting spacecraft. When it drops to a lower orbit, increasing its speed relative to the speed it had obtained in a higher orbit conserves its momentum.

Law III: *The square of planets' orbital periods is proportional to the cubes of the semi-major axes of their orbits.*

Another way to say this is that the square of the time it takes a planet to complete one orbit around the sun (orbital period) is proportional to the planet's average distance from the sun cubed. This may be expressed as: $P^2 = (AU)^3$, where *P* is the time it takes a planet to complete one revolution around the sun in years, and *AU* is the astronomical unit, which is equal to the average distance between Earth and sun, or about ninety-three million miles. Kepler developed his third law while attempting to devise a mathematical basis for musical-type harmony as related to his first two laws.

Kepler tried to apply mathematics to Plato's concept of five regular solids and from this derive mathematical harmony for Plato's model of the universe. He continued to apply his mathematics to achieve harmony with Copernicus' concept of a sun-centered solar system. After leaving his home in Germany, Kepler secured a position in Prague with Tycho Brahe, a firm proponent of Ptolemy's Earth-centered universe who never accepted the Copernican heliocentric theory. Tycho assigned his new assistant Kepler the task of observing and measuring the orbit of Mars. Kepler at first thought it would be a simple task, but it took him over eight years to complete it. Tycho died soon after Kepler joined his staff, leaving reams of data from his own extensive observations, which Kepler put to good use. Many historians give Tycho, not Kepler, credit for discovering that Mars' orbit is elliptical because this discovery was partly due to Tycho's sizable records that aided Kepler's accomplishment. This work led to Johannes Kepler's three laws of planetary motion, which remain valid.

See also Brahe; Galileo; Newton

> Kepler's laws have applications far beyond the orbits of planets. They apply to all kinds of bodies found in the universe that are influenced by gravity: moons orbiting their mother planets, other solar systems, binary stars, and artificial satellites orbiting Earth, for example. Kepler was also interested in other sciences beside astronomy. From his study of optics and vision, he developed a theory stating that light from a luminous body is projected in all directions, but when the human eye viewed this light, only the rays that enter the pupil of the eye were refracted, ending up as points on the retina. This is much closer to today's concept of electromagnetic radiation as related to vision than the ancient Greek idea of the eye sending out a signal to the object, which was then reflected back to the eye. After Galileo developed his telescopes, Kepler explained how the lenses of the instruments worked. He did the same for the new eyeglass spectacles. He was very supportive and complimentary of Galileo's work, even when others ridiculed Galileo's observations. It seems Galileo either did not understand or appreciate Kepler's mathematical contributions to astronomy because Galileo ignored his publications.

KERR'S THEORY OF QUADRATIC ELECTRO-OPTIC EFFECT (I.E., KERR EFFECT): Physics: *John Kerr* (1824–1907), Scotland.

As the electric field slowly varies with the voltage across a material, the material becomes birefringent at different indexes of refraction.

The Kerr effect, also know as the "quadratic electro-optical effect" or QEO, is a change in the refractive index of a material in response to an electric field, also known as **birefringence**. A more complete definition of the Kerr effect is the nonlinear interactions of a light beam within a medium (transparent solid or liquid) with an instantaneous response as related to the nonlinear electronic polarization.

The QEO theory can be expressed by a series of complex equations for linear and nonlinear relationships where the polarization varies with the electric field involved. Different equations are derived for different types of materials, including nonsymmetric media, such as liquids, where the refractive index is changed in the direction of the electric current. For instance, some liquids such as nitrobenzene ($C_6H_5NO_5$) have large Kerr constants that are exhibited by a glass cell (jar) containing these types of liquids. It is called a Kerr cell and is used to modulate light by quickly responding to changes in an electric field. Some applications of the Kerr cell are to measure the speed of a beam of light and for use with super-high-speed camera shutters. The Kerr effect can determine the change in the reflective index of an electric field that is the variation in the index of refraction, which is proportional to the local irradiance of the light beam. This effect is greatest when associated with very intense laser light beams. The Kerr effect can accomplish the same on magnetic fields, which is called the "magneto-optic effect."

KERST'S THEORY FOR ACCELERATING NUCLEAR PARTICLES: Physics: *Donald William Kerst* (1911–1993), United States.

> *Using a process for "stacking beams" by means of radio frequencies, it is possible to achieve center-of-mass energies through the use of colliding beams, thus greatly increasing the Me-V energies of accelerated particles.*

In 1929 John Cockcroft and Ernest T.S. Walton constructed a low-energy linear-type of "atom smasher" that used a voltage multiplier to build up a voltage high enough to accelerate alpha particles beyond the speeds they naturally obtained from radiation. These early linear accelerators built in a straight line did not provide the accelerated ions (charged particles) the energy required to enable them to smash into the nuclei of atoms to produce smaller subnuclear particles for study.

Dissatisfied with the low energies of particles achieved in the straight-line accelerators, Ernest O. Lawrence produced a design for a new type of accelerator. He believed that if the charged particles could be made to follow a circular path in a spiral, they could be influenced by a strong magnet during each path at every revolution. This "kick" at the beginning of each revolution would increase their speed and thus the energies necessary to interact with heavy nuclei of atoms. Lawrence's idea for a circular particle accelerator was based on the theory that the "bullet" particles, such as electrons, positrons, or heavier beta ions, could be continually pushed around the circular path building up to very high speeds, resulting in electron volts (eV) adequate to "smash" the nuclei of atoms. He constructed a small device with two "D" shaped semicircular units facing each other with a four-inch gap between them. This formed a circular unit he called a "cyclotron (*see* Figure F3).

Applying high-frequency fields to particles that have a charge (ions) will send them round-and-round in the two Ds. Thus, particles receive a "push" every time they pass the gap and reach very high energies as they follow the circular path formed by the two Ds.

While a professor at the University of Illinois in 1940, Donald Kerst developed the "betatron" to provide beams of electrons that exhibited much higher energy than the particles could achieve in the cyclotron. To increase the speed (and thus energy) of electrons to almost the speed of light, Kerst designed a torus-shaped (semicircular convex) vacuum tube that was associated with a transformer. Because the mass of all particles increases at relativistic speeds as they approach the speed of light, they become extremely massive—meaning that particles with mass can never reach the speed of light because they would become infinitely heavier. Therefore, it requires great energies to accelerate particles to near the speed of light. Cyclotrons became less efficient as they could only generate energies of a few electron volts (eV). However, by the 1950s the betatrons could generate over 310 MeV. At these levels high-energy beams of electrons, when directed at metal plates, produce X-rays and gamma rays that are useful in industrial and medical applications, such as radiation for cancer treatment. Today, the next generation of accelerators, called synchrotrons, achieve even higher MeVs to explore the basic nature of particles. Other more powerful particle accelerators are planned.

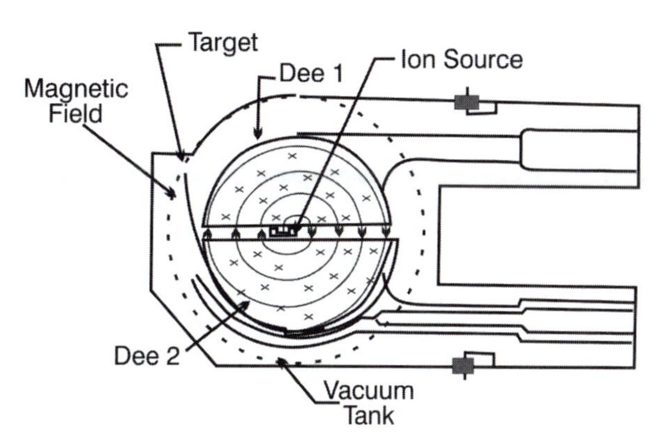

Figure K3. The cyclotron is a circular particle accelerator that was designed by E. O. Lawrence. It was based on the concept of two "D" shaped rings that could bring the particles to very high energies.

Donald Kerst was a widely recognized and honored physicist who held many honorary degrees and awards. He began his academic career in the Midwest at the Universities of Illinois and Wisconsin in the late 1930s. During World War II he worked at Los Alamos, New Mexico, from 1943 to 1945. Colleagues remember him as an influential, hard-working researcher whose development of the betatron was only one of the contributions he made to the field of nuclear physics.

See also Higgs; Lawrence; Van de Graaff

KHORANA'S THEORY OF ARTIFICIAL GENES: Biochemistry: *Har Gobind Khorana* (1922–), India, United States. H. Gobind Khorana shared the 1968 Nobel Prize in Physiology or Medicine with Robert W. Holley and Marshall Warren Nirenberg.

There is a biological language code common to all living organisms that is spelled out in three-letter words, and each set of three nucleotides is a code for a specific amino acid which can also be produced artificially.

Har Gobind Khorana was born in West Punjab (now in Pakistan) and educated in Indian schools and universities. In 1945 he entered the University of Liverpool in England where he received his PhD in 1948. After a brief time in India in the fall of 1949, he returned to England and the University of Cambridge where he became interested in research involving proteins and nucleic acids. In 1960 he moved to the University of Wisconsin where he started his work on understanding and possibly unraveling the genetic code. By combining different methods of analysis Khorana was able to overcome many obstacles to the chemical synthesis of polyribonucleotides. He proved that **codons** are triplets of mRNA (messenger RNA) that carry codes for each amino acid.

He synthesized each of the sixty-four nucleotide triplets that make up the code. This meant that three nucleotides specify an amino acid that determines the direction in which the information in messenger RNA can be read and that these codons (tri-nucleotide sequences) do not overlap. This also demonstrated that RNA is involved in translating the sequence of nucleotides in DNA into the sequences for the amino acids that make up proteins. Dr. Khorana's research added to the work already done in this field by American biochemists Marshall Nirenberg (1927–) at the National Institutes of Health and Robert Holley (1922–1993) at Cornell University and the Salk Institute in La Jolla, California. All three shared in the 1968 Nobel Prize for Physiology or Medicine.

This work with the genetic code led to Khorana's major achievement in 1970 when he synthesized oligonucleotides (strings of nucleotides), that is, the first artificial gene. After moving to the Massachusetts Institute of Technology he announced in 1976 the invention of a second artificial gene that was capable of living and operating in other living cells. This discovery opened the doors for research in various areas involving genetics. One possibility of practical use is in developing a cure for hereditary (genetic) diseases and artificially producing a source of proteins such as insulin.

See also Crick; Franklin (Rosalind); Pauling; Watson (James)

KIMURA'S NEO-DARWINIAN THEORY FOR MUTATIONS: Biology: *Motoo Kimura* (1924–1994), Japan.

Genetic mutations at the molecular level can increase within a population without being affected by Darwinian natural selection.

Motoo Kimura developed data indicating there are certain types of mutations that multiply in a given population without resulting in these mutations being selected out, as proposed by Darwin's concept of natural selection. By using chemical means, he identified several molecules for mutant genes that proved not to be harmful to the individual. In fact, some of these seemed to adapt better than nonmutated genes. He concluded that evolutionary changes might be caused by a normal drift of selected genes that may have mutated. There is recent evidence of exceptions to his theory, which is based on findings that several mutations at the molecular level are selective (not random) and do cause evolutionary changes. One exception is the mutated gene affecting human hemoglobin in the blood. Another is the genetic decoding study conducted for the population of Iceland. Over the centuries, Icelanders were isolated and their numbers often reduced by natural disasters. Recent researchers found that an ancient survivor carried a mutated gene that was missing five units of DNA. The researchers concluded that the 275,000 modern Icelanders are somewhat inbred, as evidenced by similar characteristics (blue eyes, blond hair). Fortunately, Icelanders have maintained excellent genealogical and health records, which assisted in this genetic research. The descendants of this one person, who now compose a large portion of the present population, carry this mutation, which causes a high risk of breast cancer for men and women. This seems to be an example of a mutated gene at the molecular level that has drifted and was selected according to evolutionary theory, thus disproving Kimura's theory.

See also Darwin; Haldane

KIMURA'S THEORY FOR VARIATIONS IN EARTH'S LATITUDES: Astronomy: *Hisashi Kimura* (1870–1943), Japan.

> *Slight differences in latitudes at different geographic locations are due to an uneven distribution of mass around Earth's axis.*

In 1765 Leonhard Euler observed that there were slight variations in **latitude** measurements at different geographic locations on Earth, but he had no understanding of why this might be so.

Hisashi Kimura spent his career as an astronomer studying and measuring the variations in Earth's latitudes. He was the director of the International Latitude Observatory in Mizusawa, Japan, that became one of the six observatories selected by the International Geodetic Association in 1899 for the study of latitudes around the world on a line of 39°08′ North. Kimura knew that a perfect symmetrical sphere with an even distribution of internal mass would have a stable spin. But it was also known for some years that Earth's mass is not distributed evenly around its axis. There are variations on both its surface and internally of quantities of mass at various geographic locations. Therefore, Earth is not symmetrical, and the axis of rotation and the axis of the moment of inertia do not match up. This discrepancy results in a slight wobble with a periodicity of about 14 months. However, the distance is small, and over this fourteen-month period Earth's poles drift about 63 feet (this spin is somewhat like the wobble of a child's top as it spins). This wobble is known as "**precession**" and was first discovered in 1891 by American astronomer Seth Carlo Chandler, Jr. (1846–1913). Today it is known as the "Chandler wobble." This precession is evident as Earth wobbles and its axis sweeps in a cone-shaped path at the poles. It takes about twenty-six thousand years to complete one cycle. In the year 2100 Earth's axis will point toward the star Polaris and then change its direction and thirteen thousand years later (one-half the twenty-six thousand period), the axis will point to the star Vega (*see* Figure K4).

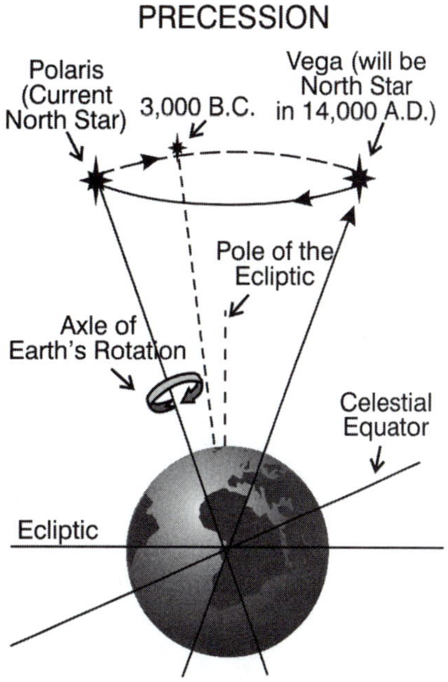

The six observatories in the latitude study compared their latitude measurements at their locations and found discrepancies in latitudes of the six different observatories due to motion of Earth's poles. In 1902 Kimura also came up with a new term to describe some of these differences in latitude that was not dependent on precession. More recently this phenomena has been considered as one of the possible natural variables influencing climate change and the slight global warming (average about 1°C over the last century) of Earth due to changes in the direction of Earth's axis over long periods of time and also possibly due to natural phases of the sun's output of heat, as well as possible effects of modern civilization.

See also Euler

PRECESSION

Polaris (Current North Star) — 3,000 B.C. — Vega (will be North Star in 14,000 A.D.) — Pole of the Ecliptic — Axle of Earth's Rotation — Celestial Equator — Ecliptic

Figure K4. Since the Earth's axis of rotation and the axis of the moment of inertia are not the same, and the Earth's mass is not evenly distributed around its axis, the amounts of mass are not the same at various geographic locations. This results in a slight wobble (precession), both short term and over thousands of years.

KIPPING'S THEORY OF INORGANIC-ORGANIC CHEMISTRY: Chemistry: *Frederick Stanley Kipping* (1863–1949), England.

It is possible to form organic compounds with interactions of inorganic elements.

Born and educated in Manchester, England, Frederick Kipping left England in 1885 and entered Munich University in Germany where he received his PhD in 1887, thus began a forty-year career studying organo-silicon compounds.

At the beginning of the twentieth century many chemists, including Kipping, experimented with combining a variety of inorganic elements with silicon to form simple molecules such as SiH_4, $SiCl_4$, $SiFl_4$, $SiBr_4$, and SiI_4. Because these compounds are very reactive, they were used to combine other elements including metals to form *organometallic* compounds such as diethyl-zinc and diphenyl-mercury. These types of compounds are referred to as *organic silanes* that do not react with water and are very inert and resist chemical change and withstand high temperatures. The production of chemicals, such as chlorofluorocarbons used to produce Freon and related refrigerants, have been reduced due to their stability and possible long-term effects on the ozone layer in Earth's atmosphere.

One theory for the origin of organic molecules is that inorganic elements and compounds, such as hydrogen, water, oxygen, methane (CH_4), and ammonia (NH_3) found in the atmosphere of early Earth mixed with rainwater and reacted, forming simple organic molecules that in some way replicated. Silicon is the second most abundant element on Earth (oxygen is the first). Silicon is considered an inorganic element and is a major element in the formation of rocks and sand. The atom of silicon is somewhat similar in the shape, as well as the chemical and physical properties, to the carbon atom in that both have a valence of 4 with the bonds forming a tetrahedron (*see* Figure V3 under Van't Hoff). Today organic chemical compounds all contain the element carbon, whereas carbon itself is considered a unique inorganic element because of its natural atomic structure that is capable of forming many different large molecular compounds with other inorganic elements that make up living tissue and the products of living plants and animals. In other words, organic chemistry may be defined as carbon chemistry. If one looks at the **Periodic Table of Chemical Elements**, carbon is located at the top of the Group 4 elements (indicating four valence electrons). Silicon is located just below carbon in Group 4, indicating that silicon has many similar characteristics and attributes of carbon, including four valence electrons and the tetrahedron structure. The atoms of both of these two elements are capable of forming compounds with the atoms of many other elements. Following are two examples of simple silicon molecules:

```
           Cl  Cl
            \ /
SiCl4 =     Si              SiO2 = O=Si=O
            / \
           Cl  Cl
```

Carbon in particular can form thousands of compounds with other inorganic atoms (mainly hydrogen) to form long chains and ring structures of organic molecules that make up proteins, and so forth, of living plant and animal organisms, including the pricey "organic" vegetables, fruits, and meats marketed today. The term "organic" as used for these products is a misnomer because some of the molecules found in all food contain atoms of carbon, and thus all foods, no matter how they are raised, are organic (except for some seasonings such as some salts and vitamins). When the majority of other atoms that share valence bonds with carbon are hydrogen atoms, hydrocarbons are formed. There are also many other types of carbon molecules, most of which, but not all, are organic molecules. Examples of nonorganic carbon molecules are cyanide (CN) and carbon dioxide (CO_2).

Although each silicon atom has four valence electrons, only two of these are used to combine with an oxygen atom, which has two valence electrons. Silicon can bond with many other types of atoms to form a large group of what are known as "inorganic/organic silicones." These silicones are used in the manufacture of synthetic silicon-type rubber, plastics, high-temperature silicon greases, spray lubricants, hydraulic fluids that can withstand high temperatures, water repellants, and many other related products. Despite his four decades of work and research on organo-silicons, Kipping was unable to appreciate the commercial uses for his discoveries.

See also Kekule; Pauling

KIRCHHOFF'S LAWS AND THEORIES: Physics: *Gustav Robert Kirchhoff* (1824–1887), Germany.

Kirchhoff's electrical current and voltage laws: *1) The sum of all the current flowing in the direction of a point is equal to the sum of all the currents flowing away from that point,* $\Sigma\ 1 = 0$. *2) At any given time the algebraic sum of a voltage increase through a closed network loop will be equal to the algebraic sum of any voltage drop,* $\Sigma\ IR = \Sigma\ V$.

These two laws are important in the analysis of electrical circuits and for solving problems related to complex electrical networks. They are extensions to Ohm's law (*see also* Ampère; Ohm).

Kirchhoff's law of radiation: *A hot body in equilibrium radiates energy at a rate equal to the rate that it absorbs energy. Both the absorbed and radiated energy have the same given wavelength (black body radiation).*

Gustav Kirchhoff's law of radiation states that a perfect black body will absorb all light and other forms of radiation that are not reflected. A black body may be thought of as the perfect radiator where the maximum energy obtained per unit of time is due to the temperature of the "radiator." A radiator that is a perfect absorber is also a perfect reflector. Such a device is never found in nature. When heated, black bodies emit all the different wavelengths of light. This raised the question of how the different wavelengths of light were actually given off and how the individual wavelengths changed with temperature. Important theories related to energy, in particular, the visible light spectra followed, which led to the unification theory of electricity, magnetism, and light, referred to as the *electromagnetic spectrum*. The concept of black-body radiation also provided Max Planck the idea needed to develop his "quanta" theory (*see also* Maxwell; Planck).

Kirchhoff's theory for the use of the light spectrum in chemical analysis: *Each chemical element, when heated, exhibits a different line or set of lines in the spectrum of visible light.*

Sir Isaac Newton was the first to use a glass prism to split white light from the sun into a spectrum of colored lights. The English chemist and physicist William Hyde Wollaston, using an improved prism, divided these colors into seven distinct divisions (*see* Figure F8). Fraunhofer developed a "diffraction grating" made of fine wires to substitute for a glass prism, which enabled him to detect almost six hundred individual lines in the sun's spectrum. These devices made it possible for Kirchhoff to identify specific lines in the spectra of elements, thus identifying them by their particular colored light patterns. Kirchhoff and Robert Bunsen collaborated in the development of the spectroscope, which used the diffraction grating developed by Fraunhofer. The spectroscope consisted of a tube with a thin vertical slit on the front end, followed by a light-collecting lens and either a prism or diffraction grating. An eyepiece lens for viewing the specimen was located at the other end of the tube-shaped spectroscope. A small sample of the element to be analyzed was heated on a glass bead by Bunsen's new type of gas burner, which produced a flame hotter than an alcohol burner. When viewed through the slit/prism, or later through an improved diffraction grating, the light separated into specific spectra lines, unique for each element. These lines were then measured on a scale. Later, photographic plates were used that produced spectrophotographs for chemical analysis. By using this method Kirchhoff and Bunsen discovered two new metals: cesium and rubidium. The viewed image of lines of a specific element could be a bright spectrum, thus showing the element's spectrum as bright lines. Conversely, if the light from the heated element (including sunlight) was passed through a gas similar in wavelength to the element's wavelength, a dark line spectrum was visible. Spectroanalysis now can be used for analyzing the electromagnetic spectrum beyond the range of visible light, in the infrared and ultraviolet ranges, as well as X-rays from outer space. The technique has been invaluable in the analysis of the chemical makeup of all types of objects in the universe (e.g., the chemical composition of the sun and other stars, as well as the atmospheric composition of the planets). Today, spectroanalysis of the electromagnetic waves (radiation) given off by heated elements is an important analytical tool used by all fields of science.

See also Bunsen; Fraunhofer; Maxwell

KIRKWOOD'S ASTEROID GAP THEORY: Astronomy: *Daniel Kirkwood* (1814–1895), United States.

> *The asteroids located between the orbits of Mars and Jupiter are separated by gaps that correspond to their orbital periods and are integer fractions of Jupiter's orbital period.*

From the days of Galileo, chunks of matter smaller than planets yet large enough to be seen with a telescope were found in the region between Mars and Jupiter. It is estimated that over one hundred thousand objects, ranging from less than one mile in diameter to over 500 miles in diameter are located in this asteroid belt. About two hundred of these objects have been given names. These near-Earth bodies are called meteors and asteroids, whereas other bodies known as comets may spend some time in this asteroid belt. They usually are located at a greater distance from the sun. Comets have very large elliptical orbits, and some have paths that bring the comet's body closer to Earth than do the other objects in the asteroid belt. All of these objects are smaller than planets. Although asteroids are generally considered larger than meteors, both could cause great damage if they impacted Earth. The smallest meteoroids are about the size of a grain of sand, and when entering Earth's atmosphere at great speeds, they are heated to the point of incandescence thus producing a flash across the sky. When

Earth passes through a large number of these tiny meteors, a meteor shower is produced. When a small chunk of a meteor lands on Earth, it is called a meteorite.

Daniel Kirkwood detected a tugging on these asteroids/meteors by Jupiter's great gravitational force. This force created a resonance phenomenon (reverberation) in the smaller orbiting bodies. Asteroids that were drifting randomly were then pulled toward this region between Mars and Jupiter, thus forming their own orbits around the sun (*see* Figure K5).

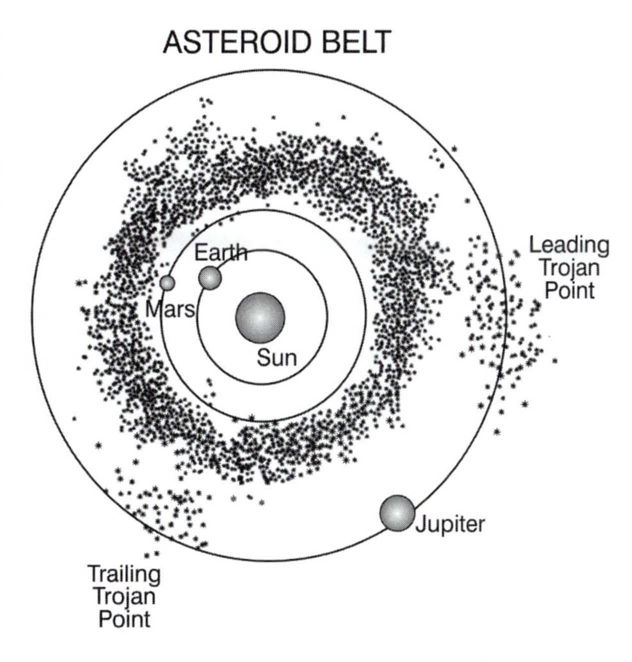

Figure K5. The diagram depicts the asteroid belt located between the orbits of Mars and Jupiter. Asteroids are leftover matter from the formation of the solar system and are too small to be considered planets. Jupiter's greater gravity most likely prevented them from coalescing into larger planets. This region is the source of most of the meteors and asteroids that intersect Earth's orbit.

This resulted in gaps among the thousands of asteroids revolving around the sun that appear as a number of "bands" void of asteroids. Kirkwood attributed this gap in the "rings" of asteroids to the influence of Jupiter's uneven gravity. The areas of depleted asteroids are called the "Kirkwood gaps" and are the consequence of their orbital separations related to their resonance motions with Jupiter. Kirkwood stated that asteroids in this gap would be forced into other orbits by perturbations caused by the bulge in Jupiter's shape, resulting in an uneven mass causing an uneven gravitational effect on the asteroids. Kirkwood also explained a similar theory for the separation of the Cassini divisions between the rings of Saturn. These divisions or separations between the rings of Saturn are partially caused by its moon, Mimas. The inner rings for the asteroid belt and Saturn's rings follow Kepler's laws of planetary motion. Because the inner sections of the ring must travel faster than the outer sections, density waves or ripples are formed that are called the *spiral density waves*.

See also Cassini; Kepler; Kuiper; Oort

KLITZING'S THEORY FOR THE QUANTIZATION OF THE HALL EFFECT:

Physics: *Klaus von Klitzing* (1943–), Germany. Von Klitzing was awarded the 1985 Nobel Prize in Physics.

> *"Hall's resistance" resulting from a two-dimensional electron gas at very low temperatures is a continuous linear function of an applied magnetic field, but it can be measured in quantized steps by varying the steps in the applied magnetic field.*

Edwin Hall was the first to note that when an electric current is flowing in a conductor, it is influenced by a magnetic field that is located perpendicular to the surface

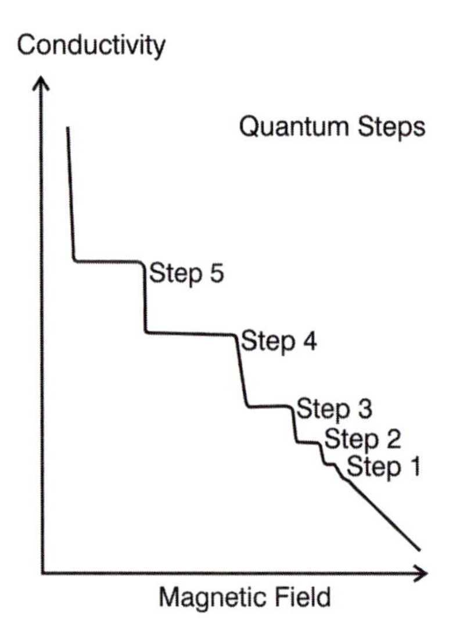

Figure K6. The tiny quantum changes in the ratio between the resistance of electrical flow and strength of a magnetic field was more of a small step-wise change than a continuous linear function.

of the conductor and produces what is known as the "Hall voltage." Hall voltage is at right angles to the direction in which the current flows and also the direction of the magnetic field. Thus, *Hall resistance* is determined by dividing the Hall voltage by its current (R = V/I). When measured and graphed, Hall resistance is linear in nature when it is influenced by a magnetic field.

The German physicist Klaus von Klitzing found that as the strength of the magnetic filed was increased, so did the Hall resistance in a linear manner, up to a certain point. It then leveled off. As the magnetic field was further increased, the Hall resistance would also increase and again level off, and so forth, which resulted in stepwise increases, rather than as a linear function (*see* Figure K6).

Klitzing devised an experiment to analyze this phenomenon. He used a very thin sheet of silicon to limit the movements of electrons to just two dimensions rather than three as normally occurs. Additional experiments were conducted using very powerful magnets kept at near absolute zero temperatures to verify the results. This setup demonstrated that the Hall effect becomes quantized as the changes in the magnetic field induced electrical changes in the silicon, only in certain distinct steps, thus not as a linear function. The result was that Klitzing discovered that the quantum steps had a value that proved to be a fundamental constant divided by an integer number. The National Institute of Standards and Technology Reference on Constants recognized Klitzing's constant as 25812.807449(86) ohms. Klitzing's discovery is extremely important to the physics community and is one of the few examples of the quantum effect being directly observed in a laboratory. He was recognized for this accomplishment with the 1985 Nobel Prize. The discovery of the quantized Hall effect gave a boost to the study of particle physics as well as providing a more precise meaning to the concept of ohms (electrical resistance).

See also Hall; Ohm; Planck

KOCH'S GERM-DISEASE POSTULATE: Biology: *Heinrich Hermann Robert Koch* (1843–1910), Germany. Robert Koch was awarded the 1905 Nobel Prize for Physiology or Medicine.

A set of specific conditions must be met before it can be established that a specific germ has caused a specific disease.

Robert Koch arrived at his postulate after many years of trying to isolate and identify specific bacteria that caused a great variety of diseases for which there was no known cause or effective cure. These included anthrax, tuberculosis, sleeping sickness, bubonic plague, malaria, cholera, and other "germ-caused" diseases. He developed techniques for growing and isolating bacteria in **agar** cultures, staining them for identification and using inoculations to test the suspected germs. His methods and postulate are responsible for bacteriology's becoming a respected science.

Robert Koch's postulate states:

1. *The bacteria (germ) must be found in all cases of the disease that have been examined.*
2. *The bacteria causing the disease must be prepared, cultured, isolated, and maintained in a pure culture that has not been contaminated by other organisms.*
3. *Bacteria grown for several generations removed from the original specimen must still be able to produce the same infection (disease).*

Koch's most famous application of his postulate was his culturing, identifying, and maintaining the rod-shaped bacillus responsible for tuberculosis, a widespread disease that killed hundreds of thousands of people in Europe and Asia. Isolating this particular bacterium was difficult because it was so much smaller than others with which he had previously worked. Koch spent most of his life traveling the world in an attempt to identify the causes of and develop eventual cures for a number of diseases.

See also Pasteur

KOHLRAUSCH'S LAW FOR THE INDEPENDENT MIGRATION OF IONS:
Physics: *Friedrich Wilhelm Georg Kohlrausch* (1840–1910), Germany.

> *The electrical conductivity of ions in a solution increases as the dilution of the solute increases in the solvent.*

Definitions for some terms follow:

- *Electrical conduction* is the passage of electrons or ionized atoms through a medium that in itself is not affected (wires, solutions, etc.)
- *Ions* are atoms or molecules that contain positive or negative charges.
- A *solution* is a mixture composed of two parts: the *solvent* (usually a fluid that effects the dissolving) and the *solute* (usually a soluble solid, such as sugar or salt—an electrolyte.
- *Electrolytes* are chemicals that, when molten or dissolved in a solvent, conduct an electric current.

Kohlrausch measured the electrical resistance of electrolytes, which are the dissolved substances that conduct electricity by transferring ions in solution. This means that when some substances, such as salt or acids, dissolve in water, an electric current can pass from the negative to the positive electrodes placed in the solution. The reason the current "flows" through the solution is that the electrolyte has formed ions (charged atoms) that carry the current. When direct current (**DC**) is used, there is polarization at the electrodes that interferes with the resistance of the current, thus making measurements difficult. To overcome this obstacle, Kohlrausch used alternating current (**AC**) instead of DC. This enabled him to measure accurately the conductivity of different electrolytes in solution. To some it may seem strange that the weaker the solution, the greater the flow of ions (current), but this is what Kohlrausch's law states. His theory is correct only over a limited range of dilutions because if the dilution increases to the extent that no ions are available in the solution, no electric current will be conducted between the electrodes (*see* Figure A7 under Arrhenius). In other words, absolutely pure water will not conduct electricity.

See also Arrhenius

KREBS CYCLE: Chemistry: *Sir Hans Adolph Krebs* (1900–1981), England. Sir Hans Adolf Krebs shared the 1953 Nobel Prize for Physiology or Medicine with Fritz Albert Lipmann.

> *All ingested foods undergo a chemical breakdown while going through a cyclic sequence of reactions where sugars are broken down into lactic acid, which is metabolized into citric and other acids and further oxidized into carbon dioxide and water, releasing a great deal of energy.*

The Krebs cycle, also known as the citric acid cycle or the tricarboxylic acid cycle, is a sequence of enzyme reactions involving the process of oxidation in the metabolism of food, resulting in release of energy, carbon dioxide, and water in the body (*see* Figure K7).

Prior to his work with the citric acid cycle, Hans Krebs worked with the German biochemist Kurt Henseleit (1907–1973) in 1932 on the Krebs–Henseleit Cycle, which is another cycle that occurs during the metabolism of food that supplies energy in animals. It is also known as the *urea cycle* where amino acids eliminate nitrogen from the body in the form of urea that is expelled from the body in the urine. The result was that the purified amino acids (which make up proteins) now provide energy to animals by other metabolic paths. This urea cycle and the Krebs citric acid cycle are responsible for the source of energy in animals that take in oxygen in the processes of respiration and the metabolism of digested food.

A number of scientists studied the metabolism of carbohydrates and the various forms of organic carbon molecules, but they were unsure as to how these chemical reactions all came together in living organisms. Food is first broken down into smaller carbon groups that combine with a four-carbon compound called citric acid. This citric acid molecule loses its carbon atoms, which frees up four electrons that produce energy in the form of

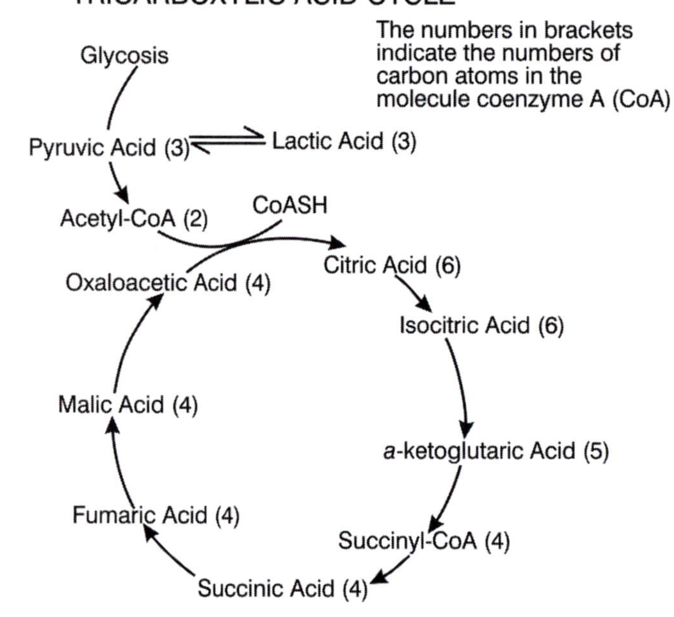

Figure K7. The Krebs cycle is also known as the Tricarboxylic Acid Cycle and is a sequence of oxidation processes involved in the metabolism of food resulting in the release of energy in the body.

adenosine triphosphate (ATP). Next, another energy-type molecule, guanosine triphosphate (GTP), is formed in the cycle. Still later in the cycle, the original molecules are regenerated and act as catalysts to start the whole cycle over again. Hans Krebs located two organic acids that contained six-carbon atoms, as well as other acids already known to contain five- and four-carbon atoms (*see* Figure K7). From this information he devised the cycle, with the carbon dioxide molecules exiting at the other end of the cycle. The process ends with the release of energy to the organism followed by an uptake of more carbohydrates (sugars), enabling the whole cycle to start all over again. Krebs was aware of the significance of this cycle of chemical reactions as the main path for the breakdown of all foods eaten by animals. This important process is both the source of energy for our living cells, as well as the process that is responsible for **biosynthesis**.

KROTO'S "BUCKYBALLS": Chemistry: *Harold Wiliam Kroto* (1939–), England. Harold Kroto, Richard E. Smalley, and Robert F. Curl shared the 1996 Nobel Prize in Chemistry for their discovery of fullerenes.

> *Graphite subjected to laser beams produces sixty carbon atoms bunched together in the form of a twenty-sided polyhedron.*

Harold Kroto collaborated with the American professor of chemistry, physics, and astronomy Richard E. Smalley (1943–2005) who used laser beams to vaporize metals that, when cooled, formed compact masses or groups of one-of-a-kind metallic atoms.

> Applied scientists and engineers, who study the chemical and physical nature of various materials and how these materials can be utilized in practical ways, are exploring some exciting avenues. In addition to macro and micro technologies that rely on materials of a relatively large scale, nanotechnology is a subclassification of technology in colloidal science, biology, physics, chemistry, and other fields that involve manipulations of materials at the nano level. This level is far smaller than micro, or even submicro levels. It deals with particles at the 10^{-9} size (one-billionth of a meter), which is at the size of molecular and atomic particles of matter. There are two types of nanotechnology—the bottom-up approach that builds up materials atom by atom and the top-down method where nano-size particles are removed from larger substances to form nano-sized materials. In the computer business new and faster chips that use macro materials are nearing the physical limits of their semiconducting materials. According to Moore's Law (stated in 1965 by Gordon Moore of Fairchild Semiconductors) the number of transistors that could be placed on a chip would double every eighteen months. This has been pretty much true over the past decades, but there are limits to this continued improvement, and those limits are physical limitations at the molecular and atomic levels. This is a challenge for technologists to make nano-sized computer circuits at the single molecule or atomic levels. At this nano level, there are indeterminacy problems related to the quantum behavior of particles, in other words, where a particle's position and momentum cannot be determined at the same time. There is great potential for nanotechnology, which has an exciting future, but there are a few dangers. One is the physical law: *the smaller the size of a particle the larger is its surface area compared to its volume.* This means that nano particles have different properties than normal-sized particles and could be harmful to living organisms, tissues, and cells. Such small particles can enter the body by swallowing or breathing; they are even small enough to pass through the pores of the skin. Nano particles may also be harmful to the environment. Further study still needs to be done in this exciting field as it promises more advantages than disadvantages in the future.

Kroto suggested graphite (a form of carbon) could be used to create new structures for groups of carbon atoms. Before finishing their research, they exhausted their research funds. A few years later, several other scientists repeated the experiment and produced an amount of material adequate for analyzing and verifying its structure. The result was a ball-like molecule that resembled a field soccer ball or a geodesic dome, similar to the ones designed by American architect Buckminster Fuller (*see* Figure C7 under Curl.) Thus, the C_{60} molecule was given the name *buckminsterfullerene* or, more commonly, "buckyballs." The unique sixty atoms, twenty-side polyhedron group of carbon molecules continues to be explored for properties that may improve the structure and functions of other materials. Since the discovery of C_{60}, other polyhedrons with more than sixty atoms have been produced experimentally. This discovery was the beginning of the material-sciences of *nanotechnology*.

See also Curl; Van't Hoff

KUIPER'S THEORY FOR THE ORIGIN OF THE PLANETS: Astronomy: *Gerard Peter Kuiper* (1905–1973), United States.

The planets evolved from condensing gas clouds distinct from the gas that formed the sun.

Gerard Kuiper discovered a new satellite for Uranus and another for Neptune. He also determined that carbon dioxide gas is present in the thin atmosphere of Mars and that Titan, the largest of Saturn's satellites, has an atmosphere composed of methane. Kuiper theorized some comets originated beyond the Oort cloud, at a distance of more than one hundred thousand AU (the AU is the unit used to measure distances in the solar system that is equal to the average distance between Earth and the sun. One AU equals approximately ninety-three million miles). He also postulated that the space beyond Pluto contains many comets with solar orbital periods of hundreds of years. In 1992 the first objects were found at about 120 to 125 AU in the Kuiper belt, thus verifying the theory he had advanced forty years earlier. All of these ideas relate to the formation of comets, asteroids, and planets from gas clouds. Like the planets, the objects in the Kuiper belt are primitive remnants left over from the formation of our solar system. Kuiper also theorized that some of our short-term comets originate in the Kuiper belt, whereas other long-term comets originate in the Oort cloud.

Kuiper proposed two other theories that are not generally accepted. One was his measurement of the planet Pluto. Based on his estimates of the perturbations of Pluto on Uranus, he estimated Pluto's size to be about half Earth's diameter (it later proved to be only about one-fourth Earth's diameter). Second, his estimate of the mass of Pluto was also much too large. These errors were later corrected. Other astronomers no longer accept Kuiper's theory that the planets formed from their own gaseous clouds. Before his death, Kuiper was involved in several of the early NASA space missions.

See also Oort

KUSCH'S THEORY FOR THE MAGNETIC MOMENT OF THE ELECTRON: Physics: *Polykarp Kusch* (1911–1993), United States. Polykarp Kusch shared the 1955 Nobel Prize for Physics with Willis Lamb.

The electron, like all other charged particles, will possess a magnetic moment (field) due to the motion of its electric charge.

Polykarp Kusch used the interacting principle of electromagnetism to demonstrate that the electron, which has an electrical charge, is affected by a magnetic field. His measurements, which were extremely accurate, were based on the structure of different energy levels of several elements. He discovered that all particles containing an electrical charge do exhibit a turning effect in a strong magnetic field. Later, it was determined that the neutron, which has zero electrical charge, will also be affected by a strong magnetic field due to its internal structure, which involves a polarity distribution of positive and negative charges (A neutron is basically composed of a negative electron and a positive proton, or it might be thought of as three quarks held together by gluons.) This polarity is why nuclear magnetic resonance (NMR) and magnetic resonance imaging (MRI) work. The magnetic field around the patient causes nuclei in the different tissues to resonate (oscillate at different rates) with the applied field and is thereby detected by the MRI instrument, producing an X-ray type image of the tissue. Kusch's work also led to the concept of quarks and the field of quantum electrodynamics.

See also Gell-Mann